T0419523

ADVANCES IN MATERIALS SCIENCE RESEARCH

VOLUME 26

ADVANCES IN MATERIALS SCIENCE RESEARCH

Additional books in this series can be found on Nova's website under the Series tab.

Additional e-books in this series can be found on Nova's website under the e-book tab.

ADVANCES IN MATERIALS SCIENCE RESEARCH

ADVANCES IN MATERIALS SCIENCE RESEARCH

VOLUME 26

MARYANN C. WYTHERS
EDITOR

New York

Library of Congress Cataloging-in-Publication Data

ISBN: 978-1-53610-059-4
ISSN: 2159-1997

Published by Nova Science Publishers, Inc. † New York

CONTENTS

PREFACE

This book provides readers with the latest developments in materials science research. Chapter One discusses advantages for optical applications with transparent polycrystalline ceramic materials. Chapter Two reviews $K_{0.5}Na_{0.5}NbO_3$-based lead-free transparent ceramics. Chapter Three describes the operation principles and requirements for photodetector (PD) devices, followed by the latest developments in materials and device architectures. Chapter Four studies applications of infrared thermography to the analysis of steel welded junctions. Chapter Five discusses Suspension Plasma Spraying (SPS) as a method that uses liquid feedstock. Chapter Six reviews the recent studies on the structural, electronic, optical and mechanical properties of silicene-derivative structures. In Chapter Seven, the correlations between experimental and theoretical results have been established in order to gain complete insight into structural features which enable modeling and synthesis of novel APD molecules with potential usage in various industries. Chapter Eight considers the analytical capabilities of two methods for studying the composition of gold-bearing samples.

Chapter 1 - Optical materials cover a broad spectrum of organic and inorganic, amorphous and crystalline materials.

To be applied in general primary function related material characteristics (refractive index, Abbe number, transmission), secondary material characteristics/combinations of material characteristics (mechanical strength, thermal and chemical stability) and economically relevant/technologically relevant material characteristics or criteria (material costs/machinability/ processing costs/machining time) have to be considered.

There is no one optical material with optimal characteristics for all application-relevant aspects. It is hence necessary to find the particular optima for each demand. Crystalline hard materials are interesting and promising alternatives to conventional optical glass materials for optical components, which have to withstand extremely high loads during their application (mechanical loads, high temperatures, chemical influences).

One requirement for a good optical performance is the surface quality, generated by grinding and polishing processes. Various studies have been carried out with mono- and polycrystalline varieties of materials (materials with the same chemical composition but different structure: Sapphire/alumina; Spinel/$MgAl_2O_4$- spinel-ceramics). As the result of the anisotropy of mechanical properties in the case of monocrystalline materials some typical influences on the surface formation are noticeable.

An interesting question with respect to practical applications of the newly developed polycrystalline transparent hard materials is whether high temperatures affect the performance or function-related properties of the optical components.

Within the experimental examinations, several optical component samples with spherical surfaces were manufactured from $MgAl_2O_4$- spinel-ceramics by grinding/lapping/polishing. Both the geometrical parameters (shape accuracy and roughness) and the optical material characteristics (in-line transmission, light scattering) were measured before and after a simulated thermal load at comparatively high temperatures.

The results of these experiments highlight the potential of polycrystalline transparent materials for applications under extreme and complex loads and may be considered as a reference for developers of optical systems concerning the imaginable application conditions of transparent high performance ceramics.

Chapter 2 - Hot-press sintering has been used to fabricate lead-free transparent ceramics $(K_{0.5}Na_{0.5})_{1-x}Li_xNb_{1-x}Bi_xO_3$ (KNNLB-x, $0.05 \leq x \leq 0.09$) based on alkali metal oxalates as the raw chemicals. Due to the co-modification of Li^+ and Bi^{3+}, the ceramics possess a dense and fine-grained structure with cubic-like symmetry. Owing to the polar nano-regions, the ceramics have a diffuse phase transition, showing the relaxor-like characteristics. These can reduce the light scattering by the grains, at the grain boundaries and the domain walls, thus improving the optical transmittance of the ceramics. The optical transmittance (T) of the ceramics reaches a high value of ~60% in the near-IR (NIR) region. The transparent ceramics can be considered as a good electro-optic (EO) material because of a large effective linear EO coefficient (r_c) in the range of 120-200 pm/V.

After improving the sintering technique and selecting alkali metal carbonate as the raw chemicals, KNNLB-x transparent ceramics have been successfully prepared by pressureless sintering. At x = 0.04-0.08, the ceramics exhibit high transmittances (T = 60-70%) in the NIR region, and strong linear EO response (r_c = 120-200 pm/V) comparable to the hot-press sintered ceramics. The good optical transparency should be attributed to the cubic-like crystal structure as well as the relaxor-like characteristics, which have been attested through the temperature dependences of dielectric constants and polarization-electric field hysteresis loops. Together with the relatively large piezoelectric coefficient (50-90 pC/N), high dielectric constant (~1400) and low dielectric loss (< 0.03), the pressureless sintered KNNLB-x transparent ceramics should have potential in the applications of multifunctional devices.

Chapter 3 - Ultraviolet (UV) photodetector (PD) has attracted extensive attention owing to their broad application in digital imaging, missile plume detection, optical communications, and biomedical sensing. Conventional silicon (Si) based UV PD requires costly high pass optical filters and phosphors to tune its spectral response and thus poses extreme challenge to achieve a UV detection system with miniature size, high energy-efficiency and low-cost. To overcome the limitations inherent from the narrow bandgap of Si, UV PDs consisted of wide-bandgap materials, such as gallium nitride (GaN), silicon carbide (SiC), and transition metal oxides (TMOs) such as zinc oxide (ZnO), nickel oxide (NiO) etc. have recently emerged and are being rapidly developed. The application of TMOs in the UV PDs mainly rely on their intrinsic properties, including optical transparency in the near-UV and visible spectra, low impact to environment, thermal stability and feasibility of constructing band-tunable heterojunctions. Here, the authors provide a perspective on the recent developments in UV photodetection based on TMO materials including thin films and one-dimensional nanostructures. The chapter begins with describing the operation principles and requirements for the PD devices, followed by the latest developments in materials and device architectures. This chapter is concluded with future challenges and outlook in this area.

Chapter 4 - The thermographic methodology based on the analysis of the response of the materials after an external stimulation can be used as a tool for the detection, characterization and evaluation of surface cracks in steel welds. These types of cracks are usually difficult to detect and their dangerousness could be a problem for the integrity of welded elements in machines and structures. Several studies have demonstrated the usefulness of the thermographic technique for the assessment of composites and metallic materials. However, the study of cracks in welds is a topic which has only been addressed in recent times. The thermographic technique, in combination with different processing algorithms allows the efficient detection of surface cracks in welds. If the procedure is complemented with the geometric calibration of the infrared camera, the accurate measurement of the surface dimensions of the cracks can be performed. What is more, the generation of the depth-prediction model of the cracks is possible through the correlation of the thermographic data with the three-dimensional data obtained using the macro-photogrammetry technique.

Chapter 5 - One of the most frequent topic of thermal spraying research field in recent years concerns the use of liquid feedstock instead of dry powder. In particular, the application of suspensions allowed introducing nanometric or submicrometric particles into flames and jets and enabled obtaining finely grained coatings having modified microstructure and interesting properties. This development may lead to create a fair competition with regard to vapor deposition methods in many industrial applications.

This chapter discusses Suspension Plasma Spraying (SPS) as a method that uses liquid feedstock. The present review should help to understand which are the advantages and drawbacks of this method. The suspension preparation and its influence on the deposition process and on the coating microstructure is discussed. Then, the suspension injection into flame or jet using nozzles or atomizers is presented. Moreover, the different types of plasma torches which can be used in the SPS processes having axial and radial injectors are discussed. The different microstructures of SPS coatings are presented. The particular attention is paid to the grain size and their orientation as well as to the phases' composition. The build-up mechanisms of various types of coatings morphologies called *columnar* and *two-zones-microstructure* is discussed. Finally, two important applications of SPS zirconia coatings, namely Thermal Barrier Coatings (TBC) and Solid Oxide Fuel Cells (SOFC), are discussed in more detailed way.

Chapter 6 - Successful isolation of graphene from graphite opened a new era for material science and condensed matter physics. Due to this remarkable achievement, there has been an immense interest to synthesize new two dimensional materials and to investigate their novel physical properties. Silicene, form of Si atoms arranged in a buckled honeycomb geometry, has been successfully synthesized and emerged as a promising material for nanoscale device applications. However, the major obstacle for using silicene in electronic applications is the lack of a band gap similar to the case of graphene. Therefore, tuning the electronic properties of silicene by using chemical functionalization methods such as hydrogenation, halogenation or oxidation has been a focus of interest in silicone research. In this paper, the authors review the recent studies on the structural, electronic, optical and mechanical properties of silicene-derivative structures. Since these derivatives have various band gap energies, they are promising candidates for the next generation of electronic and optoelectronic device applications.

Chapter 7 - Arylazo pyridone dyes (APDs) represent disperse dyes commonly used in textile industry due to their good technical features: excellent coloration properties, simplicity of preparation, good light and wash fastness properties. They find application in inkjet printing, liquid crystal displays and inks for heat-transfer printing. The main characteristic of APDs bearing –OH group in *ortho-* position to azo group, is existence of an azo-hydrazone tautomerism. Generally, APDs are present in the hydrazone form in the solid state and most of the solvents, while in highly dipolar solvents there is hydrazone-anion equilibrium with the predominance of the hydrazone form.

Subtle structural changes of APD molecules lead to their different physico-chemical properties and the knowledge of the origin of these changes is essential for their further application. Significant progress has been made in the area of experimental investigations of APDs using various techniques including UV-Vis, FT-IR and NMR spectroscopy, as well as X-ray crystallography. Furthermore, the investigation of the geometry and the relative stabilities of the APD structures, as well as the charge transfer through the molecule and the possibility of forming intramolecular hydrogen bonds can be obtained using quantum chemical calculations. The transmission of electronic effects of the substituents through the molecule can be interpreted by linear free energy relationships (LFER) analysis of spectroscopic data (UV-Vis, NMR, FT-IR). Specific and nonspecific interactions (solvent-solute), which strongly influence the coloration of the APD solutions, most commonly are analyzed by linear solvent energy relationships (LSER) concept. So, in this chapter, the correlations between experimental and theoretical results have been established in order to gain complete insight into structural features which enable modeling and synthesis of novel APD molecules with potential usage in various industries.

Chapter 8 – This chapter considers the analytical capabilities of two methods for studying the composition of gold-bearing samples: arc scintillation atomic emission spectrometry and automated mineralogy using the microprobe X-ray spectrometry. The authors report the results for two certified reference materials (Sukhoi Log deposit and flotation tailings gold sulfide ore) obtained by these methods. The reasons of obtaining incompatible information are discussed. The scopes of economically viable application of each method are listed.

In: Advances in Materials Science Research
Editor: Maryann C. Wythers
ISBN: 978-1-53610-059-4

Chapter 1

ADVANTAGES FOR OPTICAL APPLICATIONS WITH TRANSPARENT POLYCRYSTALLINE CERAMIC MATERIALS

Uwe Reichel[1,*], Gunther Notni[2], Angela Duparré[2] and Volker Herold[3]

[1]Fraunhofer IKTS, Fraunhofer Institute for Ceramic Technologies and Systems, Hermsdorf, Germany
[2]Fraunhofer IOF, Fraunhofer Institute for Applied Optics and Precision Engineering, Jena, Germany
[3]Friedrich-Schiller-University Jena, Otto Schott Institute for Materials Research, Jena, Germany

ABSTRACT

Optical materials cover a broad spectrum of organic and inorganic, amorphous and crystalline materials.

To be applied in general primary function related material characteristics (refractive index, Abbe number, transmission), secondary material characteristics/combinations of material characteristics (mechanical strength, thermal and chemical stability) and economically relevant/technologically relevant material characteristics or criteria (material costs/machinability/processing costs/machining time) have to be considered.

---

* Corresponding Author: uwe.reichel@ikts.fraunhofer.de.

There is no one optical material with optimal characteristics for all application-relevant aspects. It is hence necessary to find the particular optima for each demand. Crystalline hard materials are interesting and promising alternatives to conventional optical glass materials for optical components, which have to withstand extremely high loads during their application (mechanical loads, high temperatures, chemical influences).

One requirement for a good optical performance is the surface quality, generated by grinding and polishing processes. Various studies have been carried out with mono- and polycrystalline varieties of materials (materials with the same chemical composition but different structure: Sapphire/alumina; Spinel/$MgAl_2O_4$- spinel-ceramics). As the result of the anisotropy of mechanical properties in the case of monocrystalline materials some typical influences on the surface formation are noticeable.

An interesting question with respect to practical applications of the newly developed polycrystalline transparent hard materials is whether high temperatures affect the performance or function-related properties of the optical components.

Within the experimental examinations, several optical component samples with spherical surfaces were manufactured from $MgAl_2O_4$-spinel-ceramics by grinding/lapping/polishing. Both the geometrical parameters (shape accuracy and roughness) and the optical material characteristics (in-line transmission, light scattering) were measured before and after a simulated thermal load at comparatively high temperatures.

The results of these experiments highlight the potential of polycrystalline transparent materials for applications under extreme and complex loads and may be considered as a reference for developers of optical systems concerning the imaginable application conditions of transparent high performance ceramics.

INTRODUCTION

These days, ceramic materials reliably perform tasks for different applications.

The potential of typical characteristics takes effects especially at complex requirements, whereas other materials often predestinated for only one particular requirement type.

Extreme cases for innovative applications with complex material requirements shown at transparent polycrystalline optoceramic materials [1-3].

Compared to usual optical materials (like optical glass, single crystal, transparent polymer) transparent ceramic materials offer explicit higher specific values for mechanical and thermal stability. In combination of typical ceramic and optical characteristics, transparent ceramic materials show attractive options compared to well-known optical materials and can offer an enormous innovation potential.

COMPARISON OF THE MATERIAL PROPERTIES OF TRANSPARENT HARD MATERIALS

For optical materials a classification can be applied according to the following overview (Figure 1: Classification of optical materials by their structure).

In many cases amorphous materials /optical glasses are used. There are also applications of partially crystalline materials like glass ceramics. For applications with special requirements also crystalline materials are important.

The publication is focused on some facts concerning crystalline, especially polycrystalline materials in the form of transparent high performance ceramics which could be suitable for special future applications.

The material characteristics of optical materials can be divided in

a) primary function related material characteristics, which define the primary function
b) secondary function related material characteristics or combinations of material characteristics, which could restrict the application or operation conditions and
c) economically relevant/technologically relevant material characteristics or criteria which affect manufacturing costs or time consumption (Figure 2: Application-relevant aspects of optical materials)

The comparison of selected groups of optical materials shows the generally known fact, that there is hardly any optical material, which fulfills all possible criteria.

Glasses have some advances concerning the suitability for a wide range of applications and the availability of long-time as well as comprehensive application experiences. Crystalline materials have a good transmission and thermal stability/mechanical characteristics but there is the potential disadvantage of the anisotropy of their characteristics.

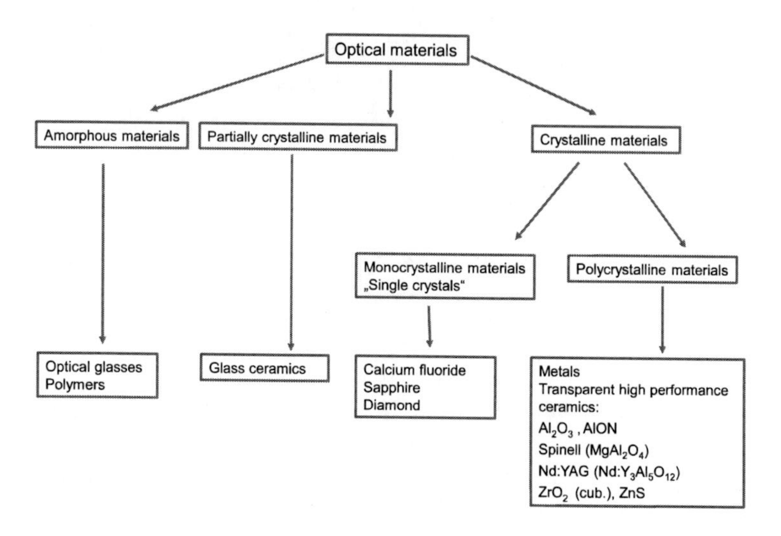

Figure 1. Classification of optical materials by their structure and respective examples.

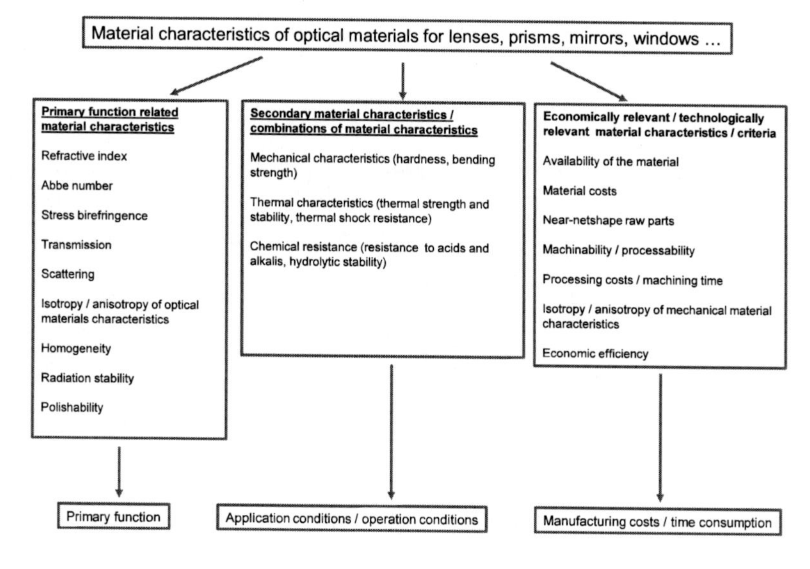

Figure 2. Application-relevant aspects of optical materials.

Polycrystalline transparent materials seem to be a good compromise concerning some characteristics for special applications. Some important material characteristics of selected transparent hard materials are listed in the following table.

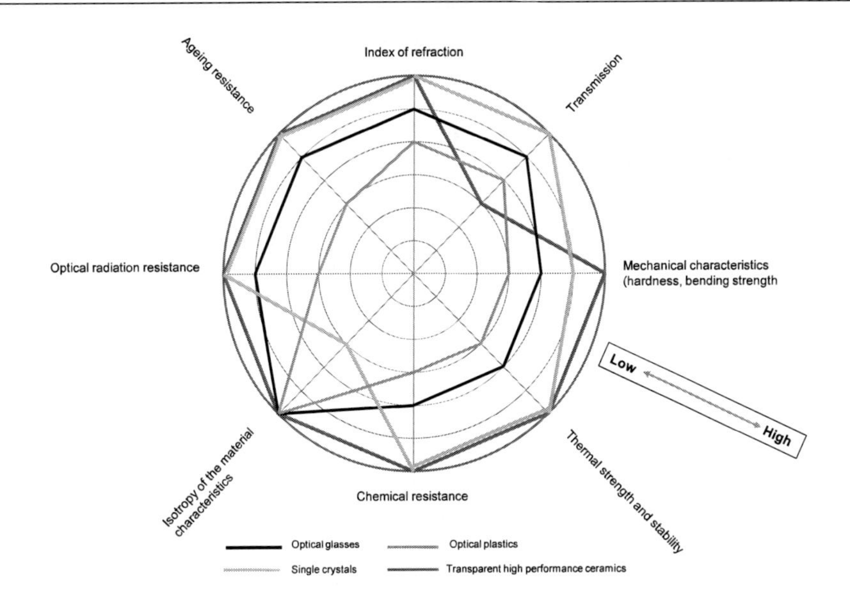

Figure 3. Comparison of selected groups of optical materials.

Table 1. Material characteristics of selected transparent hard materials

Selected materials	Material characteristics		Optical characteristics		Mechanical characteristics		Thermal characteristics		Chemical characteristics
Transparent Materials	Crystal / Structure	Materials density [g/cm³]	Refractive Index	Trans-mission Range / Wavelength [nm]	Hardness (Knoop) [kg/mm²]	Bending strength [MPa]	Maximum of Working-temperatur [°C]	Thermal Expansion (50-400°C) [10^{-6} K^{-1}]	Chemical durability Acid and alkali resistant (Solubility)
Crystal growth materials:	**Single-crystal**								
Sapphire	Hexag.-rhomb.	3,99	n=1,76 II *) @590nm	150 – 5500	2200 II	400	2000	6,2 II	0 (to 300 °C)
Spinel	Cubic	3,61	n=1,727 @589nm	150 - 6000	1175-1380	150 - 200	2100	5,9	0 (RT)
Ceramic materials:	**Poly-crystal**								
Al_2O_3	Hexag.-rhomb.	3,99	n_0=1,760/1,768 (Birefringence!)	400 - 6000	2800	700	1800	7,5	Slightly soluble
$MgAl_2O_4$	Cubic	3,61	n_0=1,72	200 – 6000	1700	200 – 300	1400	5,9	
for comparison:									
Fused silica IR	amor-phous	2,21	n=1,40 @3,7µm	300 - 3500	460	50	1000	0,54	Soluble in HF
BK 7	Glass	2,52	n=1,50 @1,5µm	350 - 2500	400	20 - 40	500	3,3	Only in HF, concentrated phosphoric acid and strong alkali

Previous Developments and Applications of Transparent Ceramic (Polycrystalline) Materials

Optical materials cover a broad spectrum of organic and inorganic, amorphous and crystalline materials. There is no optical material with optimal characteristics for all application-relevant aspects and it is hence necessary to find the particular optima for each demand.

At the authors affiliations Fraunhofer IKTS, Fraunhofer IOF and Friedrich-Schiller-Universität Jena there is a longstanding expertise in the development of materials and technologies for optical applications, their processing and inspection [4-8]. Crystalline hard materials are interesting and promising alternatives to conventional optical glass materials for optical components, which have to withstand extremely high loads during their application.

Optoceramic materials are also able to address new optical positions outside of the glass area, with different refractive indices and Abbe Numbers. As a result of further development in the field of high performance ceramics, transparent polycrystalline materials based on different material systems are available with potential technical and/or economic benefits compared to the already used monocrystalline material sapphire.

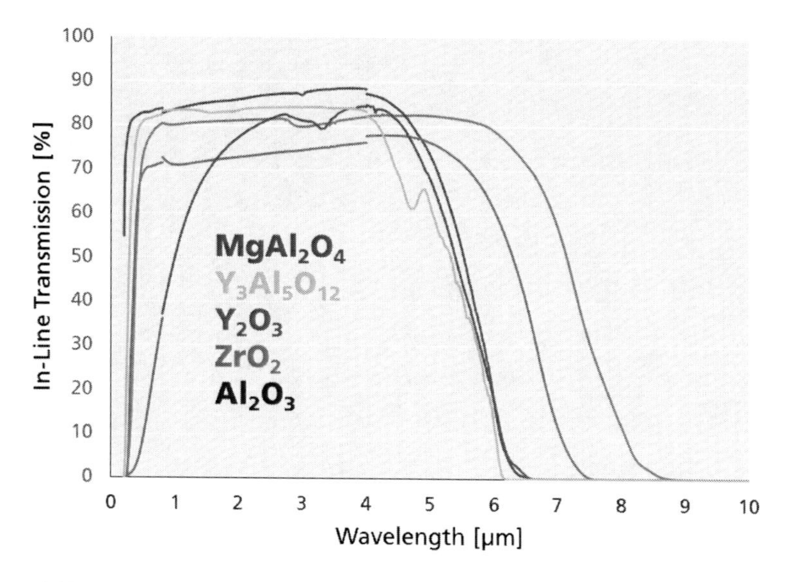

Figure 4. Transparency of oxide ceramics from UV-VIS to IR (Thickness ~ 3 mm).

Figure 5. Optoceramics with Phosphore dopants while UV-lighting (right side).

Figures 4 and 5 shows different transmission behavior of transparent ceramic materials. Remarkable material characteristics (either used individually or in combination with others) of the transparent high performance ceramics are:

- isotropy of the material characteristics
- high index of refraction
- transmission in a wide wavelength range (UV to IR); as a function of the chemical and crystalline compositions
- mechanical material characteristics (hardness, long-term scratch resistance, abrasion resistance, bending strength)
- thermal strength and stability
- optical radiation resistance

Compared to monocrystalline materials (such as sapphire), transparent high performance ceramics offer substantial technical advantages (such as isotropy of the material characteristics) or improved economic efficiency of the component manufacturing process (such as cost-effective near net shape raw parts by dry pressing) [9].

These characteristics highlight the potential of polycrystalline transparent materials for applications under extreme and complex loads and can be considered as a reference for developers of optical systems concerning the imaginable application conditions of transparent high performance ceramics.

Innovative applications for transparent high performance ceramics in the field of "micro photonics" could be (for example) special optics for optical metrology (especially IR) and medical devices (surgery, photodynamic therapy), optics for special laser systems (laser emission, laser mirrors) and lighting. Figures 6 and 7 shows transparent ceramics for optical systems (lens) and laser emission (dopants for special transmission/absorption) [10].

Highest optical performances of LEDs such as optimized luminous efficiency are posing a challenge for phosphor materials. Usual the reached properties of phosphor materials are optimized starting at their synthesis methods (e.g., sol-gel synthesis, hydrothermal synthesis or solid-state reaction) to meet the modern requirements on high performance materials. These phosphor powders with enhanced optical performance and optimized particle size distribution are used to develop optical ceramics with different clarity and variable chromaticity coordinates (see figure 5).

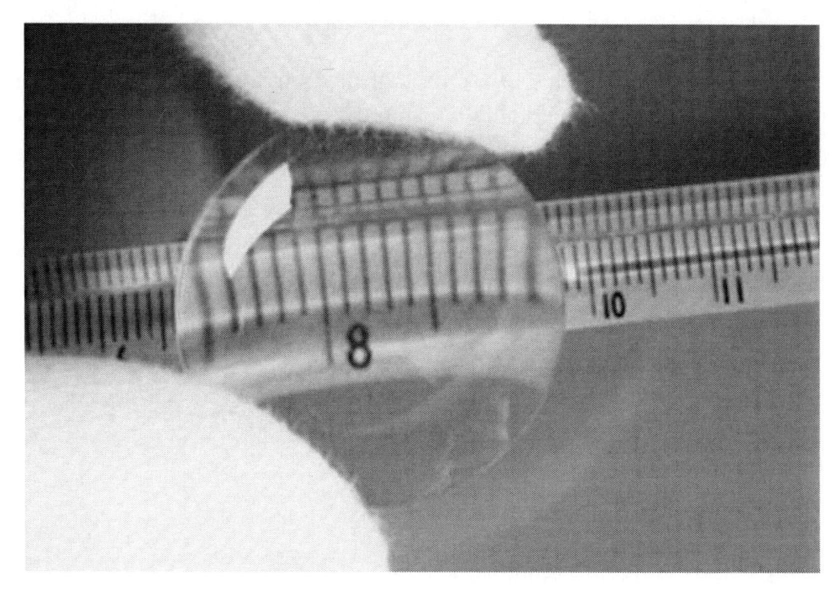

Figure 6. Biconvex lens from transparent $MgAl_2O_4$-Spinel (n_D = 1.73).

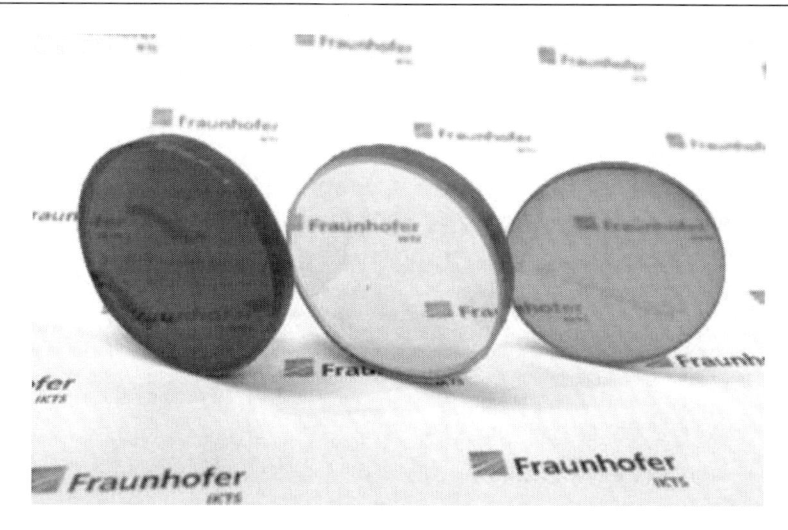

Figure 7. Transparent ceramics with different dopants (blue: $MgAl_2O_4$:Co; rose: $MgAl_2O_4$:Cr; yellow: ZrO_2:Ce).

Optical ceramics are combining the properties of phosphor and ceramic materials such as high quantum yield and increased thermal conductivity compared to common phosphor-silicone composites.

New technologies such as high-performance LEDs or laser-based head lamps will be realized prospectively with optical ceramics.

Basic Conditions for Promising Applications of Transparent High Performance Ceramics (Transparent Polycrystalline Hard Materials)

Basic conditions for promising applications of transparent high performance ceramics (transparent polycrystalline hard materials) are:

- the demands on primary optical material characteristics can be fulfilled (index of refraction, transmission, scattering …)
- it relates to components with very high mechanical and/or thermal and/or chemical loads, where conventional optical materials fail
- compared to monocrystalline materials (such as sapphire) transparent high performance ceramics offer substantial technical advantages (such as isotropy of the material characteristics) or improved

economic efficiency of the component manufacturing process (such as cost-effective near net shape raw parts by dry pressing)

Transparent High Performance Ceramics: Remarkable Material Characteristics/Combinations of Material Characteristics in Comparison to Monocrystalline Materials

Remarkable material characteristics of transparent polycrystalline hard materials (transparent high performance ceramics) are

- isotropy of the material characteristics
- high index of refraction
- transmission in a wide range of wave length (UV to IR); in dependence of chemical and crystalline composition
- mechanical material characteristics (hardness, long-lasting scratch resistance, abrasion resistance, bending strength)
- thermal strength and stability
- optical radiation resistance

Due to their special material characteristics or combinations of material characteristics for transparent high performance ceramics an extension of the application possibilities of optical materials for optical parts with extraordinary profiles of requirements are conceivable.

EXPERIMENTAL INVESTIGATIONS

Mechanical Machinability and Surface Quality

In practice for shaping of brittle optical materials mainly mechanically based machining processes (various grinding processes, lapping, polishing) are used. The machinability with these machining processes is an important aspect for the economic efficiency of parts manufacturing, whereas the generation of the surfaces decisively codetermines the optical performance of the surfaces. Comparison of the surface generation between monocrystalline and polycrystalline materials was performed on the base of grinding experiments. Transparent hard materials (monocrystalline materials (crystals) and polycrystalline materials (ceramics)) are difficult to machine materials. The commonly used grinding processes are characterized by

- high machining forces
- (local) high machining temperatures
- excessive tool wear
- subsurface damage because of typical brittle mode of material removal

Impacts of the anisotropy on the machining process (especially the grinding process) arise from locally different mechanical material characteristics and lead to:

- locally different grindability
- locally different grinding forces (possibility of self-excited chatter vibrations, potential effects on shape accuracy)
- locally different mechanisms of material removal (ductile mode / brittle mode)
- locally different surface generation (qualitatively: surface structures, quantitatively: roughness parameters with possible effects on subsequent machining processes (e.g., smoothing in the polishing process)

In the experimental investigations samples with the same material composition (alumina) but with different structures (monocrystalline structure: sapphire with various crystallographic orientations; polycrystalline structure: alumina ceramics) were machined by cylindrical grinding.

After grinding the circumferential surface of the sample was measured by different methods. Roughness profiles (2D-measurement) in axial direction were recorded by a tactile profilometer in angular steps of 2.5° and evaluated by the parameters Ra and Rq. Furthermore local surface topographies (3D-measurement) by means of the circumference were detected by a white light interference microscope. Figure 8 shows experimental results on the Influence of anisotropy on surface generation (quantitative evaluation):

Variation of surface roughness around the work surface circumference for sapphire parts with different crystal orientation.

The distributions of roughness values both for Ra and Rq show typical patterns depending on the crystallographic orientation of the sapphire samples.

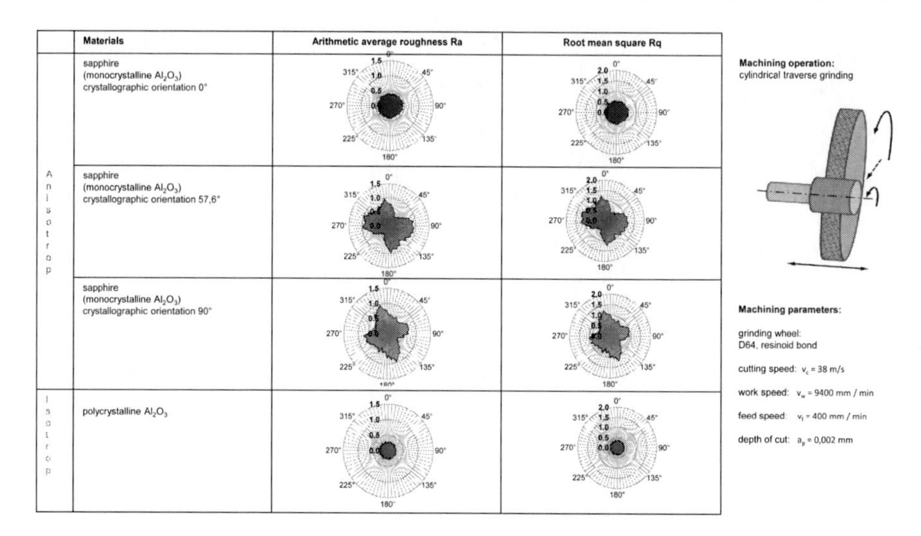

Figure 8. The Influence of anisotropy on surface generation (quantitative evaluation): Variation of surface roughness around the work surface circumference for sapphire parts with different crystal orientation.

In comparison, on polycrystalline alumina samples roughness values only with random variations around the cylindrical surface were measured.

The qualitative evaluation of the Influence of anisotropy on surface generation shows characteristic variations of the surface structure on locations with roughness maxima and roughness minima (Figure 9).

A probable cause of these different surface structures can be seen in locally different ratios of ductile mode and brittle mode of material removal.

An indication for predominant ductile mode are scratches of the abrasive grains, whereas surfaces consisting of random craters are an indication for brittle mode of material removal.

High Temperature Effects on Geometrical Features and Optical Material Characteristics of Polycrystalline Spinel Parts

If the new developed transparent high performance ceramics meet the requirements imposed on the optical materials for special applications there is on question among others how do change geometrical and optical characteristics of the optical components in the case of high thermal loads.

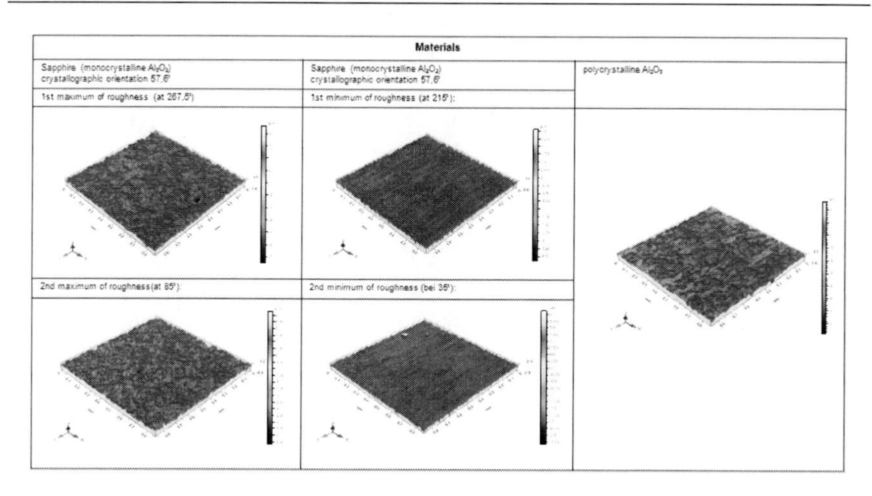

Figure 9. The Influence of anisotropy on surface generation (qualitative evaluation): Typical variations of surface structure on locations with roughness maxima and roughness minima.

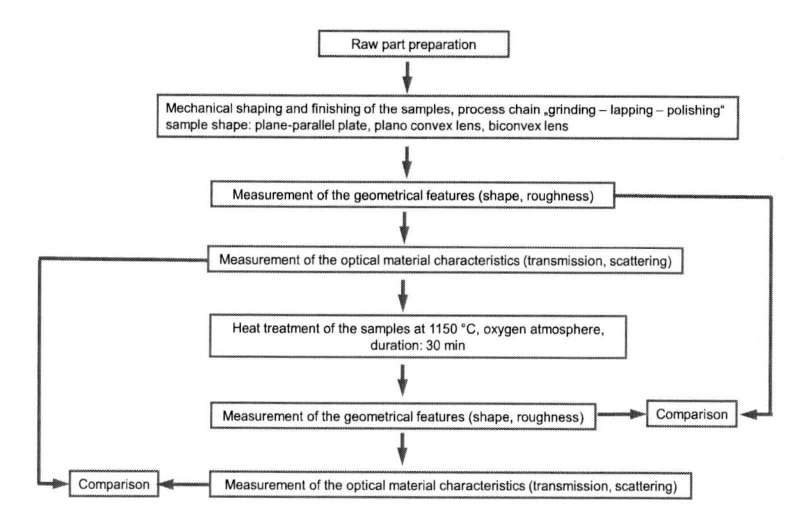

Figure 10. Flow chart of the sample preparation and analysis.

Some samples were prepared with optical function surfaces and exposed to high thermal load. The radii of curvature, the shape accuracy, the roughness and surface structure of the optical surfaces have been used for the characterization of the geometrical properties.

The transmission and scattering were selected as important optical material characteristics.

The Flow chart of the sample preparation and analysis are explained in Figure 10.

The samples were manufactured by uniaxial pressing, pressureless sintering on air, T = 1500°C and redensification by Hot Isostatic Pressing (HIP), p > 100 MPa, T = 1500°C.

After machining of the raw parts the geometrical features and optical material characteristics were measured before and after a thermal treatment at 1150°C, oxygen atmosphere with a duration of 30 min and then compared.

Table 2 contains some informations about the sample material and the raw part manufacturing process.

Table 2. Sample material/sample manufacturing

Materials characterization:

- Powder: Baikowski High Purity Spinel S30CR
- Chemical composition: $MgAl_2O_4$ 99,9 %
- Crystal composition (XRD): 98,3% Mg-Al-Spinell, 1,4% Korund (Al_2O_3), 0,4% Periklas (MgO)

Samples manufacturing:

- Moulding: uniaxial pressing
- Sintering: pressureless on air, T = 1500 °C
- Redensification: Hot Isostatic Pressing (HIP), p > 100 MPa, T = 1500 °C

Sample	Press granulate	Sinter density	Density after HIP	Thermal treatment before machining
A	< 100 µm	3,48 g/cm³	3,60 g/cm³	unannealed
B	< 100 µm	3,48 g/cm³	3,60 g/cm³	2h annealed
C	< 100 µm	3,48 g/cm³	3,60 g/cm³	2h annealed
D	< 315 µm	3,55 g/cm³	3,60 g/cm³	2h annealed

The Sample geometry was according a plano convex lens with a radius of curvature of 60 mm (Figure 11).

In order ensure the angular positions of the surface measurements before and after the thermal treatment a flat was ground on the circumference of the samples.

After CNC-controlled grinding the samples were machined by lapping and polishing (Figure 12).

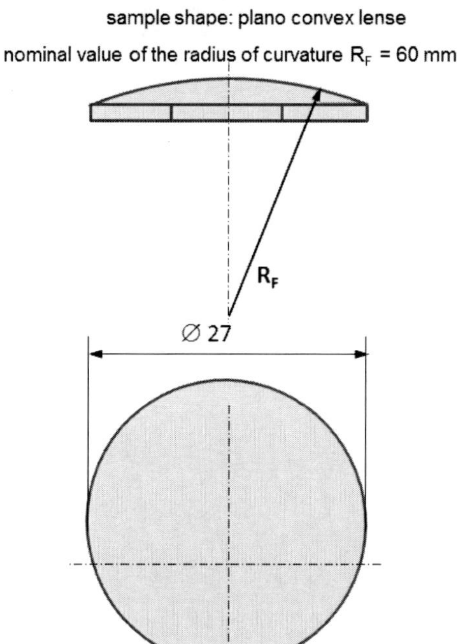

Figure 11. Sample geometry.

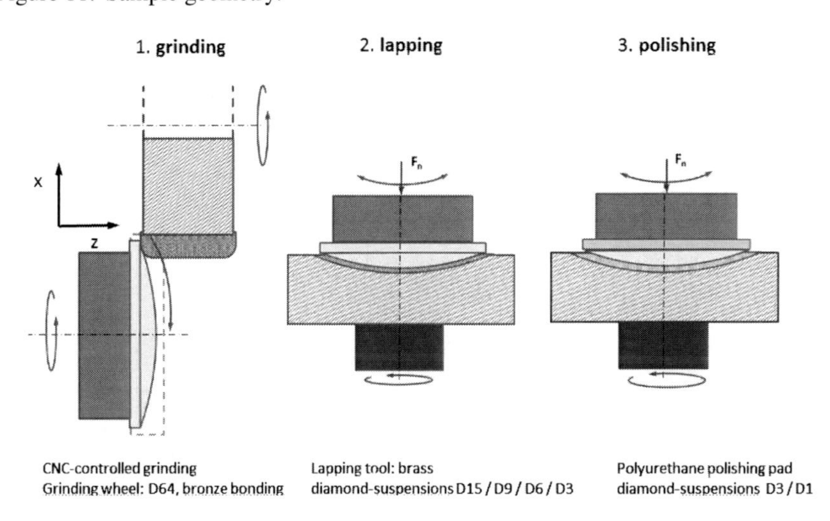

Figure 12. Mechanical shaping and finishing of the samples: process chain grinding – lapping – polishing.

At first the comparison of the radii of curvature before and after thermal treatment was done (Table 3).

Table 3. Comparison of the radii of curvature before and after thermal treatment

	Radii of curvature / mm						
sample identification	before heat treatment		average value	after heat treatment		average value	average difference of radii
	measuring site 1	measuring site 2		measuring site 1	measuring site 2		
A	59,3112	59,3110	59,3111	59,3200	59,3162	59,3181	0,0070
B	61,1856	61,2537	61,2197	61,1837	61,2535	61,2186	-0,0010
C	61,4297	61,4308	61,4303	61,4254	61,4272	61,4263	-0,0040
D	59,4914	59,4936	59,4925	59,4879	59,4914	59,4897	-0,0029

Measurement by tactile profilometry.

The measurements were performed as 2D-measurements with a tactile profilometer “FormTalysurf” (AMETEK Taylor Hobson) on two sites of every lens. The measurement results show a very low difference between the radii, which are probably practically negligible for the most possible applications.

The shape accuracy of the spherical optical surfaces was evaluated on the basis of both the tactile 2D-measurements with the profilometer and optical 3D-measurements with a digital interferometer (GPI XP/Zygo).

Figure 13 shows the comparison of the shape accuracy before and after heat treatment, The difference of the profiles is in the range of the measurement uncertainty.

Similar results were obtained by interferometric shape measurement (Figure 14).

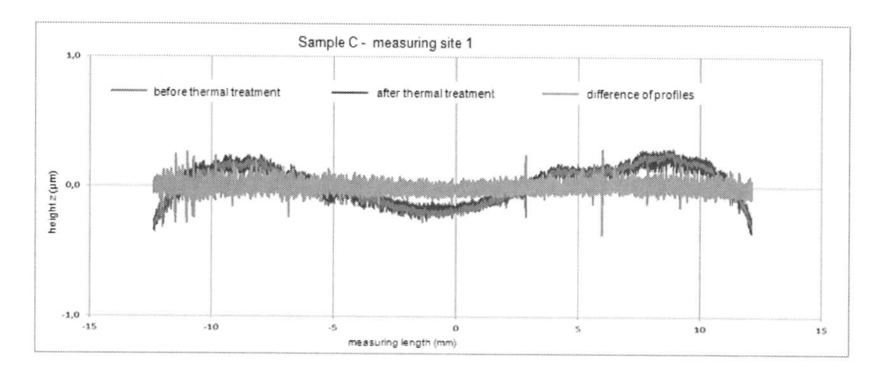

Figure 13. Comparison of the shape accuracy before and after heat treatment, measurement by tactile profilometry, elimination of the least square circle, presentation of the primary profile.

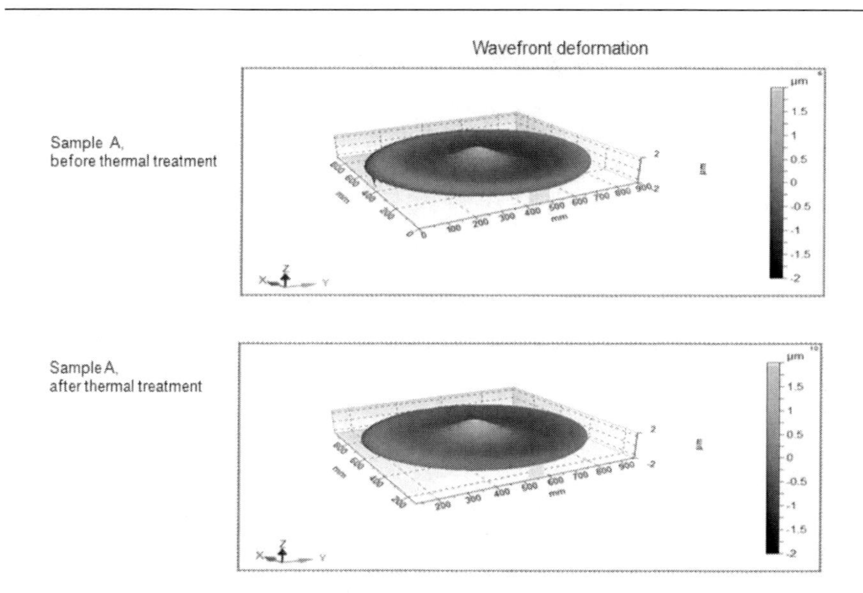

Figure 14. Interferometric shape measurement, elimination of the best fit sphere.

The Comparison of surface roughness before and after thermal treatment is presented in Figure 15. The surface measurements were performed with a white light interference microscope Talysurf CCI (AMETEK Taylor Hobson). As a result there was an increasing roughness measured.

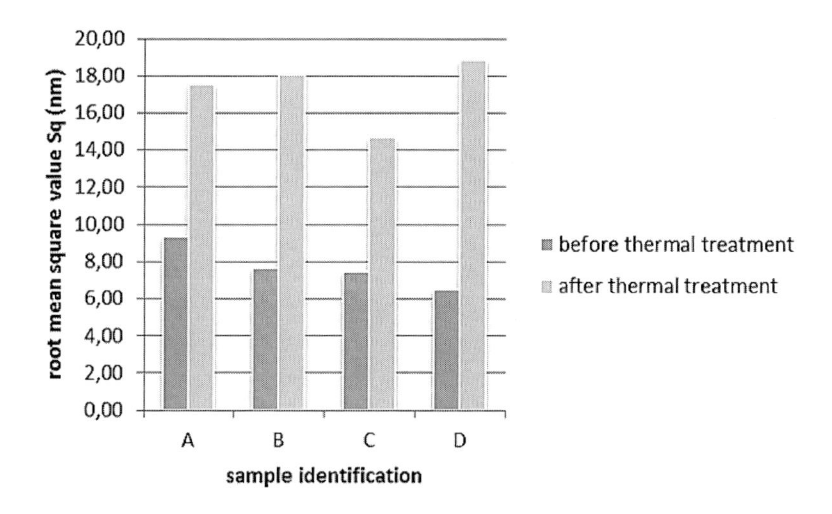

Figure 15. Comparison of surface roughness before and after thermal treatment, measurement by white light interference microscope.

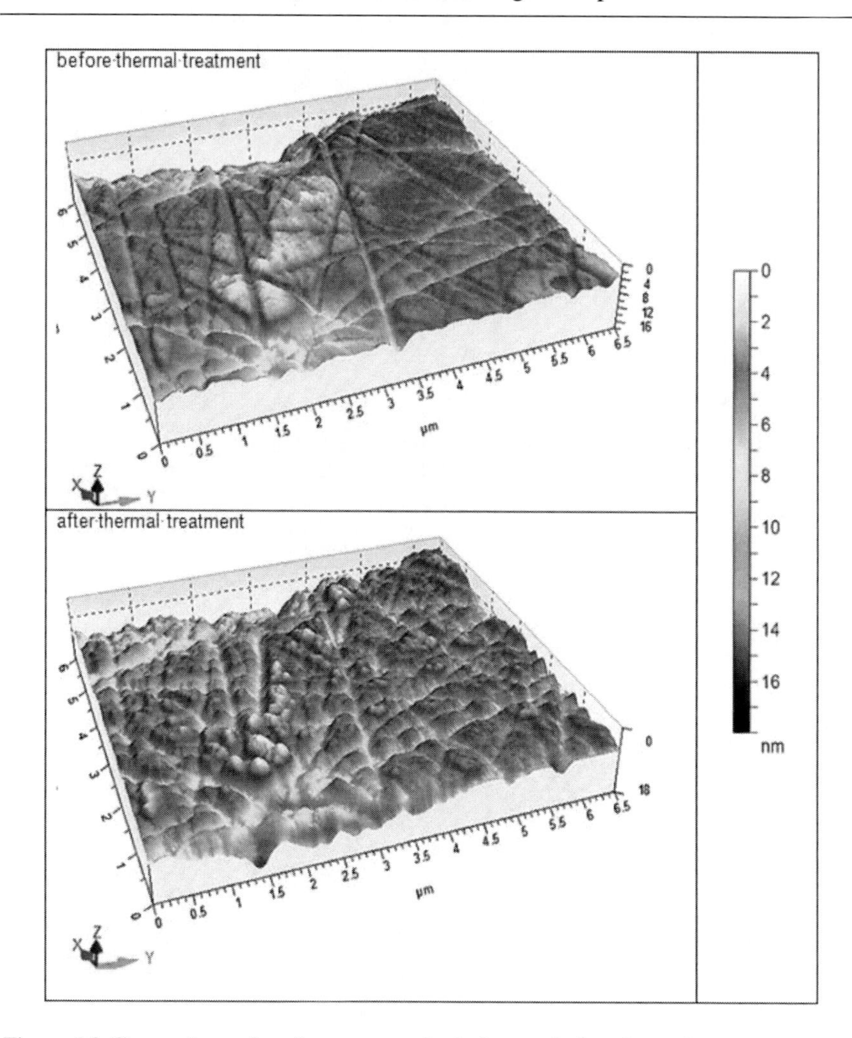

Figure 16. Comparison of surface topography before and after thermal treatment, measurement by Atomic Force Microscope.

On the basis of surface measurements with an Atomic Force Microscope "Dimension Icon" Bruker AXS S.A.S. with correspondingly high lateral and vertical resolution it becomes apparent which changes of the surface structure lead to increased roughness parameters after thermal treatment.

On the one hand pronounced fissuring has to be stated on the plateaus of the surface topography after thermal treatment (Figure 16).

A plan view on the surface structure at the same place shows furthermore sculptured grain boundaries as a result of the thermal treatment (Figure 17).

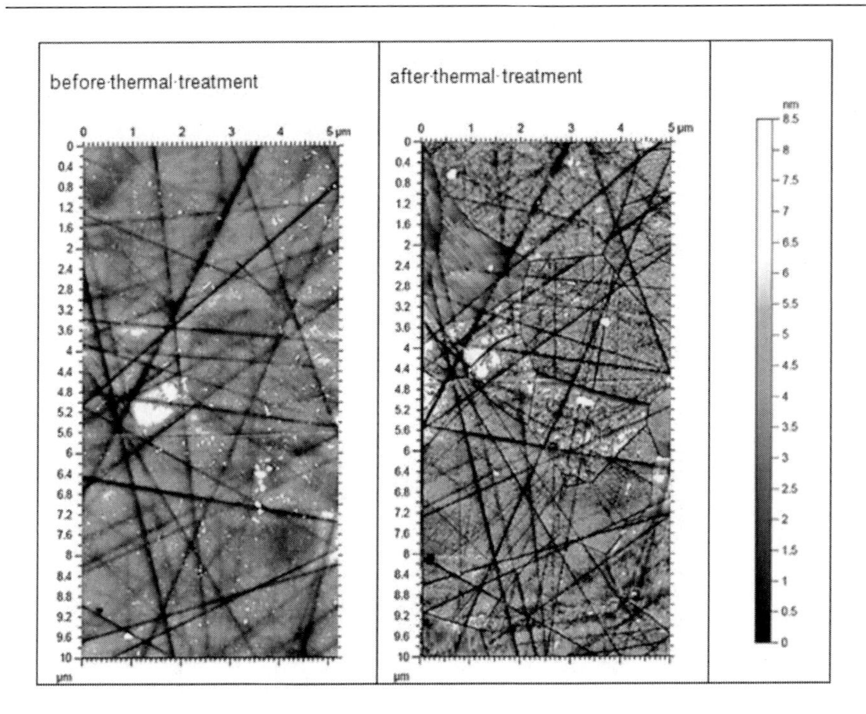

Figure 17. Comparison of the surface topography before and after thermal treatment, measurement by Atomic Force Microscope.

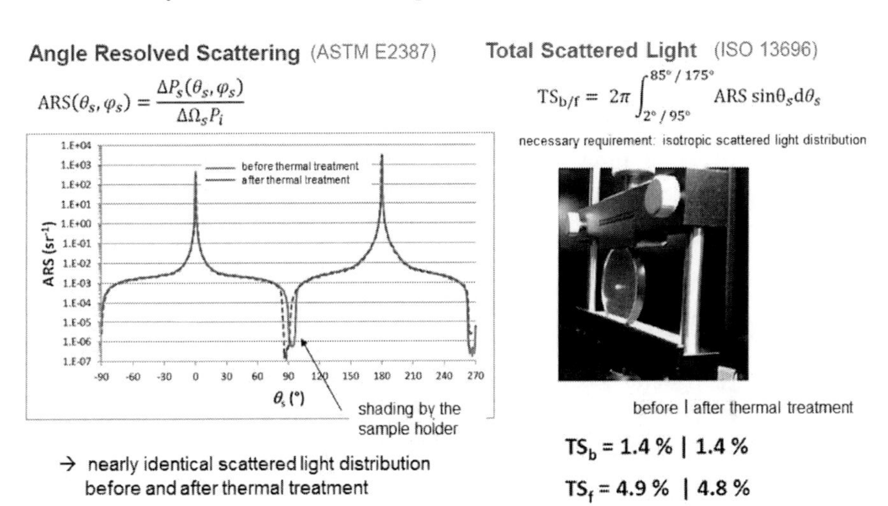

Figure 18. Comparison of light scattering before and after thermal treatment, measurement by ALBATROSS TT.

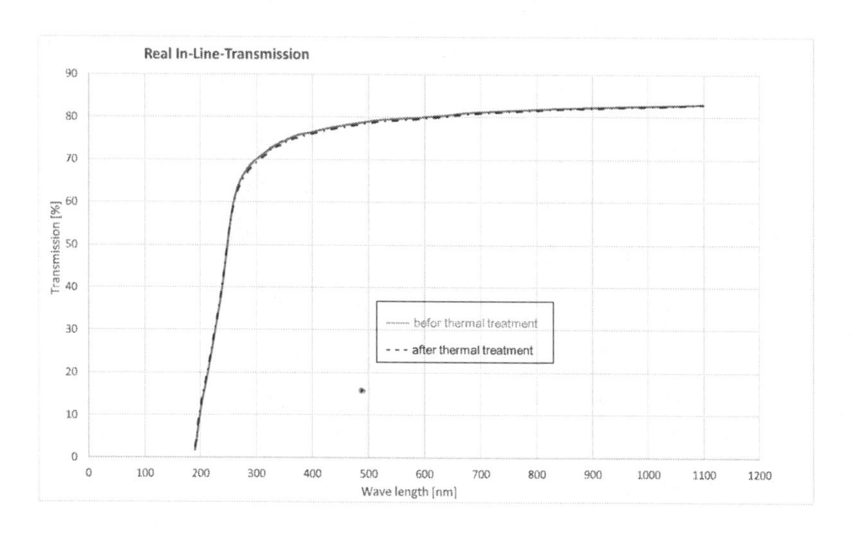

Figure 19. Comparison of real in-line transmission before and after thermal treatment, measurement by SPECORD 200 (Analytik Jena AG).

The light scattering and the transmission were measured as optical material characteristics before and after thermal treatment [11].

The comparison of the light scattering before and after thermal treatment shows a nearly identical scattered light distribution (Figure 18).

The comparison of the real in-line transmission before and after thermal treatment shows also nearly identical curve characteristics (Figure 19).

These curves are the result of multiple measurements before and after thermal treatment at this probe (plane-parallel surfaces/sample thickness 3 mm) [12].

CONCLUSION

Geometrical features and optical characteristics of polycrystalline spinel samples with spherical functional surfaces were evaluated.

Apart from a minor increase of the roughness of the samples the measurements revealed practically unchanged shape, transmission and scattering after 30 min at 1150°C in an oxygen atmosphere. Compared to monocrystalline materials transparent advanced ceramics can offer some substantially advantages because of the isotropy of their material characteristics.

Because of their optical material characteristics in combination with mechanical strength, thermal and chemical resistance transparent high performance ceramics represents an alternative solution or complement for conventional optical materials.

For special components with requirements concerning combinations of materials characteristics potential applications may be resulted.

These results can be seen as basic informations concerning application conditions of transparent advanced ceramics.

REFERENCES

[1] Wang, S. F. et al. Transparent ceramics: Processing, materials and applications. *Progress in Solid State Chemistry*, 41 (2013) 20 – 54.

[2] Sanghera, J. et al. Ceramic laser materials: Past and present. *Optical Materials*, 35 (2013)12, 693-699.

[3] Ramisetty, M.; Goldman, L.; Nag, N.; Balasubramanian, S. and Sastri, S. SURMET CORP.: Transparent Ceramics Enable Large, Durable, Multifunctional Optics. *Photonica Spectra*, June 2014, 58 – 63.

[4] Reichel, U.; Ludwig, H. Transparent $MgAl_2O_4$ Ceramics by automated uniaxial Pressing. Fraunhofer IKTS Annual Report 2011/12 (2012) 36-37.

[5] Krell, A.; Waetzig, K.; Klimke, J. Influence of the structure of MgO*n*Al_2O_3 spinel lattics on transparent ceramics processing and properties. *J. Europ. Ceram. Soc.*, 32 (2012) 2887-2898.

[6] Reichel, U.; Kemnitz, E. Optoceramic materials with nano-scaled metal-fluorine antireflective coatings. Fraunhofer IKTS Annual Report 2013/14 (2014) 98-99.

[7] Klimke, J.; Krell, A. Gelcasting – a shaping method for particularly defect-free ceramic parts. Fraunhofer IKTS Annual Report 2013/14 (2014) 36-37.

[8] Klimke, J.; Krell, A. Optical Ceramics with specifically adjusted Spectral Transmission. Fraunhofer IKTS Annual Report 2014/15 (2015) 90–91.

[9] Reichel, U.; Müller, F. A.; Notni, G.; Duparré, A.; Claussen, I.; Herold, V. Comparative studies of monocrystalline and polycrystalline transparent hard materials for optical applications. 2nd European Seminar

on Precision Optics Manufacturing, April 14th-15th 2015, Teisnach, Germany.
[10] Reichel, U.; Schubert, R. Innovative concepts of ceramic materials and processes for new applications in optical and electronic industry. Keram. Z. 67 (2015) [5–6] 292-295.
[11] von Finck, A.; Hauptvogel, M.; Duparré, A. Instrument for close-to-process light scatter measurement of thin film coatings and substrates. *Appl. Opt.*, 50 (2011) C321-C328.
[12] Reichel, U.; Notni, G.; Duparré, A.; Müller, F.; König, St.; Herold, V. High temperatures stability of optical relevant performances of transparent $MgAl_2O_4$ Spinel ceramic. Handbook „Technical Ceramic Materials“ (HvB Verlag/150. completion/January 2016).

Biographical Sketch

Uwe Reichel

Affiliation: Fraunhofer Institute for Ceramic Technologies and Systems IKTS, Hermsdorf Branch, Hermsdorf, Germany

Education: Technical University „Bergakademie Freiberg“, Freiberg, Germany: Study of „Materials and Technologies of Silicates“ (Technology of Glass, Ceramic and Construction Materials); finished with certificate of academically engineer (Dipl.-Ing.); also there scientific dissertation in the area of materials science / ceramic materials; finished with certificate as a Doctor Engineer (Dr.-Ing.)

Research and Professional Experience:

Professional Experience: Silicate materials (Glass/Ceramic), Advanced Ceramic, Nano structured materials (Ceramic), Technique for joining parts (ceramic/glass – metal);

Focus of Research: development of materials and technologies for Advanced Ceramics, Nano structured materials and Transparent Ceramics

Professional Appointments: over 20 years executive positions in Technical and R&D management in different companies of Advanced Ceramic; since 2007 scientific officer and project manager at the Hermsdorf Institute for

Technical Ceramic, by now Fraunhofer Institute for Ceramic Technologies and Systems IKTS, Hermsdorf, Germany.

Publications Last Three Years:

Reichel, U.; Kemnitz, E.: Optoceramic materials with nano-scaled metal-fluorine antireflective coatings. Fraunhofer IKTS Annual Report 2013/14, 98-99.

Reichel, U.; Kinski, I.: Optische Funktionsschichten/Antireflex-Beschichtung auf transparenten Keramiken. mo Magazin für Oberflächentechnik, 69 (2015) [1-2] 16-19.

Reichel, U.; Schubert, R.: Innovative concepts of ceramic materials and processes for new applications in optical and electronic industry; Keram. Z. 67 (2015) [5–6] 292-295.

Reichel, U.; Notni, G.; Duparré, A.; Müller, F.; König, St.; Herold, V.: High temperatures stability of optical relevant performances of transparent MgAl2O4 Spinel ceramic.

Handbook „Technical Ceramic Materials“ (HvB Verlag/150. completion/ January 2016).

In: Advances in Materials Science Research ISBN: 978-1-53610-059-4
Editor: Maryann C. Wythers

Chapter 2

$K_{0.5}Na_{0.5}NbO_3$-BASED LEAD-FREE TRANSPARENT CERAMICS

Xiao Wu,[1,2] Faliang Li[1,3] and K. W. Kwok[1,*]
[1]The Hong Kong Polytechnic University, Kowloon, Hong Kong, China
[2]College of Materials Science and Engineering, Fuzhou University, China
[3]WuHan University of Science and Technology, China

ABSTRACT

Hot-press sintering has been used to fabricate lead-free transparent ceramics $(K_{0.5}Na_{0.5})_{1-x}Li_xNb_{1-x}Bi_xO_3$ (KNNLB-x, $0.05 \leq x \leq 0.09$) based on alkali metal oxalates as the raw chemicals. Due to the co-modification of Li^+ and Bi^{3+}, the ceramics possess a dense and fine-grained structure with cubic-like symmetry. Owing to the polar nano-regions, the ceramics have a diffuse phase transition, showing the relaxor-like characteristics. These can reduce the light scattering by the grains, at the grain boundaries and the domain walls, thus improving the optical transmittance of the ceramics. The optical transmittance (T) of the ceramics reaches a high value of ~60% in the near-IR (NIR) region. The transparent ceramics can be considered as a good electro-optic (EO) material because of a large effective linear EO coefficient (r_c) in the range of 120-200 pm/V.

[*] Corresponding Author address: The Hong Kong Polytechnic University, Kowloon, Hong Kong, China. Email: apkwkwok@polyu.edu.hk.

After improving the sintering technique and selecting alkali metal carbonate as the raw chemicals, KNNLB-x transparent ceramics have been successfully prepared by pressureless sintering. At x = 0.04-0.08, the ceramics exhibit high transmittances (T = 60-70%) in the NIR region, and strong linear EO response (r_c = 120-200 pm/V) comparable to the hot-press sintered ceramics. The good optical transparency should be attributed to the cubic-like crystal structure as well as the relaxor-like characteristics, which have been attested through the temperature dependences of dielectric constants and polarization-electric field hysteresis loops. Together with the relatively large piezoelectric coefficient (50-90 pC/N), high dielectric constant (~1400) and low dielectric loss (< 0.03), the pressureless sintered KNNLB-x transparent ceramics should have potential in the applications of multifunctional devices.

Keywords: transparent ceramics, KNN, electro-optic, photoluminescence

INTRODUCTION

Hot-Press Sintered KNNLB-*x* Transparent Ceramics

Electro-optic (EO) materials usually possess good optical transparency and strong EO effect, exhibiting great potential in the optoelectronic applications, such as optical filters, switches, modulators, deflectors and attenuators [1]. $LiNbO_3$ single crystal is considered as the industry benchmark of EO materials used in the telecoms market. However, it has some drawbacks limiting the widespread applications, containing low EO effect (r_c = 19.9 pm/V), high temperature dependency, small available size and high fabrication cost [2]. Compared to the $LiNbO_3$ single crystal, EO ceramics usually have many advantages, such as better EO response, higher ruggedness, easier fabrication, lower cost and larger available size [3]. In 1971, Haertling and Land reported the first transparent EO ceramics lanthanum-modified lead zirconate titanate [$(Pb_{1-x}La_x)(Zr_yTi_{1-y})_{1-x/4}O_3$, or PLZT] [1], in which $Pb_{0.91}La_{0.09}(Zr_{0.65}Ti_{0.35})_{0.9775}O_3$ is a relaxor ferroelectric with a cubic-like structure, exhibiting a large quadratic EO effect. Recently, the other two lead-based EO ceramics, $(1-x)Pb(Mg_{1/3}Nb_{2/3})O_3-xPbTiO_3$ ($0.10 < x < 0.35$) (PMN-PT) and $(1-x)Pb(Zn_{1/3}Nb_{2/3})O_3-xPbTiO_3$ ($0.15 < x < 0.35$) (PZN-PT), have been investigated and applied in various optoelectronic devices [2]. Nevertheless, these EO ceramics contain a large amount of lead (> 65 wt%), which show detrimental effect on both human bodies and environment. From the perspective of environmental protection, it is urgently needed to explore alternative lead-free ceramics.

Potassium sodium niobate $K_{0.5}Na_{0.5}NbO_3$ (KNN) is one of the most promising lead-free piezoelectric ceramics because of their high Curie temperature (T_c = 420°C), high planar coupling (k_p ~ 45%), moderate dielectric constant (ε_r = 470), and large remnant polarization (P_r = 33 μC/cm^2) [4]. Owing to the high volatility of alkaline metal elements (K and Na), dense KNN ceramics with good piezoelectric properties are difficult to be fabricated. Different ions, such as Li^+, Ta^{5+}, Bi^{3+} and Sb^{5+}, were commonly doped in the KNN host to improve the sinterability and piezoelectric properties [5, 6]. In addition, the co-modification of Li^+ and Bi^{3+} can also influence the optical performance of material.

The $(K_{0.5}Na_{0.5})_{1-x}Li_xNb_{1-x}Bi_xO_3$ (KNNLB-x, $0 \leq x \leq 0.09$) ceramics were fabricated by the hot-press sintering method [7], which is a promising technique and has been used to prepare dense and high performance transparent ceramics [8]. Alkali metal oxalates ($K_2C_2O_4$ and $Na_2C_2O_4$) and lithium acetate ($CH_3COOLi \cdot H_2O$) were utilized and dissolved in de-ionized water. Then oxide-based chemicals (Nb_2O_5 and Bi_2O_3) were added and mixed thoroughly. After that, the mixture was dried and calcined at 850°C, and then ball milled in ethanol for 24 h. The dried powders were mixed with a polyvinyl alcohol (PVA) binder solution and pressed into disks. Subsequently, the disk samples were sintered at 1060°C for 4 h under a uniaxial pressure of 10 MPa. Finally, the sintered ceramics with a diameter of 10 mm were polished to obtain smooth surfaces.

The optical transmittances (T) measured in the range of 300-900 nm of the KNNLB-x ceramics are shown in Figure 1. And a photograph of the typical ceramics is given in the inset of Figure 1. The ceramics with $x \leq 0.02$ are almost opaque, while those with $x \geq 0.05$ are optically transparent. The observed T increases considerably with increasing x from 0.02 to 0.05, and then become almost unchanged with further increasing x to 0.09. For each ceramic with $x \geq 0.04$, the observed T increases rapidly from zero as the wavelength increases from ~390 nm, and becomes almost saturated in the near-IR region (~900 nm). Among the transparent ceramics, the KNNLB-0.05 ceramic possesses the highest transmittance (T ~60% at ~900 nm). As the transmittance of material is affected by the polishing technology, it can be anticipated that the transmittance in the visible region will be further improved through optical-grade polishing. The reflection loss mainly occurs at the air-ceramic interface and can be estimated by:

$$R = (n - 1)^2 / (n + 1)^2 \qquad (1)$$

where R is the reflectance and n is the refractive index (2.257 for the KNNLB-x transparent ceramics). If proper antireflection coating is used to eliminate the reflection loss at the air-ceramic interfaces of the transparent ceramics, the transmittance can be increased to a higher value (~90%).

Due to the interband transition, the observed T for the KNNLB-x transparent ceramics decreases to zero at wavelengths shorter than ~390 nm. Based on the Tauc equation [9], the optical band gap energy (E_g) can be estimated from the absorption spectra. For direct transition, the relationship between E_g and the absorption coefficient (α) is given as:

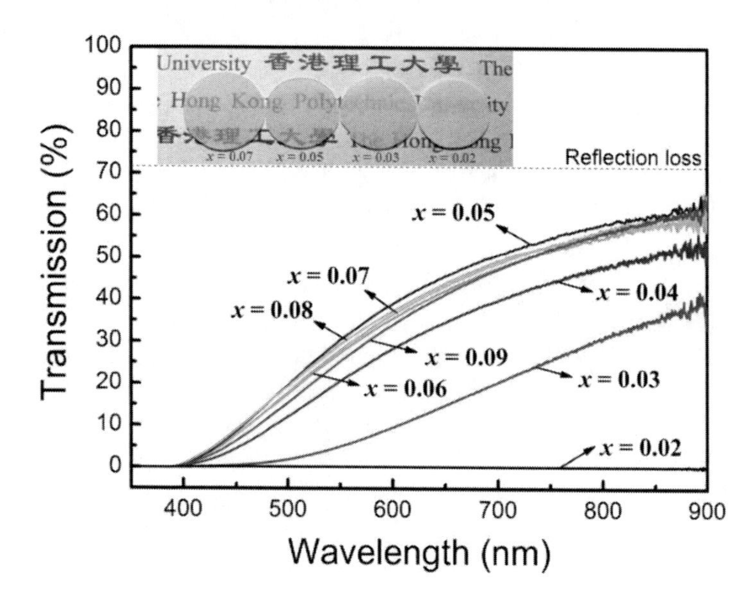

Figure 1. Optical transmittance spectra of the KNNLB-x ($0.02 \leq x \leq 0.09$) ceramics. The inset is photograph of the typical ceramics (x = 0.02, 0.03, 0.05 and 0.07). The thickness of ceramics are all 0.5 mm.

$$(\alpha h\nu)^2 = A(h\nu - E_g) \qquad (2)$$

where ν is the photon frequency, h is Planck's constant and A is a constant. Then α can be calculated from T by [10]:

$$\alpha = \frac{1}{t}\ln(\frac{1}{T}) \tag{3}$$

where t is the thickness of the sample (i.e., 0.5 mm for the KNNLB-x transparent ceramics). Accordingly, E_g can be obtained by plotting $(\alpha h\nu)^2$ versus $h\nu$ and estimating the linear portion of the curve to zero, as shown in Figure 2. The E_g values for these ceramics are listed in Table 1.

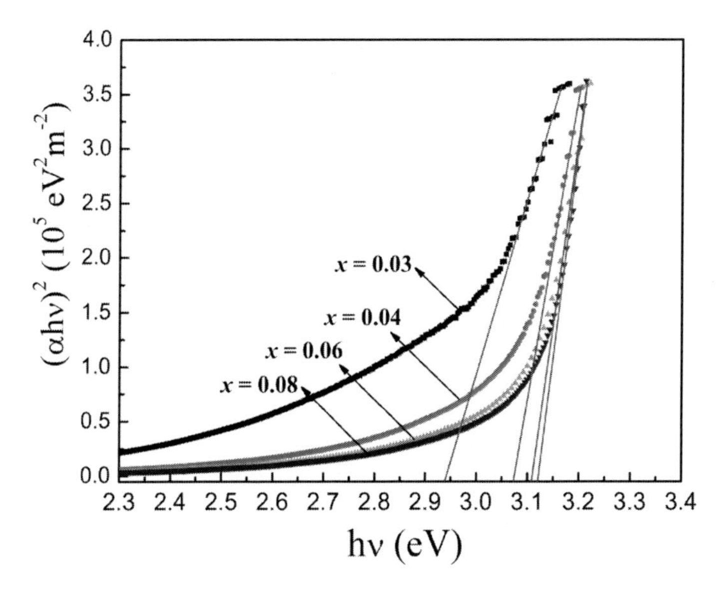

Figure 2. Plots of $(\alpha h\nu)^2$ versus $h\nu$ for the typical KNNLB-x ceramics, from which E_g is obtained by extrapolating the linear portion of the curve to zero.

Table 1. Optical band gap energy (E_g) for the KNNLB-x ceramics

x	0.03	0.04	0.05	0.06	0.07	0.08	0.09
E_g (eV)	2.93	3.08	3.09	3.11	3.12	3.12	3.12

The observed E_g increases slightly from 3.08 to 3.12 eV with increasing x from 0.04 to 0.09, suggesting that the effects of modifiers (Li^+ and Bi^{3+}) on the band gap of the ceramics are insignificant. Probably due to the low transmittance for the KNNLB-0.03 ceramic, the linear portion of the curve is short and the observed E_g value (2.93 eV) may not be accurate enough (Figure 2). Nevertheless, the E_g values of the ceramics with $x \geq 0.04$ are close to that of KNN nanorods (3.09 eV) [11]. The band gap of KNN corresponds to the transition from the top of the valence bands occupied by O_{2p} electron states to the bottom of the conduction bands dominated by the empty Nb_{4d} electron states.

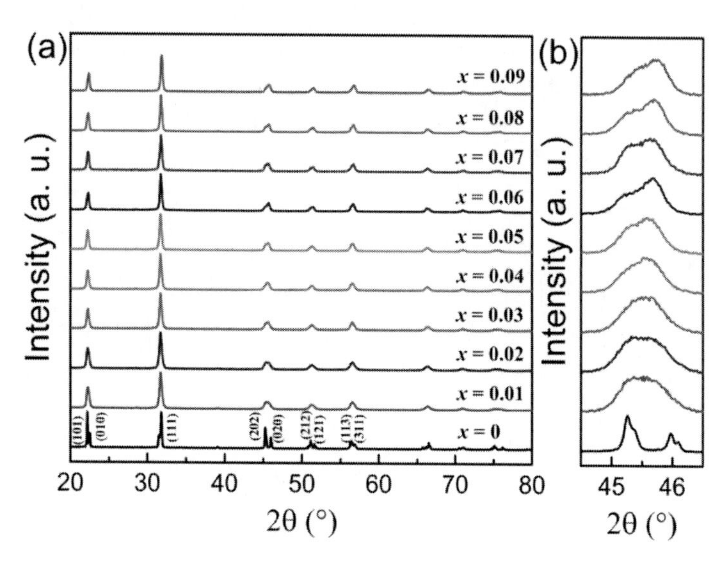

Figure 3. (a) XRD patterns of the KNNLB-x ceramics, (b) enlarged XRD patterns near 45.5°.

The X-ray diffraction (XRD) patterns of the KNNLB-x ceramics are shown in Figure 3. All the ceramics possess a single phase perovskite structure (Figure 3a), suggesting that Li^+ and Bi^{3+} have diffused into the KNN lattice, i.e., Li^+ have entered the A-sites ($(K_{0.5}Na_{0.5})^+$ sites) and Bi^{3+} have occupied the B-sites (Nb^{5+} sites). For the ceramics with $x > 0$, the two diffraction peaks (202) and (020) merge together (Figure 3b), meaning that the lattice constants a, b and c become close to each other. And the ceramics may transform into another phase, e.g., pseudocubic or cubic-like. Similar results have been observed for other KNN-based ceramics, such as Ta-modified KNN [12] and KNN-$BaTiO_3$ [13]. The Rietveld method was used to refine the lattice constants of the ceramics according to the diffraction peaks (Figure 3a). In the MAUD software, different crystal models, including tetragonal *P*4*mm*, orthorhombic *Amm*2 and monoclinic *Pm*, have been used for the refinements. Probably due to the cubic-like crystal structure, similar results with reasonably good fits have been obtained for all the models. The calculated lattice parameters (a, b, c and β) together with the reliability factor (R_{wp}) and goodness-of-fit indicator (S) are given in Table 2. The low R_{wp} (<15%) and S

(<2) values denote a good fit between the observed and calculated patterns. With increasing x, c remains almost unchanged, but a decreases and b increases, and both become saturated at high x. The difference between c and b decreases as x increases. With $x \geq 0.03$, the difference is very small (~0.04 Å), suggesting that the ceramics possess a cubic-like crystal structure with minimal optical anisotropy. Therefore, the light scattering is reduced and the optical transparency is improved (Figure 1).

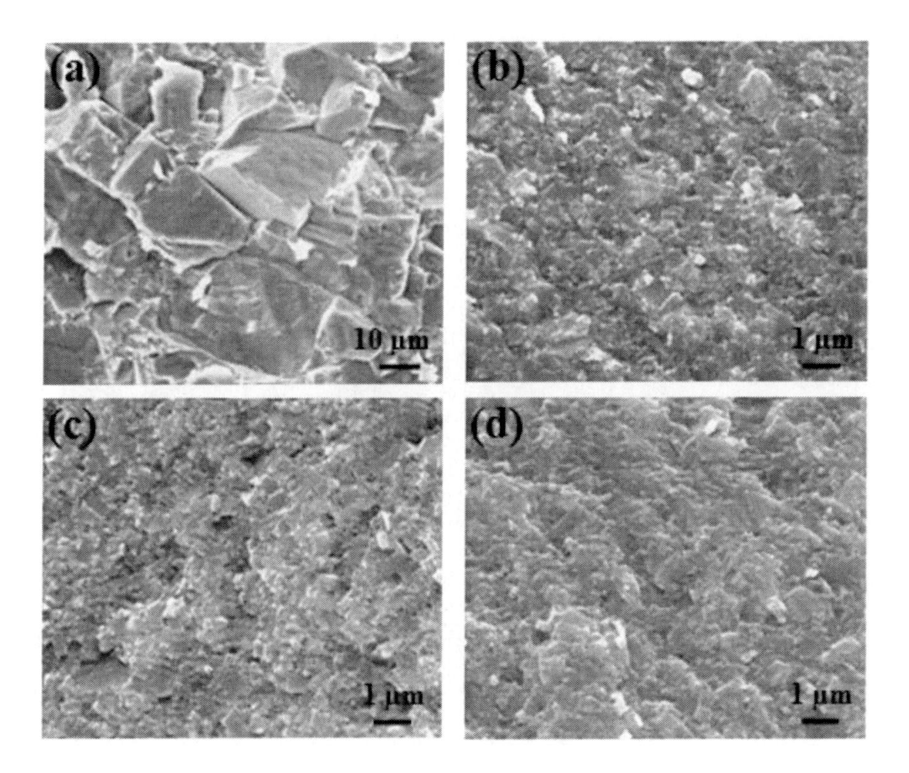

Figure 4. SEM micrographs of the fracture surfaces of the KNNLB-x ceramics, (a) x = 0; (b) x = 0.03; (c) x = 0.05; (d) x = 0.07.

Table 2. Lattice parameters *a*, *b*, *c* and *β* for the KNNLB-*x* ceramics

x	0	0.01	0.02	0.03	0.04	0.05	0.06	0.07	0.08	0.09
a (Å)	3.999	3.981	3.981	3.978	3.974	3.975	3.970	3.973	3.968	3.970
b (Å)	3.942	3.956	3.955	3.960	3.960	3.958	3.960	3.958	3.955	3.957
c (Å)	4.002	4.001	4.002	3.999	3.998	3.998	3.999	3.996	3.996	3.994
β (°)	90.03	90.29	90.31	90.31	90.34	90.33	90.43	90.34	90.40	90.35

Figure 4 shows, as examples, the scanning electronic microscopy (SEM) micrographs of the fracture surfaces of the KNNLB-x ceramics with $x = 0$, 0.03, 0.05 and 0.07. Similar results have been observed for the other ceramics. All the ceramics possess dense structures with the cuboid morphologies. Table 3 lists the densities, relative densities and average grain sizes of the KNNLB-x ceramics. As x increases, the densification is improved and the average grain size decreases. Compared to the pure KNN ceramics with the large grain size (> 20 μm), the grain sizes of the KNNLB-x ($x > 0$) ceramics are substantially smaller (< 1 μm) because of the co-modification of Li^{+} and Bi^{3+}. Similar results have been reported that Bi^{3+}-doping can significantly suppress the grain growth in perovskite ceramics [14, 15].

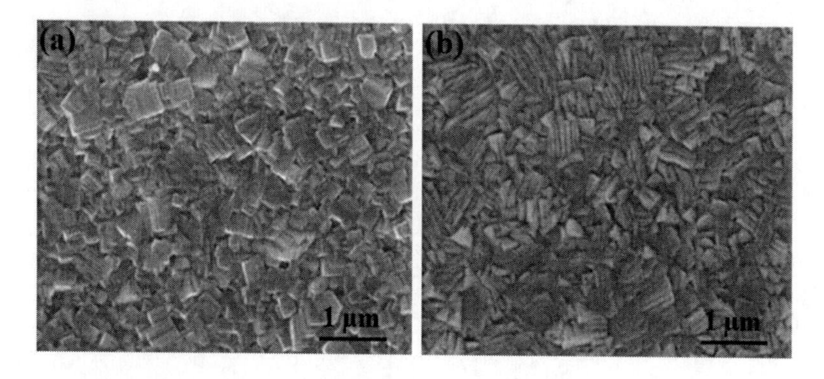

Figure 5. SEM micrographs of the thermally etched surfaces of the KNNLB-x ceramics, (a) $x = 0.04$ and (b) $x = 0.06$.

Table 3. Density and average grain size of the KNNLB-*x* ceramics

x	0.01	0.02	0.03	0.04	0.05	0.06	0.07	0.08	0.09
Density (g/cm^3)	4.30	4.33	4.38	4.43	4.57	4.61	4.62	4.66	4.68
Relative density (%)	94.3	94.5	95.0	95.5	98.1	98.3	97.9	98.1	98.1
Grain size (μm)	0.76	0.72	0.70	0.69	0.67	0.67	0.66	0.66	0.62

In the KNNLB-x ceramics, Bi^{3+} may concentrate near the grain boundaries and decrease their mobility during the densification. As a result, the mass transportation is weakened, inhibiting the grain growth. For the ceramics with $x \geq 0.04$, the grains are even smaller, generally < 0.7 μm (Figure 5). The transparency of ceramics is generally influenced by three main factors, grain size, grain-boundary phase and pores [16]. For the hot-press sintered lead-based EO transparent ceramics (such as PLZT), the microstructure is dense and grains are large (> 5 μm), thus the grain-boundary becomes thin and the area of the grain-boundary becomes small. Hence, the optical scattering is reduced and the optical transmittance is high.[4] Unlike the lead-based EO ceramics, the good optical transparency of the KNNLB-x ceramics should be partly derived from the reduced scattering of visible (Vis) and infrared red (IR) lights by the fine grains [17].

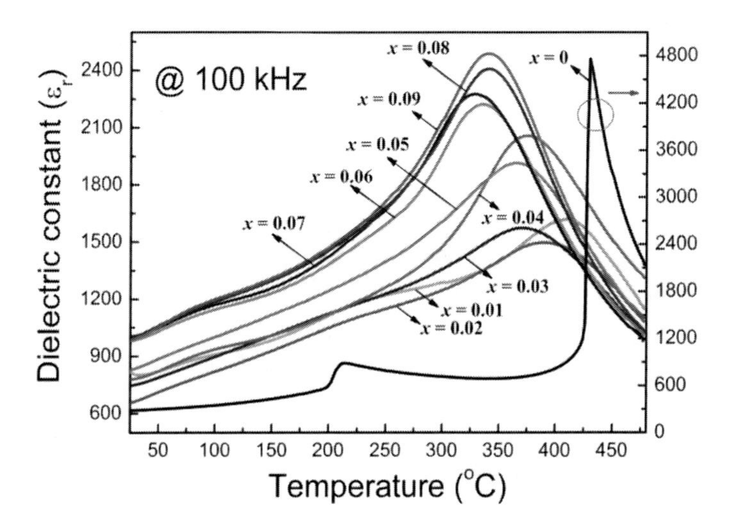

Figure 6. Temperature dependences of ε_r for the KNNLB-x ceramics.

The temperature dependences of dielectric constant (ε_r) at the frequency of 100 kHz for KNNLB-x ceramics are shown in Figure 6. The pure KNN ($x = 0$) ceramic exhibits two transition peaks: the ferroelectric orthorhombic to ferroelectric tetragonal phase transition (at ~226°C) and the ferroelectric tetragonal to paraelectric cubic phase transition (at ~425°C). With $x > 0$, all the ceramics exhibit a broadened cubic-tetragonal transition peak at lower temperature, suggesting that diffuse phase transitions are induced in these ceramics. As x increases, the observed T_m (the temperature of maximum ε_r) decreases from 425°C to 330°C (from $x = 0$ to 0.06), and then remains almost unchanged (from $x = 0.06$ to 0.09). In addition, the orthorhombic to tetragonal phase transition peak is unobvious at $x \geq 0.02$, which may be arisen from the cubic-like crystal structure.

It is well known that for a normal ferroelectric, the ε_r above the Curie temperature (T_c) should obey the classical Curie-Weiss law:

$$\varepsilon = \frac{C}{T - T_C} \tag{4}$$

where C is the Curie-Weiss constant. Deviation from the Curie-Weiss law can be defined by:

$$\Delta T_m = T_B - T_m \tag{5}$$

where T_B is the temperature from which ε_r starts to follow the law. At $T < T_B$, the paraelectric phase transforms into the ergodic relaxor state and forms the polar nano-regions [18]. Figure 7 shows the plots of $1/\varepsilon_r$ versus T for the typical KNNLB-x ceramics. For the KNNLB-0 ceramic, the observed T_B is the same as T_m, indicating that it is a classical ferroelectric. With increasing x (at $x \geq 0$), the ε_r begins to follow the Curie-Weiss law at higher temperature ($T > T_m$). The observed T_B, T_m and ΔT_m values for the KNNLB-x ceramics are shown in Figure 8. As x increases, the calculated ΔT_m increases, exhibiting a diffuse phase transition with an enhanced diffuseness.

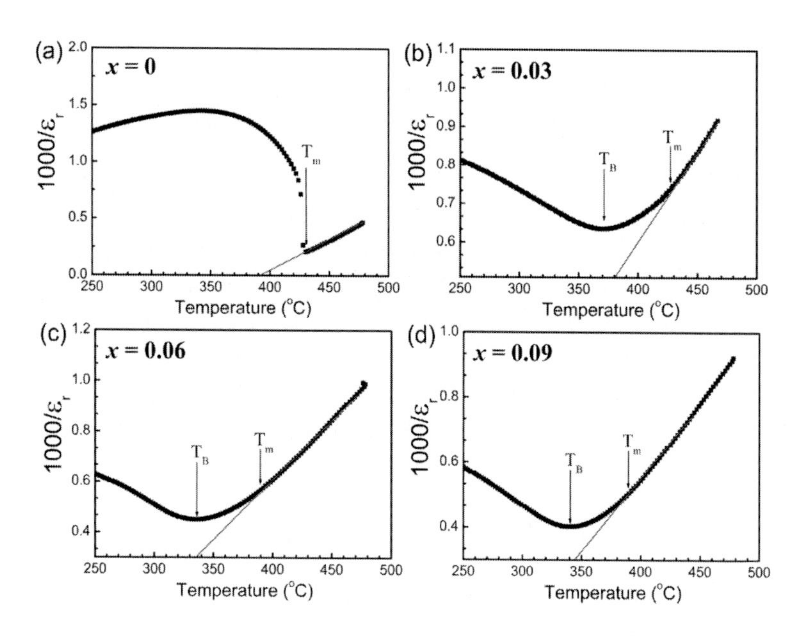

Figure 7. Plots of $1/\varepsilon_r$ versus T for the KNNLB-x ceramics, (a) $x = 0$, (b) $x = 0.03$, (c) $x = 0.06$, (d) $x = 0.09$. The symbols denote the experimental data and the solid lines denote the least-squares fitting lines to the Curie-Weiss law.

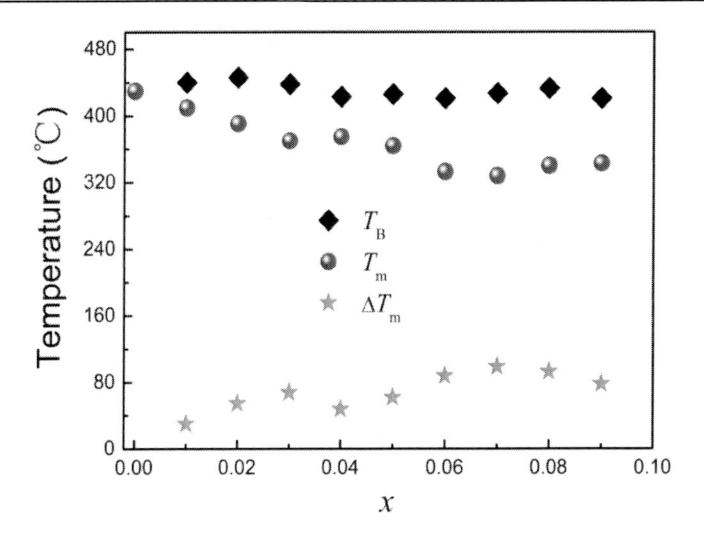

Figure 8. Variations of T_B, T_m and ΔT_m with x for the KNNLB-x ceramics.

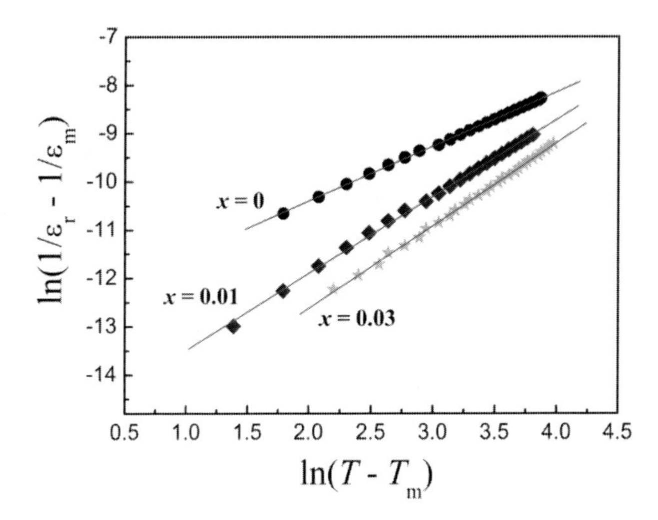

Figure 9. The plots of $\ln(1/\varepsilon_r - 1/\varepsilon_m)$ as a function of $\ln(T - T_m)$ for the KNNLB-x ceramics ($x = 0$, 0.01 and 0.03).

Uchino and Nomura suggested a modified Curie-Weiss law [19] to determine the diffuseness of phase transition in a relaxor ferroelectrics:

$$\frac{1}{\varepsilon_r} = \frac{1}{\varepsilon_m} + \frac{(T - T_m)^\gamma}{C_1} \tag{6}$$

where ε_m is the maximum value of ε_r at T_m, C_1 is the Curie-like constant and γ is the degree of diffuseness ranging from 1 (a normal ferroelectric) to 2 (an ideal relaxor ferroelectric). The plots of $\ln(1/\varepsilon_r - 1/\varepsilon_m)$ as a function of $\ln(T - T_m)$ for the typical KNNLB-*x* ($x = 0, 0.01, 0.03$) ceramics are shown in Figure 9. By fitting the experimental data to the modified Curie-Weiss law, the γ values of are obtained from the slopes of the fitting curves, as listed in Table 4. The observed γ values of the ceramics with $x \geq 0.03$ are in the range of 1.73 - 1.87, suggesting that they are more close to a relaxor ferroelectric.

Table 4. Degree of diffuseness (γ) for the KNNLB-*x* ceramics

x	0	0.01	0.02	0.03	0.04	0.05	0.06	0.07	0.08	0.09
γ	1.04	1.67	1.72	1.87	1.80	1.81	1.81	1.87	1.74	1.73

EO properties of the KNNLB-*x* ceramics were measured with a transverse geometry using a modified Sénarmont method, of which the details can be found elsewhere [20]. The ceramics were polarized under a DC electric field of 5 kV/cm at 150°C for 30 min, and the measurements were carried out one day after the poling at room temperature. The field-induced birefringence $\delta(\Delta n)$ as a function of electric field (E) for the ceramics was measured, the effective linear EO coefficients r_c were calculated according to the general expression for the linear EO effect (Pockels effect) [7]:

$$\delta(\Delta n) = \frac{1}{2} n_e^3 r_c E \tag{7}$$

where n_e is the refractive index for light with the vibrating electric-field parallel to the optical axis (i.e., the polarization axis of the sample). A n_e value of 2.257 (approximated by the value of the KNNLB-0 ceramic) was used in the calculation of r_c and the results are listed in Table 5. The observed r_c increases with increasing x, exhibiting quite large values (120 - 200 pm/V) at $x \geq 0.04$. Although they are not as large as that of $Pb_{0.92}La_{0.08}(Zr_{0.65}Ti_{0.35})_{0.98}O_3$ transparent ceramics (612 pm/V), they are about 6-9 times larger than that of lead-free $LiNbO_3$ single crystals (19.9 pm/V) [21], and comparable to that of $0.67Pb(Mg_{1/3}Nb_{2/3})O_3$-$0.33PbTiO_3$ single crystals (182 pm/V) [20].

As compared with the Bi-modified KNN ceramics, the KNNLB-*x* ceramics exhibit larger r_c and ε_r, for example, $r_c = 160$ pm/V, $\varepsilon_r = 1000$ for the KNNLB-7 ceramic and $r_c = 43$ pm/V, $\varepsilon_r = 716$ for the

$K_{0.45}Na_{0.52}Nb_{0.925}Bi_{0.075}O_3$ ceramic [22, 23]. Probably due to the introduction of Li^+ in the KNN lattices, the $[BO_6]$ octahedra distort greatly [24], resulting in the enhancement of r_c and ε_r in the KNNLB-x ceramics [25]. Based on the theoretical calculations, Shih and Yariv reported that a distorted octahedral structure was a promising mechanism for a strong EO response [24]. By considering the ferroelectric phase as a quadratic phase biased with a spontaneous polarization P_s, r_c can be related to ε_r and P_s via a general equation [25, 26]:

$$r_c = 2\varepsilon_o\varepsilon_r g_c P_s \tag{8}$$

where g_c is the polarization-related quadratic EO coefficient and ε_0 is the vacuum dielectric constant. Eq. 8 is also available for many oxygen-octahedra materials, such as $LiTaO_3$, $0.88Pb(Zn_{1/3}Nb_{2/3})O_3$-$0.12PbTiO_3$, $NaBa_2Nb_5O_{15}$ and $KSr_2Nb_5O_{15}$ [27]. For the KNNLB-3 ceramic, the Li^+ content is not high enough to bring large lattice distortion, leading to a relatively weak EO response with small r_c value.

Table 5. Effective linear EO coefficient r_c for the KNNLB-x ceramics

x	0.03	0.04	0.05	0.06	0.07	0.08	0.09
r_c (pm/V)	44	133	121	155	160	168	198

Pressureless Sintered KNNLB-*x* Transparent Ceramics

After selecting appropriate raw chemicals and optimizing the fabricating process, KNNLB-x transparent ceramics have been successfully prepared by pressureless sintering. The starting raw chemicals are K_2CO_3 (99.9%), Na_2CO_3 (99.5%), Nb_2O_5 (99.99%), Li_2CO_3 (99%) and Bi_2O_3 (99.9%). These powders were weighed according to the stoichiometric ratio of the compositions. The mixing and grinding processes are achieved by ball-milling using zirconia balls for 12 h in ethanol as medium. Then the mixture was baked to remove moisture and calcined at 850°C for 4 h. After that, the calcined powders were ball-milled again for 8 h, baked, and sieved through an 80-mesh screen. Subsequently, they were mixed thoroughly with PVA binder solution and pressed into disk samples under ~300 MPa of pressure. To completely burn out the binder and remove pores, the samples were heated to 800°C at a slower rate of 0.5°C/min. After soaking at 800°C for 2 h, the disk samples were sintered at 1060-1100°C for 4 h in air for densification. All the sintered KNNLB-x ($0.01 \leq x \leq 0.10$) ceramics were polished to a thickness of 0.3 mm.

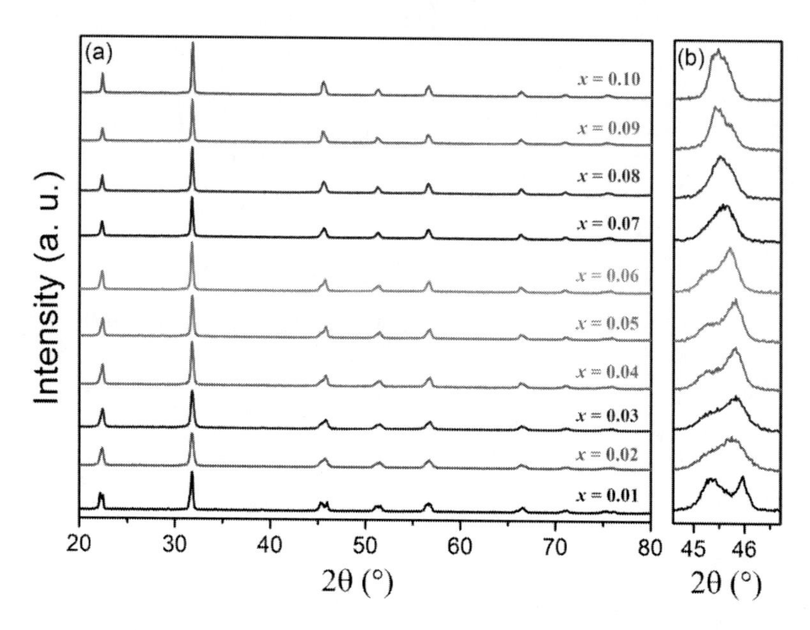

Figure 10. (a) XRD patterns of the pressureless sintered KNNLB-x ceramics, (b) enlarged XRD patterns near 45.5°.

Figure 10 shows the XRD patterns of the KNNLB-x ($0.01 \leq x \leq 0.10$) ceramics. All the ceramics possess a single-phase perovskite structure without any impurity phases (Figure 10a), suggesting that Li^+ have entered the $(K_{0.5}Na_{0.5})^+$ sites and Bi^{3+} ions have diffused into the Nb^{5+} sites according to the designed formula $(K_{0.5}Na_{0.5})_{1-x}Li_xNb_{1-x}Bi_xO_3$. The enlarged (220) and (002) peaks of 2θ around 45.5° of the KNNLB-x ceramics are shown in Figure 10b. As the increase of x, the two peaks gradually merge together to form another phase with similar lattice constants, e.g., pesudocubic or cubic-like, which is similar to that of hot-press sintered KNNLB-x ceramics. However, probably due to the different raw chemicals and fabricating processes between the two preparation methods, the shapes of XRD peaks (especially the enlarged 2θ peaks around 45.5° as shown in Figure 3b and Figure 10b) are a bit distinct, indicating that the crystal structures are slightly different in spite of the same target composition.

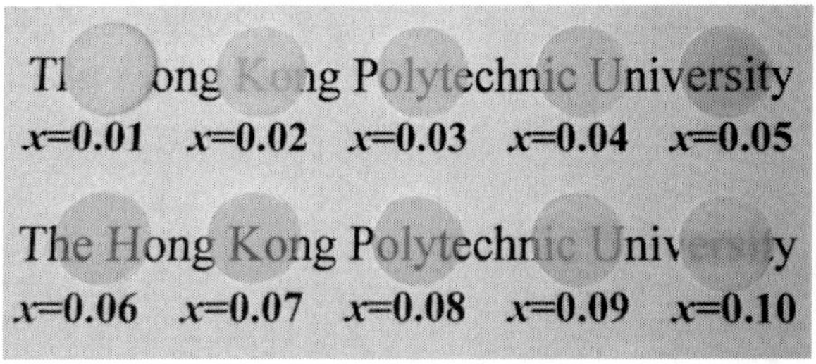

Figure 11. Photograph of the KNNLB-*x* ceramics. The thickness of them are all 0.3 mm.

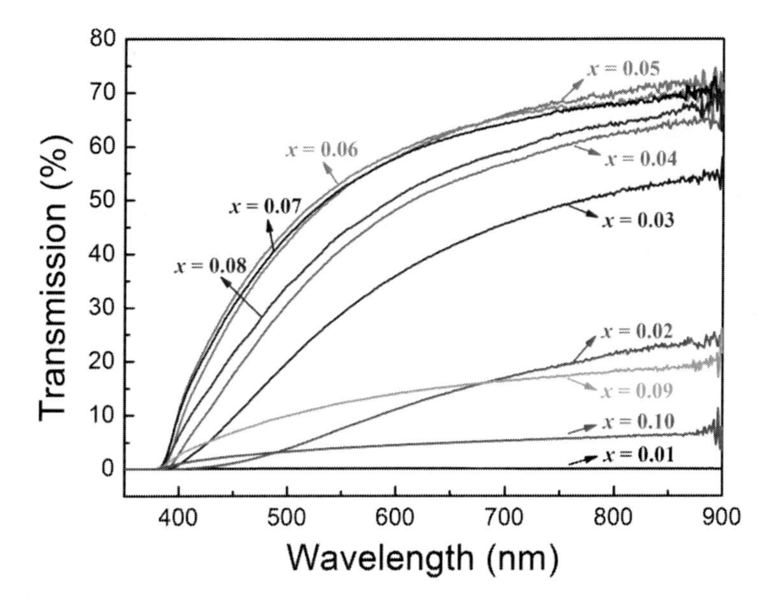

Figure 12. Optical transmittance spectra for the KNNLB-*x* ceramics. The thickness of them are all 0.3 mm.

A photograph of the KNNLB-*x* ceramics is shown in Figure 11. The ceramics are opaque or translucent with $x \leq 0.02$. As x increases from 0.03 to 0.08, the optical transparency improves and the ceramics become optically transparent. When further increasing x to 0.10, the ceramics change back to translucent. Figure 12 shows the optical transmittance spectra of the ceramics. The improvement in transparency with increasing x can also be expressed in the transmittance spectra. It can be seen that the KNNLB-*x* ceramics with x = 0.04-0.08 exhibit the highest transmittance with a high T of 60-70% in the near-IR region and become almost saturated near 900 nm. For each ceramic

($0.03 \leq x \leq 0.08$), due to the interband transition, the observed T decreases rapidly to zero at the wavelength shorter than 390 nm, which is consistent with the hot-press sintered KNNLB-x ceramics.

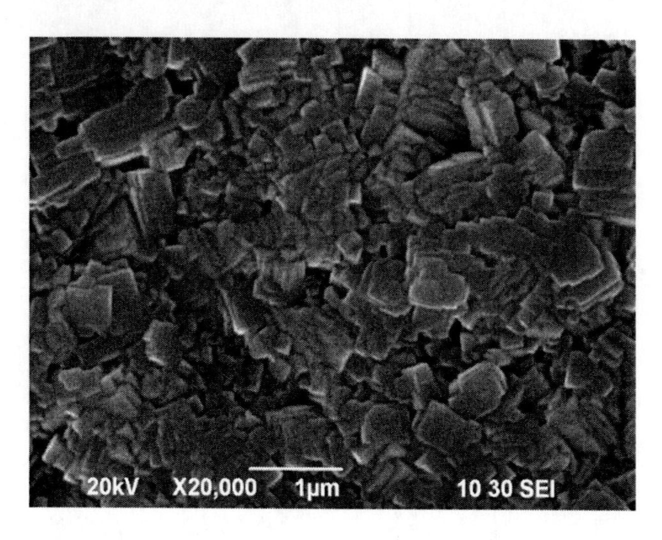

Figure 13. SEM micrograph of the free surface of the KNNLB-0.06 ceramic.

Figure 13 shows, as an example, the SEM micrograph of the KNNLB-0.06 ceramic. Similar results have also been observed for other KNNLB-x ($0.04 \leq x \leq 0.08$) ceramics. The ceramic has a dense and fine-grained structure with cuboid morphology. Probably due to the pressureless sintering, a small number of tiny pores are inevitably formed and distributed randomly in the ceramic. A theoretical study has shown that the optical transmittance of a dense ceramic under 645-nm illumination increases exponentially with decreasing the grain size from 3 to 0.3 μm. And the light scattering at the grain-boundary is insignificant when the grain size becomes smaller than the wavelength of light [17]. Therefore, the good optical transparency of the KNNLB-x ceramics should be partly ascribed to the reduced optical scattering by the fine grains. Due to the tiny size and small amount of pores, the optical scattering by them should be negligible.

The EO properties of the pressureless sintered KNNLB-x ceramics were measured, giving the r_c values in Table 6. The low r_c value of the KNNLB-0.04 may be attributed to the tetragonal crystal structure (Figure 10b). With the more cubic-like structure at $x \geq 0.05$, the observed r_c values are quite large (120-200 pm/V), comparable to that of the hot-press sintered transparent ceramics. The piezoelectric and dielectric properties of the KNNLB-x

ceramics have also been investigated, exhibiting relatively high piezoelectric coefficients (d_{33} = 50-90 pC/N), large dielectric constants (ε_r ~1400) and low dielectric losses (tan $\delta < 0.03$).

Table 6. Effective linear EO coefficients (r_c), piezoelectric coefficients (d_{33}), dielectric constants (ε_r) and dielectric losses (tan δ) for the pressureless sintered KNNLB-*x* ($0.04 \leq x \leq 0.08$) transparent ceramics

x	0.04	0.05	0.06	0.07	0.08
r_c (pm/V)	120	145	126	185	200
d_{33} (pC/N)	50	78	90	90	83
ε_r @1kHz	1250	1382	1461	1435	1410
tan δ @1kHz	0.029	0.027	0.024	0.026	0.027

Figure 14 shows the temperature dependences of ε_r measured at 10 kHz, 100 kHz and 1 MHz for the KNNLB-*x* ceramics, taking x = 0.06 as the example. Similar results have also been observed for the other ceramics ($0.04 \leq x \leq 0.08$). The ceramic expresses two broadened tetragonal-orthorhombic and cubic-tetragonal phase transition peaks, which are induced by the co-modifications with Li^+ and Bi^{3+} ions. The modification of Li^+ would not enhance the diffuseness in the phase transition because Li^+ has the same valence as K^+ and Na^+. Hence, the diffuse phase transition is attributed to Bi^{3+}, which replaces the Nb^{5+} sites and increases the degree of disorder on the B-sites and the local compositional fluctuation in the ceramics. However, unlike a typical relaxor ferroelectric, the observed T_m does not increase with increasing frequency. Although it has been shown that the change in T_m for KNN-based relaxors is not very large, only about 7°C as the frequency increases [28], the KNNLB-*x* ceramics may not be a typical relaxor and possess only relaxor-like characteristics [29].

Probably due to the relaxor-like characteristics as well as the fine grains and cubic-like symmetry, the polarization-electric field (*P-E*) hysteresis loops of the KNNLB-*x* ceramics become slant and flatted. Figure 15 shows the *P-E* loops of the KNNLB-0.06 ceramic at a frequency of 100 Hz under different electric fields at room temperature. Similar results have also been observed for the other ceramics ($0.04 \leq x \leq 0.08$). As the electric field increases, the observed remnant polarization P_r and coercive field E_c increase and almost become saturated at 5 kV/mm, with P_r = 9 μC/cm^2 and E_c = 1.5 kV/mm.

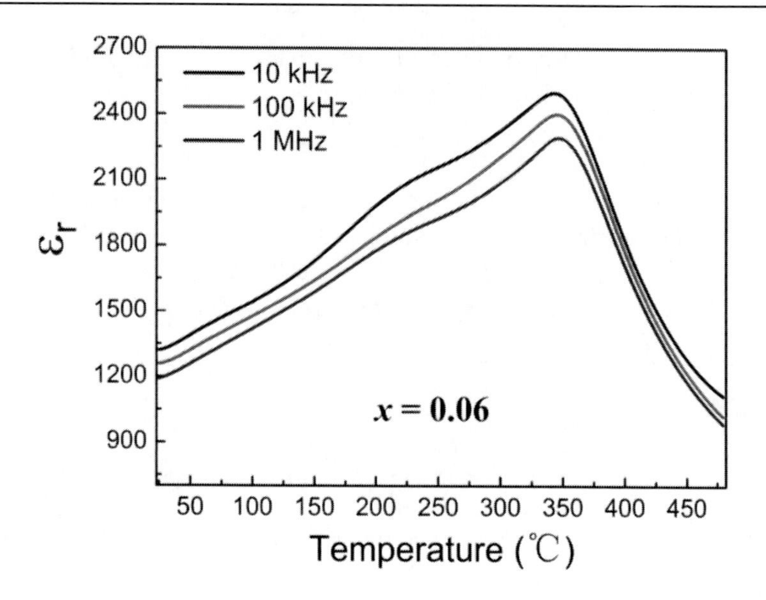

Figure 14. Temperature dependences of ε_r measured at frequencies in the range of 1 kHz to 1 MHz for the KNNLB-0.06 ceramic.

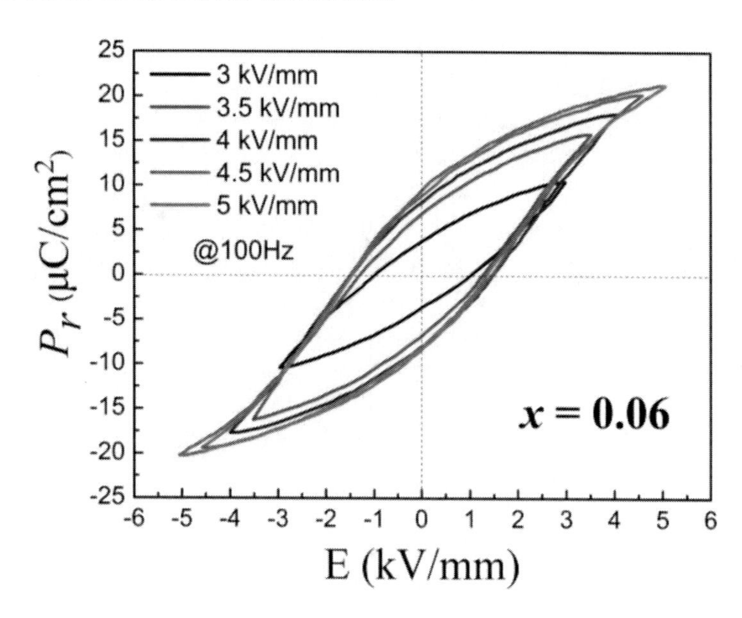

Figure 15. *P-E* loops of the KNNLB-0.06 ceramic under different electric fields at room temperature.

To further investigate the effect of LB on the ferroelectric properties, the P-E loops of the pure KNN and KNNLB-0.06 ceramics are shown in Figure 16 under the electric field of 4 kV/mm at room temperature. Compared to the pure KNN with a standard and saturated shape, the KNNLB-0.06 ceramic exhibit a slant and flattened *P-E* loop with larger E_c and smaller P_r, which is probably due to the relaxor-like characteristics induced by the co-modifications with Li and Bi.

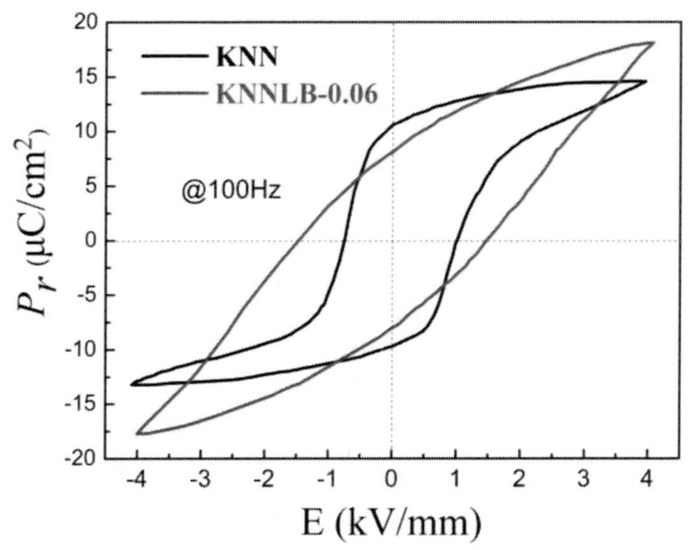

Figure 16. *P-E* loops of the pure KNN and KNNLB-0.06 ceramics.

CONCLUSION

Lead-free KNNLB-*x* transparent ceramics have been successfully fabricated by both hot-press sintering and pressureless sintering through the ordinary solid-state reaction. Due to the effective suppression of grain growth by Bi-modification, the ceramics possess fine and cubic-like crystalline grains with dense structure. The co-modification with Li^+ and Bi^{3+} induces a transformation of crystal structure from orthorhombic to cubic-like and a diffuse phase transition, making the ceramics exhibit relaxor-like characteristics. As a result, the optical anisotropy and hence the scattering of light at the grain boundaries are greatly reduced. The observed optical transmittances of the KNNLB-*x* ($0.04 \leq x \leq 0.08$) transparent ceramics reach high values of 60-70% in the near-IR region. Owing to the large distortion of [BO_6] octahedron and increase in dielectric constant due to the substitution of Li^+ in the KNN lattices, the KNNLB-*x* ceramics exhibit strong EO response, giving a large effective linear EO coefficient (120-200 pm/V).

KNNLB-*x* ($0.04 \leq x \leq 0.08$) transparent ceramics have also been fabricated by pressureless sintering. The ceramics possess a dense and fine-grain structure. The observed transmittance reaches a high value of 60-70% in the NIR region. The KNNLB-*x* transparent ceramics also exhibit strong linear EO response (r_c = 120-200 pm/V). The temperature dependences of dielectric constants and polarization-electric field hysteresis loops have been investigated, exhibiting relaxor-like characteristics. The pressureless sintered KNNLB-*x* transparent ceramics have relatively large piezoelectric coefficient (50-90 pC/N), high dielectric constant (~1400) and low dielectric loss (< 0.03), showing great potential in the applications of multifunctional electronic devices.

REFERENCES

[1] Haertling, G. H.; Land, C. E. *J. Am. Ceram. Soc.* 1971, *54*, 1-11.

[2] Jiang, H.; Zou, Y. K.; Chen, Q.; Li, K. K.; Zhang, R.; Wang, Y.; Ming, H.; Zheng, Z. Q. Transparent electro-optic ceramics and devices, *Proceedings of the SPIE* 5644, 2005, 380-394.

[3] Li, F. L.; Kwok, K. W. *J. Am. Ceram. Soc.* 2013, *96*, 3557-3562.

[4] Saito, Y.; Takao, H.; Tani, T.; Nonoyama, T.; Takatori, K.; Homma, T.; Nagaya, T.; Nakamura, M. *Nature* 2004, *432*, 84-87.

[5] Lin, D. M.; Kwok, K. W.; Chan, H. L. *J. Appl. Phys.* 2007, *102*, 034102.

[6] Lin, D. M.; Kwok, K. W.; Chan, H. L. *J. Am. Ceram. Soc.* 2009, *92*, 2765-2767.

[7] Li, F. L.; Kwok, K. W. *J. Eur. Ceram. Soc.* 2013, *33*, 123-128.

[8] Jaeger, R. E.; Egerton, L. *J. Am. Ceram. Soc.* 1962, *45*, 209-213.

[9] Tauc, J. C. Optical properties of solids. Amsterdam, North-Holland, 1972.

[10] Naser, M. A.; Zaliman, S.; Uda, H.; Yarub, A-D.; *Int. J. Nanoelectron. Mater.* 2009, *2*, 189-195.

[11] Wang, Z.; Gu, H. S.; Hu, Y. M.; Yang, K.; Hu, M. Z.; Zhou, D; Guan, J. G. *CrystEngComm* 2010, *12*, 3157-3162.

[12] Lin, D.; Kwok, K. W.; Chan, H. L. W. *Appl. Phys. A* 2008, *91*, 167-71.

[13] Lin, D.; Kwok, K. W.; Chan, H. L. W. *J. Appl. Phys.* 2007, *102*, 074113.
[14] Skidmore, T. A.; Comyn, T. P.; Milne, S. J. *J. Am. Ceram. Soc.* 2010, *93*, 624-626.
[15] Wang, X. X.; Kwok, K. W.; Tang, X. G.; Chan, H. L. W.; Choy, C. L. *Solid. State. Commun.* 2004, *129*, 319-323.
[16] Liu, M.; Wang, S. W.; Tang, D. Y.; Chen, L. D.; Ma, J. *Sci. Sinter.* 2008, *40*, 311-317.
[17] Apetz, R.; Bruggen, M. P. B. V. *J. Am. Ceram. Soc.* 2003, *86*, 480-486.
[18] Bokov, A.; Ye, Z. G. *J. Mater. Sci.* 2006, *41*, 31-52.
[19] Uchino, K.; Nomura, S. *Ferroelectr. Lett. Sect.* 1982, *44*, 55-61.
[20] Wan, X.; Wang, D. Y.; Zhao, X.; Luo, H. S.; Chan, H. L. W.; Choy, C. L. *Solid. State. Commun.* 2005, *134*, 547-551.
[21] Aillerie, M.; Fontana, M. D.; Abdi, F.; Carabatos-Nedelec, C.; Theofanous, N.; Alexakis, G. *J. Appl. Phys.* 1989, *65*, 2406-2408.
[22] Li, K.; Li, F. L.; Wang, Y.; Kwok, K. W.; Chan, H. L. W. *Mater. Chem. Phys.* 2011, *131*, 320-324.
[23] Kwok, K. W.; Li, F. L.; Lin, D. M. *Funct. Mater. Lett.* 2011, *4*, 237-240.
[24] Shih, C. C.; Yariv, A. *J. Phys. C: Solid. State. Phys.* 1982, *15*, 825-846.
[25] DiDomenico, M.; Wemple, S. H. *J. Appl. Phys.* 1969, *40*, 720-734.
[26] Burns, G.; Smith, A. *IEEE J. Quantum Elect.* 1968, *4*, 584-587.
[27] He, C. J.; Chen, H. B.; Sun, L.; Wang, J. M.; Xu, F.; Du, C. L.; Zhu, K. J.; Liu, Y. W. *Crys. Res. Technol.* 2012, *47*, 610-614.
[28] Guo, Y.; Kakimoto, K.; Ohsato, H. *J. Phys. Chem. Sol.* 2004, *65*, 1831-1835.
[29] Du, H. L.; Zhou, W. C.; Luo, F.; Zhu, D. M.; Qu, S. B.; Pei, Z. B. *J. Appl. Phys.* 2009, *105*, 124104.

BIOGRAPHICAL SKETCH

Xiao Wu

Affiliation: The Hong Kong Polytechnic University, Hong Kong, China
Fuzhou University, Fuzhou, China

Education: 2004.9-2008.6, Soochow University, Bachelor
2008.9-2011.6, Soochow University, Master

2011.9-2015.5, The Hong Kong Polytechnic University, PhD

Address: College of Materials Science and Engineering, Fuzhou University, 2#, Xueyuan Road, University Town, Fuzhou, 350116, China.

Research and Professional Experience:

2009.8-2011.5, Shanghai Institute of Ceramics, Chinese Academy of Sciences, Joint student.

2014.10-2015.3, The Hong Kong Polytechnic University, Research assistant.

2015.4-2016.4, The Hong Kong Polytechnic University, Postdoctoral Fellow.

Professional Appointments:

From 2016.4 Fuzhou University, "Qishan" scholar.

Publications Last 3 Years:

1. Xiao Wu, K.W. Kwok, and F.L. Li, Upconversion fluorescence studies of sol-gel-derived Er-doped KNN ceramics, *Journal of Alloys and Compounds,* 580, 88-92 (2013).
2. Xiao Wu, K. W. Kwok, Mid-IR to Visible Photoluminescence, Dielectric, and Ferroelectric Properties of Er-Doped KNLN Ceramics, *Journal of the American Ceramic Society*, 97 [5] 1504–1510 (2014).
3. Xiao Wu, Chi Man Lau, K.W. Kwok, Effect of phase transition on photoluminescence of Er-doped KNN ceramics, *Journal of Luminescence,* 155, 343–350 (2014).
4. Xiao Wu, Mei Lin, Chi Man Lau, K.W. Kwok, Photoluminescence, ferroelectric, dielectric and piezoelectric properties of sol-gel-derived Er-doped KNN-LN lead-free multifunctional ceramics, *Advances in Science and Technology,* 90, 19−24 (2014).
5. Xiao Wu, Chi Man Lau, K.W. Kwok, Photoluminescence Properties of Er/Pr-Doped $K_{0.5}Na_{0.5}NbO_3$ Ferroelectric Ceramics, *Journal of the American Ceramic Society*, 98, 2139–2145 (2015).
6. Xiao Wu, Tat Hang Chung, K.W. Kwok, Enhanced visible and mid-IR emissions in Er/Yb-codoped $K_{0.5}Na_{0.5}NbO_3$ ferroelectric ceramics, *Ceramics International*, 41, 14041–14048 (2015).
7. Xiao Wu, Tat Hang Chung, Hailing Sun, K.W. Kwok, Tunable photoluminescence properties of Pr^{3+}/Er^{3+}-doped $0.93Bi_{0.5}Na_{0.5}TiO_3$-

$0.07BaTiO_3$ low-temperature sintered multifunctional ceramics, *Ceramics International*, in press (2016).
8. Chi Man Lau, Xiao Wu, K.W. Kwok, Effects of vacancies on luminescence of Er-doped $0.93Bi_{0.5}Na_{0.5}TiO_3$-$0.07BaTiO_3$ ceramics, *Journal of Applied Physics*, 118, 034107 (2015).
9. Chi Man Lau, Xiao Wu, K.W. Kwok, Photoluminescence, ferroelectric, dielectric and piezoelectric properties of Er-doped BNT-BT multifunctional ceramics, *Applied Surface Science*, 336, 314–320 (2015).
10. Hailing Sun, Qiaoji Zheng, Yang Wan, Qiang Li, Yan Chen, Xiao Wu, K.W. Kwok, etc. Microstructure, electrical properties and electric field-induced transitions in $NaNbO_3$-$LiTaO_3$ lead-free ceramics, *Phys. Status Solidi A,* 211 [4] 869–876 (2014).
11. Mei Lin, Yuming Chen, Bolei Chen, Xiao Wu, Wei Lu, Helen Lai Wa Chan, and Jikang Yuan, Morphology-controlled synthesis of self-assembled $LiFePO_4$/C/RGO for high-performance Li-ion batteries, *ACS Applied Materials and Interfaces,* 6, 17556−17563 (2014).
12. Mei Lin, Bolei Chen, Xiao Wu, Jiasheng Qian, Linfeng Fei, Wei Lu, Helen Lai Wa Chan, and Jikang Yuan, Controllable synthesis of ε-MnO_2 hollow spheres/RGO Nanocomposite, *Nanoscale*, 8, 1854−1860 (2016).
13. LIN Mei, WU Xiao, CHEN Bo Lei and YUAN Ji Kang, Hydrothermal synthesis of corn cob-like $LiFePO_4$/C as high performance cathode material for lithium ion batteries, *Advances in Science and Technology,* 93, 152−157 (2014).
14. H.L. Sun, Q.J. Zheng, Y. Wan, Y. Chen, X. Wu, K.W. Kwok, H.L.W. Chan, D.M. Lin, Correlation of grain size, phase transition and piezoelectric properties in $Ba_{0.85}Ca_{0.15}Ti_{0.90}Zr_{0.10}O_3$ ceramics, *Journal of Materials Science: Materials in Electronics,* DOI 10.1007/s10854-015-3063-7 (2015).

Faliang Li

Affiliation: The Hong Kong Polytechnic University
WuHan University of Science and Technology

Education:

2009.8-2013.9 Dept. of Applied Physics, The Hong Kong Polytechnic University, Ph. D

2006.9-2009.6 Dept. of Materials Science and Engineering, ChangZhou University, M.E.

2002.9-2006.7 Dept. of Materials Science and Engineering, ZhengZhou University, B.E.

Research and Professional Experience:

From 2013.10, WuHan University of Science and Technology

Professional Appointments: Lecturer

Publications Last 3 Years:

1. Faliang Li, Kin-Wing Kwok, $K_{0.5}Na_{0.5}NbO_3$-based lead-free transparent electro-optic ceramics prepared by pressureless sintering, Journal of the American Ceramic Society, 2013, 96(11): 3557-3562.
2. Faliang Li, Fang Fu, Lilin Lu, Haijun Zhang, Shaowei Zhang, Preparation and artificial neural networks analysis of ultrafine β-Sialon powders by microwave-assisted carbothermal reduction nitridation of sol–gel derived powder precursors, Advanced Powder Technology, 2015, 26(5): 1417-1422.
3. Li Faliang, Meng Lu, Zhang Haijun, Zhang Shaowei, Carbothermal Reduction-Nitridation Synthesis of Ca-alpha-Sialon from Bauxite and Thermodynamic Study, Rare Metal Materials and Engineering, 2015, 44 (Supplement: 1): 434-437.
4. Xiao Wu, Kin-Wing Kwok, Faliang Li, Upconversion fluorescence studies of sol-gel-derived Er-doped KNN ceramics, Journal of Alloys and Compounds, 2013, 580: 88-92.
5. Xiangong Deng, Haijun Zhang, Faliang Li, Shaowei Zhang, Zhiyi Wang, Effect of SiO_2 on thermal stability and photocatalytic activity of SiO_2/TiO_2 films, Nanoscience and Nanotechnology Letters, 2015, 7(4):324-330.
6. Yingnan Cao, Haijun Zhang, Faliang Li, Lilin Lu, Shaowei Zhang, Preparation and Characterization of Ultrafine ZrB_2-SiC Composite Powders by a Combined Sol-gel and Microwave Boro/carbothermal Reduction Method, Ceramics International, 2015, 41(6):7823-7829.
7. Lilin Lu, Min Qiang, Faliang Li, Haijun Zhang, Shaowei Zhang, Theoretical investigation on the antioxidative activity of anthocyanidins: A DFT/B3LYP study, Dyes and Pigments, 2014, 103: 175-182.

8. Lilin Lu, Haijun Zhang, Shaowei Zhang, Faliang Li, A New Family of High-Efficiency Hydrogen Generation Catalysts Based on Ammonium, Angewandte Chemie International Edition, 2015, 54(32):9328-9332.
9. Lilin Lu, Shufang Zhu, Haijun Zhang, Faliang Li, Shaowei Zhang, Theoretical study of complexation of resveratrol with Cyclodextrins and Cucurbiturils: structure and antioxidative activity, *RSC Advance*, *RSC Advance*, 2015, 5: 14114.
10. Yajun Gu, Lilin Lu, Haijun Zhang, Yingnan Cao, Faliang Li, Shaowei Zhang, Nitridation of silicon powders catalyzed by cobalt nanoparticles, Journal of the American Ceramics Society, 2015, 98(6): 1762-1768.
11. Xiangong Deng, Shuang Du, HaijunZhang, Faliang Li, Junkai Wang, Wanguo Zhao, Feng Liang, Zhong Huang, Shaowei Zhang, Preparation and characterization of ZrB2-SiC composite powders from zircon via microwave-assisted boro/carbothermal reduction, Ceramics International, 2015, 41(10): 14419-14426.
12. Ying-Nan Cao, Shuang Du, Jun-Kai wang, Hai-Jun Zhang, Fa-Liang Li, Li-Lin Lu, Shao-Wei Zhang, and Xiangong Deng, Preparation of Zirconium Diboride Ultrafine Hollow Spheres by a Combined Sol-Gel and Boro/carbothermal Reduction Technique, Journal of Sol-gel Science and Technology, 2014, 72(1):130-136.
13. Xiangong Deng, JunKai Wang, Jianghao Liu, Haijun Zhang, Faliang Li, Hongjuan Duan, Lilin Lu, Zhong Huang, Wanguo Zhao, Shaowei Zhang, Preparation and characterization of porous mullite ceramics via foam-gelcasting, Ceramics International, 2015, 41(7):9009-9017.
14. Zhang Haijun, Zhao Wanguo, Li Faliang, Duan Hongjuan, Zhang Shaowei, Preparation and Erosion Resistance of $CaO-Al_2O_3-MgO-SiO_2$ Microcrystalline Glass Ceramics, Rare Metal Materials and Engineering, 2015, 44 (Supplement: 1): 277-280.

In: Advances in Materials Science Research ISBN: 978-1-53610-059-4
Editor: Maryann C. Wythers

Chapter 3

ULTRAVIOLET PHOTODETECTORS BASED ON TRANSITION METAL OXIDES

***Ting Xie*[1,2], *Albert V. Davydov*[1,*] and *Ratan Debnath*[1,3,†]**
[1]Materials Science and Engineering Division,
Material Measurement Laboratory,
National Institute of Standards and Technology,
Gaithersburg, MD, US
[2]Department of Electrical and Computer Engineering,
University of Maryland, MD, US
[3]N5 Sensors, Inc., Rockville, MD, US

ABSTRACT

Ultraviolet (UV) photodetector (PD) has attracted extensive attention owing to their broad application in digital imaging, missile plume detection, optical communications, and biomedical sensing. Conventional silicon (Si) based UV PD requires costly high pass optical filters and phosphors to tune its spectral response and thus poses extreme challenge to achieve a UV detection system with miniature size, high energy-efficiency and low-cost. To overcome the limitations inherent from the narrow bandgap of Si, UV PDs consisted of wide-bandgap materials, such as gallium nitride (GaN), silicon carbide (SiC), and transition metal

* Corresponding author: albert.davydov@nist.gov.
† Corresponding author: ratan.debnath@nist.gov, rdebnath@n5sensors.com.

oxides (TMOs) such as zinc oxide (ZnO), nickel oxide (NiO) etc. have recently emerged and are being rapidly developed. The application of TMOs in the UV PDs mainly rely on their intrinsic properties, including optical transparency in the near-UV and visible spectra, low impact to environment, thermal stability and feasibility of constructing band-tunable heterojunctions. Here, we provide a perspective on the recent developments in UV photodetection based on TMO materials including thin films and one-dimensional nanostructures. The chapter begins with describing the operation principles and requirements for the PD devices, followed by the latest developments in materials and device architectures. This chapter is concluded with future challenges and outlook in this area.

1. INTRODUCTION

Ultraviolet (UV) radiation was first discovered and studied by J. W. Ritter in the earlier 19 centuries. The discovery, followed by many other investigations at that time, revealed that UV light is identical to visible emission except for the shorter wavelength. UV light spans the wavelength range from 10 nm – 400 nm in the electromagnetic radiation spectrum. UV light is typically divided into the following four regions: UVA with light wavelength from 315 nm – 400 nm, UVB with light wavelength from 280 nm – 315 nm, UVC with light wavelength from 100 nm – 280 nm, and extreme UV (EUV) with light wavelength from 10 nm – 121 nm [1]. As atmospheric ozone absorbs solar light below 280 nm in wavelength and visible light spans from 400 nm – 700 nm, UV PDs are identified as solar-blind and visible-blind, which response to light wavelength below 280 nm and 400 nm respectively, based on their spectral selectivity [2]. UV light is an energetic ionization radiation that can activate many chemical processes and cause severe problems, such as skin cancer to human-beings. Moreover, the detection of UV can find its applications in chemical and biological analysis, flame detection, digital imaging, missile plume detection, and optical communication [3]. Therefore, the development of high-performance UV photodetector (PD) is of significant importance.

A high-performance PD is generally designed to meet the high *5S* standards, i.e., *sensitivity*, *signal-to-noise ratio*, *spectral selectivity*, *speed* and *stability* [2], as well as to achieve linear photoresponse with respect to the incident light power [4]. The priority of the above performance factors for a UV PD design depends significantly on the specific applications, for instance, a fast response time can be the primary requirement in UV imaging, in which fast signal acquisition is a necessity [4].

A wide variety of devices have been demonstrated and accomplished for UV detection, including photomultiplier tube, thermal detector, charge-coupled device (CCD), and Si-based photodiode. Photomultiplier tubes, in which a UV-sensitive photocathode ejects an electron upon UV illumination, have high gain and signal-to-noise ratio, but are restricted by their bulky and fragile structure [5]. Similar drawbacks also restrict the utility of thermal UV detectors (pyrometer and bolometer) that are frequently used for calibration purposes. Thermal devices are also generally slow and wavelength-independent [5]. In contrast, solid state device based UV PDs, such as CCD and Si-based photodiodes, are usually light-weight and energy-efficient. Due to the well-established technology and relative low-cost, Si-based UV enhanced photodiodes currently dominate the market [6]. The Si UV PDs are commonly configured as p-n junction photodiodes and charge-inversion photodiodes (similar to metal-oxide semiconductor structures) [7, 8]. However, limitations inherited from the intrinsic properties of Si exist in this type of UV PD. First of all, external optical filters are essential to block the visible and infrared photons, eliminating the long-wavelength response of the UV PD due to the narrow bandgap (1.1eV) of Si. Otherwise, the UV PD will be overwhelmed with the response to the visible and IR background signals. For instance, the spectral irradiance power over 1 cm^2 of visible and IR light under AM 1.5 condition is more than 0.1 W, which is in the same magnitude of order as a commercially available high output power UV LED. Additionally, the exposure to radiation which is much higher than the bandgap has been shown to degrade the silicon in long time operation [5]. Furthermore, the SiO_2 passivation layer typically deposited on the front surface of Si photodiodes absorbs UV radiation, thereby reducing the external quantum efficiency and it also degrading the quality of the coating over time [5].

To surmount many of the limitations listed above, a new generation of solid-state UV PDs employing wide-bandgap materials has been developed. Among the demonstrated wide-bandgap materials used for UV PDs, transition metal oxides (TMOs), represented by ZnO (and its alloy ZnMgO), NiO and titanium dioxide (TiO_2), are particularly favored for UV detection because of their visible-blindness, low dislocation density [9], high-energy radiation resistance [10] and chemical and thermal stability [11]. Thus, these oxides are the potential candidates for the next generation of UV PDs. This chapter includes an overview of the various growth techniques employed to prepare wide-bandgap TMOs with different properties and a comprehensive survey of the relevant PDs with different device architectures. We will begin the discussion with the figures of merits of PDs and conclude with some perspective and the future research work in this area.

2. FIGURES OF MERIT

In order to compare different PDs, there have been a number of figures of merits that have been introduced. For simplicity, we will define the most frequently used and important evaluation matrix for TMO based UV PDs that are enumerated below:

i. External quantum efficiency η is the ratio of the number of collected carriers to the number of incident photons, namely, yield of photocurrent per incident photon. It describes the coupling ability of the detector to the light radiation. The efficiency is defined as

$$\eta = \frac{I_{ph}/q}{\Phi_{in}}$$

where I_{ph} is the collected photocurrent, q is the electron charge, and Φ_{in} is the flux of incident photons.

ii. Photoresponsivity R_{ph} is determined by the external quantum efficiency η. It measures the input-output gain of a detector system. For photodetectors, it is usually defined as output photocurrent per input optical power at a specific wavelength,

$$R_{ph} = \frac{I_{ph}}{P_{in}} = \eta \frac{q\lambda}{hc}$$

where I_{ph}, P_{in}, η, q, λ, h, and c are the collected photocurrent, the incident light power, the external quantum efficiency, the electron charge, the wavelength of the incident light, Planck's constant, and the velocity of light, respectively.

iii. Noise equivalent power (NEP) quantifies the sensitivity of a PD. A more sensitive PD has a smaller NEP. It is defined as the input power that produces a unity signal-to-noise ratio (SNR) in 1 Hz output bandwidth [12]. With this definition, the NEP can be expressed as,

$$NEP = \frac{\sqrt{\langle i_n^2 \rangle}}{R_{ph}}$$

where $\langle i_n^2 \rangle$ is the mean squares noise current at an input power for SNR = 1 and R_{ph} is the photoresponsivity.

iv. Detectivity D^* is the parameter that evaluates the noise performance of a photodetector using the following expression (assuming that the dark current dominate the noise),

$$D^* = \frac{R_{ph}}{\sqrt{2qI_{dark}}}$$

where $R_{ph,}$ q and I_{dark} are measured responsivity under light illumination, the electron charge, and current measured in the dark, respectively.

v. Cutoff wavelength λ_c is closely related to the bandgap E_g of the semiconductor used in the PD. The incident photon energy should be larger than the bandgap to excite electron-hole pairs and consequently trigger an effective photoresponse. Therefore, the cutoff wavelength is defined as the following,

$$\lambda_c(nm) = \frac{hc}{E_g} = \frac{1240}{E_g\ (eV)}$$

where h, c and E_g are Planck's constant, the light velocity and bandgap of the semiconductor. However, photoresponse yielded by incident photons with lower energy is also observed frequently, which is due to intragap states associated with defects in the semiconductor.

vi. UV-to-visible rejection ratio $R_{UV\text{-}vis}$ quantifies the ratio of the photoresponse in the UV regime to the visible light regime. It is usually defined with the expression,

$$R_{UV-vis} = \frac{R_{ph}(UV)}{R_{400nm}}$$

where $R_{ph}(UV)$ and R_{400nm} are the measured photoresponsivities at a specific wavelength in the UV regime and 400 nm, respectively. A high UV-to-visible ratio is desirable for the photodetector since it indicates very weak response to the visible solar spectrum.

vii. Rise and fall time measure the response speed of a PD to a square-pulse light signal. Rise time represents the time interval for the PD response to increase from 10% to 90% of the peak output. Fall time is the time interval for the response to decay from 90% to 10% of the peak value.

3. Growth Methods

Various growth techniques have been developed for thin film and low-dimensional TMOs for acoustical and optical devices due to their unique properties. In this section, we will highlight some of the most important methods and the relevant material properties.

a. Sputtering

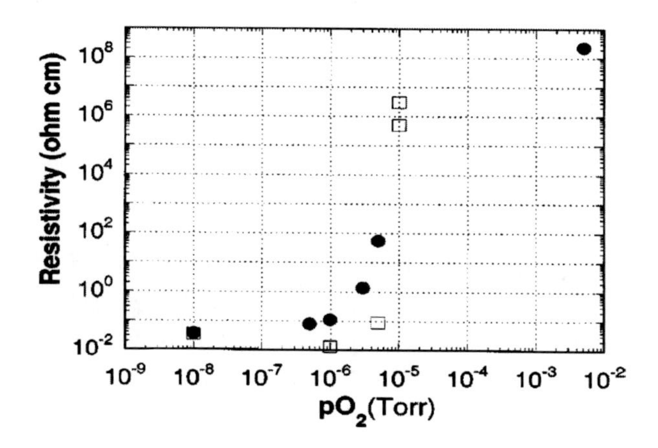

Figure 1. Dependence of the electrical resistivity of sputtered ZnO thin films on the partial pressure of O_2 in the mixture of Ar + O_2 for a total pressure of 10 mTorr (●) and 20 mTorr (□). (Adapted with permission from Ref. [13]. Copyright © 2003 AIP Publishing LLC).

Sputtering is one of the most popular growth/deposition methods for various TMOs, including ZnO, ZnMgO, NiO, TiO_2 etc. The sputtering technique is a low-cost deposition method and compatible with most microelectronic fabrication processes. To control the stoichiometry of the grown oxides, the deposition environment is generally filled with a mixture of oxygen and argon gases at a pressure of (1 – 20) mTorr. The ratio of O_2 to O_2 + Ar ranges from 0 to 1. The selection of the ratio depends on the substrate temperature, sputtering target and post-annealing conditions. In the gas mixture, Ar plasma sputters the target material and O_2 reacts with the sputtered materials. Thus, this deposition method is often called reactive sputtering. TMOs can be either deposited from a metal target or an oxide target. Depending on the conductivity of the sputtering target, metallic or insulating/semiconducting, a direct-current (DC) or radio-frequency (RF) power source is selected, respectively. The sputtering power and pressure are the primary means to control the deposition yield rate. A 5 min to 10 min pre-sputtering period is typically imposed to remove contamination from the target surface and to stabilize the deposition conditions [14].

High quality ZnO films have been obtained by sputtering on a variety of substrates such as gold, aluminum, platinum, quartz, silicon and sapphire [14, 15]. Carica et al. reported the dependence of the electrical resistivity on the partial pressure of oxygen in the Ar+O_2 mixture, in the sputtered ZnO thin films [13]. The ZnO films was sputtered from a ZnO target with low power (100 W), low cathode voltage (-100 V), and high pressure (10 and 20 mTorr). The deposition conditions were intended to mitigate the potential damage due to the bombardment of the plasma. As shown in Figure 1, the partial pressure of oxygen strongly affects the electrical resistivity of the deposited ZnO films. The results reveal an abrupt transition of the resistivity from the semiconducting state to the semi-insulating state in the oxygen partial pressure range of 10^{-6} to 10^{-5} Torr. The intrinsic ZnO films with low resistivity (< 1 ohm cm) exhibit *n*-type semiconducting with the Hall mobility of (12 – 15) cm^2/Vs. Similar oxygen pressure dependence of the resistivity was observed in the reactive sputtered ZnO film from a Zn target [16]. The hardness and modulus of the sputtered ZnO film are weakly affected by the oxygen pressure, as reported by Chen et al. [16]. The photoluminescence (PL) study of the sputtered ZnO films shows the significantly diminished visible emission in the PL spectra with the increased O_2 partial pressure [17]. Jeong et al. attributed this tendency to the mitigation of the oxygen vacancy in the films which were deposited under high O_2 pressure condition. These results indicate the improvement in the stoichiometry of the deposited ZnO under oxygen rich environment. Lu et al. investigated the effects of RF power and the post-annealing condition on the properties of the sputtered ZnO thin films [18]. The films were deposited at a substrate temperature of 400°C in Ar+O_2 gas environment and post-annealed at 600°C in Ar. The growth rate and the grain size of the deposited ZnO films increases and decreases with higher RF power, respectively. The preferred growth orientation (0002) was enhanced by the post-annealing treatment, suggesting the increase of gain size as well during the annealing process.

Highly crystalline ZnMgO thin film was reported by Kim et al., using RF magnetron sputtering technique [19]. The single-crystal ZnMgO was deposited on sapphire substrate at a substrate temperature between 500°C to 800°C. The sputtering target was prepared by mixing ZnO powders with (5 – 35) mass% MgO. As the growth temperature increases, the Mg atoms replace the Zn in the lattice, which increases the bandgap of the deposited ZnMgO films. This result corroborates with the study of high-temperature ZnMgO growth carried out by another group [20]. In that study, the researchers also demonstrated the importance of high-temperature treatment under vacuum condition to enhance the UV emission efficiency of the ZnMgO films. In addition to the sputtering from a single ZnMgO target, Han et al. used the two-source (Zn and Mg) co-sputtering method to grow high quality ZnMgO thin films [21]. The films were deposited at a substrate temperature of 600°C in a mixture of Ar and O_2. The individual sputtering power for the metal targets and the O_2 partial pressure are the determining factors for the properties of the ZnMgO films.

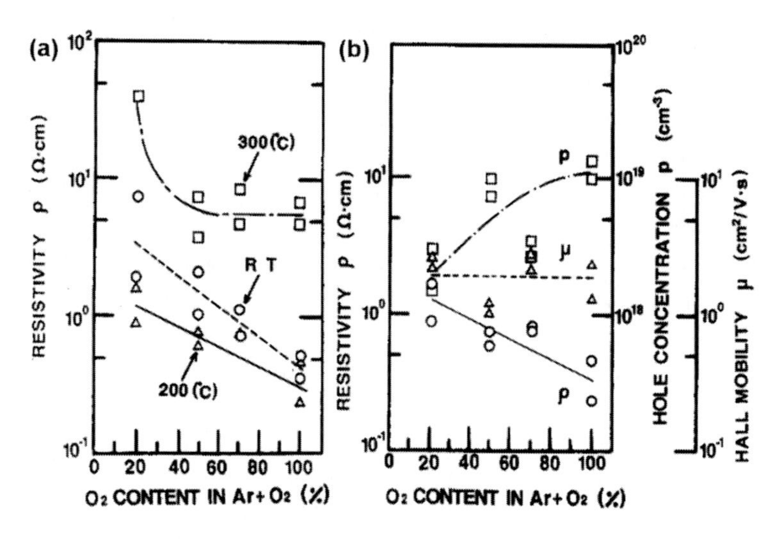

Figure 2. (a) Resistivity as a function of the O_2 content in the Ar + O_2 mixture for NiO films prepared at substrate temperature of RT (- - -), 200 (—) and 300°C (- — -). (b) Resistivity ρ (○), hole concentration p (□) and Hall mobility μ (Δ) as functions of the O2 content in the Ar + O_2 mixture for NiO films prepared at a substrate temperature of 200°C. (Adapted with permission from Ref. [22]. Copyright © 1993 Elsevier B.V.).

The unintentionally doped NiO typically exhibits a p-type semiconducting characteristic. Sato et al. reported a comprehensive study of the deposition of NiO with RF magnetron sputtering technique [22]. The thin films were sputtered from a NiO target at a low RF power of 50 W. The substrate temperature was varied at room temperature (RT), 200°C and 300°C. Note, the surface temperature of film reached 140°C during the deposition at RT. Gas mixtures of Ar + O_2 with various O_2 contents (0% to 100%) were introduced as the deposition environment. A decrease in the deposition yield rate was observed with the increase of the oxygen partial pressure. The measured optical and electrical results show that O_2 partial pressure is the dominant factor in determining the properties of the sputtered NiO films. Figure 2a shows the O_2 partial pressure dependence of the resistivity of the NiO films at various substrate temperatures. The resistivity decreases with higher O_2 content in the deposition gas, at both RT and 200°C. At 300°C, the resistivity is fairly constant for O_2 content > 50%. Figure 2b shows the resistivity, hole concentration and hole mobility as functions of the O_2 partial pressure at 200°C. The resistivity decrease with the increase in O_2 content is likely due to the increase in hole concentration since the mobility is relatively independent on O_2 (Figure 2b). The increase in hole concentration with O_2 content is attributed to the increase in Ni vacancies and/or interstitial oxygen which act as shallow acceptors. Ryu et al. investigated the preferred orientation of NiO thin films by varying O_2 partial pressure in the RF magnetron sputtering process [23]. The results show that (100) and (111) orientation dominates in the sputtered NiO films in pure Ar and pure O_2 environments, respectively. XRD studies also show that the orientation and crystallinity of NiO films change with the sputtering power and substrate temperature [24, 25].

TiO_2 has two dominant phases in sputtered films, namely rutile and anatase [27]. By investigating the substrate temperature and oxygen partial pressure, Schiller et al. revealed the dependence of the phase and grain size of the reactive sputtered TiO_2 films on these parameters [26]. The total gas ambient is Ar with O_2. For $P_{O2}/P_{tot} \geq 0.23$, only the two tetragonal TiO_2 phases were observed. The results are presented in Figure 3. Transmission electron microscopy studies relate the large crystalline size (60 nm to 200 nm) and the fine grains (less than 60 nm) to anatase phase and rutile phase, respectively. The substrate temperature dependence of the phase transition was investigated by Wicaksana et al. as well [28].

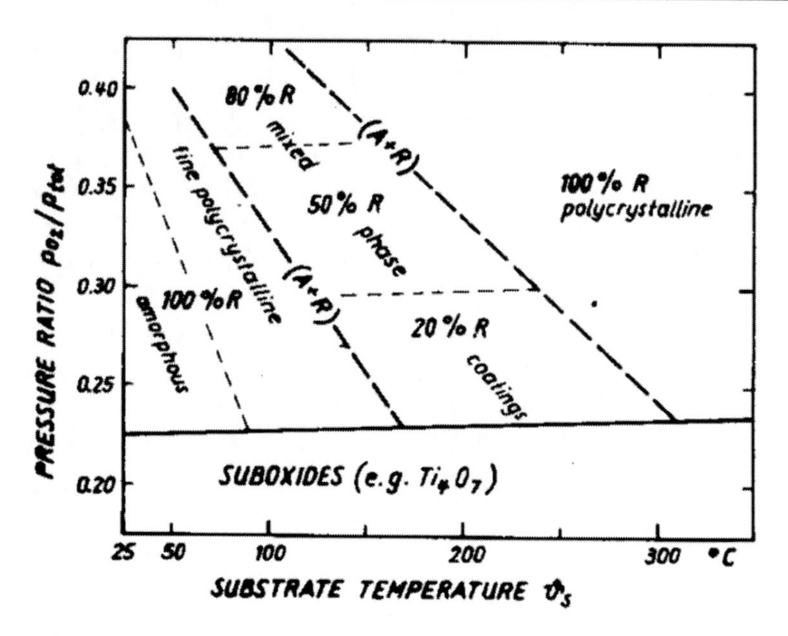

Figure 3. Phase composition of TiO2 films prepared by reactive DC sputtering as a function of the pressure ratio P_{O2}/P_{tot} and the substrate temperature ϑs (A, anatase; R, rutile; dTiO2 = (200-250) nm; substrata, silicon wafer; measuring technique, transmission electron microscopy). (Adapted with permission from Ref. [26]. Copyright © 1981 Elsevier B.V.).

b. Molecular Beam Epitaxy

Molecular-beam epitaxy (MBE) provides more precise control than other deposition processes, benefiting from the *in-situ* feedback of reflection high-energy electron diffraction [14]. For the growth of epitaxial ZnO films, Zn and O_2 are widely used as the source materials, although some groups have also reported the use of NO_2 and H_2O_2 as the source of active oxygen [29, 30]. Chen et al. demonstrated ZnO epitaxial growth on (0001) sapphire substrates by oxygen plasma assisted MBE [31, 32]. Pre-exposure of the sapphire substrate to the plasma for 30 min at 600°C enables the initial two-dimensional nucleation of the ZnO layer. By employing Zn and Mg Knudsen cells and a RF oxygen plasma cell, Ogata et al. reported the growth of ZnMgO films with a high electron mobility of 100 cm^2/Vs [33]. The broadening of the bandgap of the grown ZnMgO with increased Mg contents was observed as well.

c. Pulsed Laser Deposition

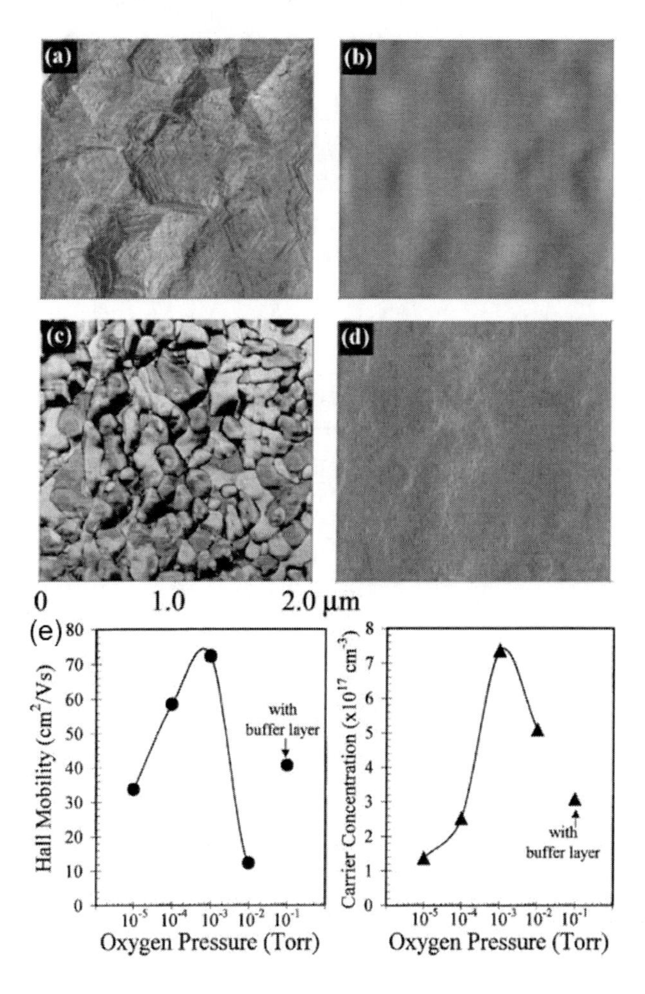

Figure 4. AFM images of the PLD ZnO films grown at various oxygen pressures (a) 10^{-4} Torr, (b) 10^{-2} Torr, (c) 10^{-1} Torr, and (d) 10^{-1} Torr with a nucleation layer of 100 Å grown at 10^{-4} Torr. (e) Hall mobility and carrier concentration vs oxygen pressure for ZnO film grown by PLD. (Adapted with permission from Ref. [34]. Copyright © 1999 AIP Publishing LLC).

Pulsed laser deposition (PLD) uses a high power laser source to evaporate materials from the targets. The plume of evaporated material deposits on the substrate and condenses at a specific substrate temperature. The growth of high quality films with PLD can be achieved at a moderate substrate temperature (200 to 800°C) [14]. Choopun et al. reported the influence of oxygen pressure on the epitaxy, surface morphology, and optical properties of PLD deposited ZnO films [34]. The films were grown on (0001) sapphire substrates at a temperature of 750°C and background oxygen pressure ranged from 10^{-5} Torr to 10^{-1} Torr. At low oxygen pressure, the deposited ZnO films were highly *c*-axis orientated with a larger lattice constant than its bulk value, which was attributed to both oxygen vacancies in the films and the compressive strain induced by the substrate. With the increase in oxygen pressure, the *c*-axis lattice constant approaches the bulk value. The surface morphology changes drastically with oxygen pressure, as shown in Figure 4a-d. The transition towards the formation of a smooth ZnO film was observed at the O_2 pressure of 10^{-2} Torr, and the oxygen pressure dependence of the Hall mobility and carrier concentration are shown in Figure 4e. The highest mobility and carrier concentration are expected from the ZnO films grown under the condition of 10^{-4} to 10^{-3} Torr O_2 pressure. Directed by these results, Choopun et al. proposed the two-stage ZnO growth procedure with an initial low-oxygen background (10^{-4} Torr – 10^{-3} Torr) and a moderate-oxygen ambient (10^{-2} Torr) for the later stage.

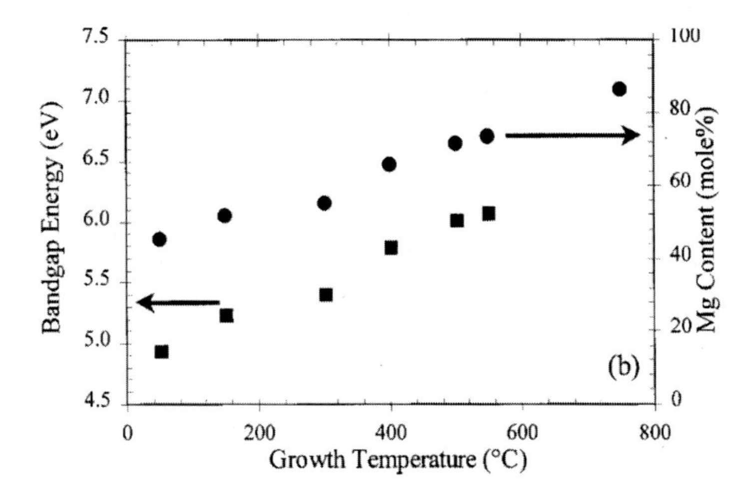

Figure 5. Band gap energy and Mg content of MgxZn1-xO films as a function of growth temperature. (Adapted with permission from Ref. [36]. Copyright © 2002 AIP Publishing LLC).

The concept of tuning the bandgap of ZnMgO using Mg content was first demonstrated by Ohtomo et al. with PLD [35]. The films were grown on (0001) sapphire substrates under ultra-high vacuum condition. The targets were prepared from ZnO and MgO using a sintering process. The films were deposited at a substrate temperature of 600°C at 5×10^{-5} Torr oxygen pressure. Later Choopun et al. reported the realization of ZnMgO films with a band gap of 5 eV to 6 eV [36]. The PLD was under oxygen environment with a pressure of 5×10^{-4} Torr and the substrate temperature was varied from RT to 750°C. The growth temperature dependence of the Mg content as well as band gap of the ZnMgO films are shown in Figure 5.

d. Chemical Vapor Deposition

Many thin film deposition techniques have been widely utilized to grow metal oxides for various applications. Among those techniques, chemical vapor deposition (CVD) is very interesting as it offers excellent crystallinity and high quality film for large scale production. When metal-organic precursors are used, the technique is called metal-organic CVD or MOCVD. Typically, vapor phase precursors are delivered in the CVD chamber by the carrier gas and the chemical reaction takes place on the substrates or deposition zone where high temperature is maintained. Although this technique is widely used for epitaxial films to realize various GaN-based optoelectronic devices, it has also been utilized for TMOs particularly for ZnO. Typically, hydride vapor phase epitaxy growth method was used i.e., hydrogen was used as a carrier gas and ZnO powder was placed in the evaporation zone [37]. The following chemical reaction takes place:

$$ZnO + H_2 \rightarrow Zn + H_2O$$

In the deposition zone, the substrate is kept at high temperature (590–610°C) and the reverse reaction occurs:

$$Zn + H_2O \rightarrow ZnO + H_2$$

Such growth method produced highly crystalline ZnO film on sapphire substrates. Other than ZnO, zinc chloride ($ZnCl_2$) [38] or zinc iodide (ZnI_2) [39] were employed as zinc sources. The ZnO grown from an iodine source shows better optical and structural properties than that grown from chloride and a higher growth rate was achieved.

For metal organic source, metal alkyls such as dimethyl zinc or diethyl zinc (DEZn) were used in MOCVD in addition to oxygen and argon or nitrogen as a carrier gas. [40, 41]. As dimethyl zinc or diethyl zinc reacts readily with oxygen and water vapor, it was very difficult to control the reaction and film degradation took place accompanied by formation of white powder. Alternatively, a stable metal source such as zinc acetylacetonate in combination with oxygen has been successfully used to achieve high-quality ZnO films on sapphire substrates [42, 43]. Due to the improvements in MOCVD technology, there has been great progress to grow high quality ZnO using DEZn and oxygen [44, 45]. Highly crystalline ZnO epilayers were grown on sapphire (0001) substrates with ZnO buffer layer.

In addition to thin film, ZnO nanostructures have also been grown using a catalyst-free method on various substrates [46-48]. Furthermore, Mg was incorporated in ZnO to grow ZnMgO nanorods and PL measurements showed a blue-shift in the bandgap suggesting the alloying effects into the system [49]. $ZnO/Zn_{0.8}Mg_{0.2}O$ nanorod single-quantum well structures had also been demonstrated [50]. A comprehensive review about ZnO growth using CVD can be found elsewhere [14].

MOCVD has been used to grow TiO_2 film on substrates including sapphire [51], glass [52], Si [53], conductive glass and quartz [54]. Titanium isopropoxide was used as a source precursor of titanium whereas oxygen and nitrogen gases were used as the oxidant gas and the carrier gas, respectively [52]. The anatase phase of as-grown TiO_2 films on glass substrate was transformed into rutile phase at an annealing temperature of 900°C. The values of optical band gap decreased from 3.5 eV to 3.25 eV at 900°C. The above mentioned precursors and gases were also used for TiO_2 film grown on sapphire substrates [51]. The morphology of the film was greatly influenced by the temperature, substrate and carrier gas flow rate although an additional oxygen flow was not essential to obtain stoichiometric TiO_2 films.

e. Sol-Gel

Sol-gel is a "*wet-chemistry*" method often used to synthesize inorganic oxides. The process is based on hydroxylation and condensation of molecular precursor in which the precursor is an aqueous solution of an inorganic salt or a metal organic salt [55]. Sol-gel processing of TMOs has grown considerably and various synthetic routes have been developed to achieve thin films as well as various types of nanostructures [56-63].

ZnO films with preferred orientation along the (002) plane were fabricated by dissolving zinc acetate in a solution of 2-methoxyethanol and monoethanolamine and a homogeneous but stable solution was prepared [62].

A dip-coating method was used to grow the film which showed different crystallographic orientation and morphology depending on the thickness. Highly concentrated ZnO wurtzite nanoclusters (crystallite sizes ~ 3-6 nm) were produced employing ultrasound that showed bright luminescence under UV light [61]. Doped ZnO films were also prepared by the sol-gel methods in which aluminum, indium and tin have been used as dopants [64]. The deposited film had stronger c-axis orientation perpendicular to the substrate and the surface morphology changed with the type of dopants used. Overall, the (002) oriented films showed higher conductivities and transmittance. For ZnO, nanostructures such as quantum dots were synthesized using zinc acetate dihydrate and diethylene glycol [65]. The average size of quantum dots has been tuned by changing the concentration of zinc precursor and size-dependent blue shifts of photoluminescence and absorption spectra revealed the quantum confinement effect. Other sol-gel chemistries were employed to synthesize undoped [66-69] and doped [70] ZnO quantum dots.

TiO_2 thin film growth by sol-gel method has been widely investigated [71-79]. As an example, transparent TiO_2 thin films were prepared on soda-lime glass substrates via the sol–gel method from TiO_2 sol solution using tetrabutylorthotitanate, diethanolamine, polyethylene glycol, ethanol and water [71, 72]. The film was annealed at 500°C to convert TiO_2 gel films into TiO_2 (anatase) crystalline films. In a similar way, TiO_2 solution was prepared by using titanium tetra-isopropoxide as starting material, hydrochloric acid as a catalyst, and isopropanol as a solvent [73]. XRD results showed that TiO_2 thin film calcined at 300°C was amorphous, transformed into the anatase phase at 400°C, and the rutile phase at 1000°C.

4. Device Configuration

As discussed in the previous section, various growth techniques have been developed to deposit high quality TMOs on various substrates. The wide-bandgap TMOs have been successfully utilized in various optoelectronic device applications such as UV photodetection. Figure 6 illustrates 5 different device configurations that have been used as UV detector: photoconductor, Schottky, metal-semiconductor-metal (MSM), *p-n*, and *p-i-n* photodiodes. Each structure has its unique advantages for UV PD application. The photoconductor consists of two ohmic contacts on semiconductors. This type of UV PD is essentially a radiation-sensitive resistor. In general, Schottky and MSM photodiodes have the common advantages of low-temperature process, fabrication simplicity, low-noise, and high response speed [6]. The *p-n* junctions are the most common and well-studied structures for UV detection. UV photodiodes are typically more sensitive than Schottky photodiodes in response to incident light but slower, due to low carrier diffusion velocity [6].

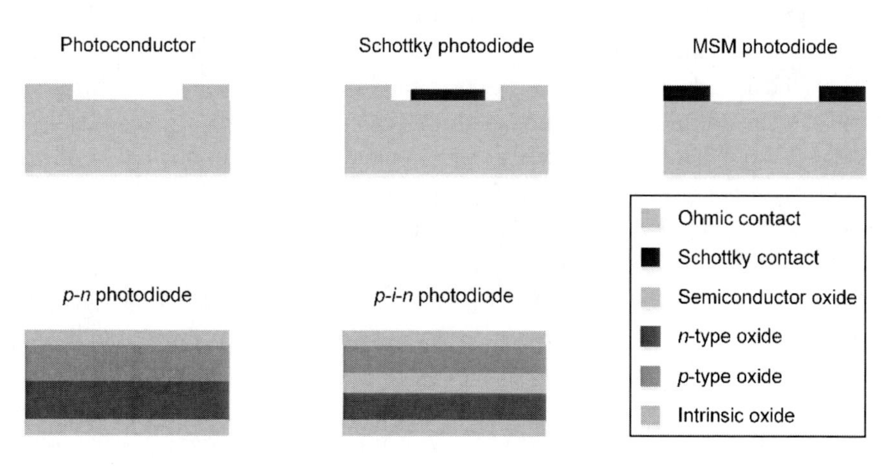

Figure 6. Schematic structures of demonstrated semiconductor oxide based UV photodetectors.

a. Photoconductor

The UV photoconductor is the simplest device structure, which consists of two ohmic contacts on a semiconductor thin film as shown in 6. Essentially, this type of photodetector is a radiation-sensitive resistor. An incident photon with energy ($h\nu$) exceeding the bandgap energy (E_g) of the semiconductor produces an electron-hole pair, generating the photocurrent throughout the device with external bias. The voltage of the applied bias and the thereby induced electrical field across the detector determines the collection efficiency of the photogenerated electron-hole pairs. This process affects the internal gain of a photoconductor with the following relation [6],

$$G = (\mu_n + \mu_p)\tau E/L = \tau/(t_n + t_p)$$

where μ_n and μ_p are the electron and hole mobility, respectively, τ is the carrier lifetime, E is the electrical field, L is the distance between the electrodes and $t_n = L/(E\mu_n)$ and $t_p = L/(E\mu_p)$ are defined as the transit time of the electrons and hole to the electrodes, respectively. This expression indicates that high internal gain can be achieved by increasing the lifetime or reducing the transit time of the photogenerated carriers. Notice that a long carrier lifetime usually results in a slow response of the UV photoconductor. The internal gain is also sub-linearly proportional to the incident UV-light power.

Table 1. Material properties of TMO for UV photodetectors

Materials	Bandgap (eV)	Electron affinity (eV)	Carrier type	Ohmic contacts
ZnO	3.37 [31]	3.7 - 4.6 [80]	n	Au [81, 82], Al [83, 84], Ni/Au [85, 86], ITO [87], Ta/Au [88], Ti/Au [89, 90], Al/Ti [91], Al/Pt [92], Ru [93], Ti/Al/Pt/Au [94], In [95]
ZnMgO	3.3 – 7.8 [36]	< 3.3	n	Cr/Au [96], Au [96], Al [97], Ni/Au [98], In [98]

Table 1 summarizes the fundamental material properties of two important TMOs for UV photoconductors that have been studied extensively due to advantages in bandgap tuning and fabrication simplicity. Takahashi et al. reported the thickness dependence of the UV photoresponse of the ZnO thin films [99]. They found that the conductivity of the ZnO films decrease with the film thickness in dark while the conductivity of the films at various thickness remain constant under UV illumination. Thereby, the thinner ZnO film results in a higher responsivity. It was also demonstrated that the photoconduction response of ZnO thin films strongly depends on the operational gas environment. Specifically, the oxidizing atmosphere, especially water vapor, expedited the decay of photoconduction when UV was turned off. Basak et al. reported the photoconductivity of sol-gel prepared 300 nm ZnO thin film with Au contacts [81]. The responsivity and the quantum efficiency of the detector reached to 0.04 A/W and 14% under 350 nm UV illumination at 5 V bias, respectively. However, this type of photoconductors showed a slow photoresponse decay. Similar device structure has been realized by Liu et al. on RF sputtered c-orientated ZnO films on SiO_2 substrate [82]. The maximum responsivity of this photoconductor was 30 A/W under

360 nm at 3 V. A much faster rise time of 20 ns and fall time of 10 μs were achieved with the sputtered device using a pulsed Nd–YAG laser (355 nm, 10 ns) as an excitation source. Figure 7 shows the obtained temporal response. The pulse shows an asymmetrical shape with a 20 ns rise time. The fall time was estimated to be 10 μs and no persistent photoconductivity was observed. The intensity decreased to zero after about 25 μs.

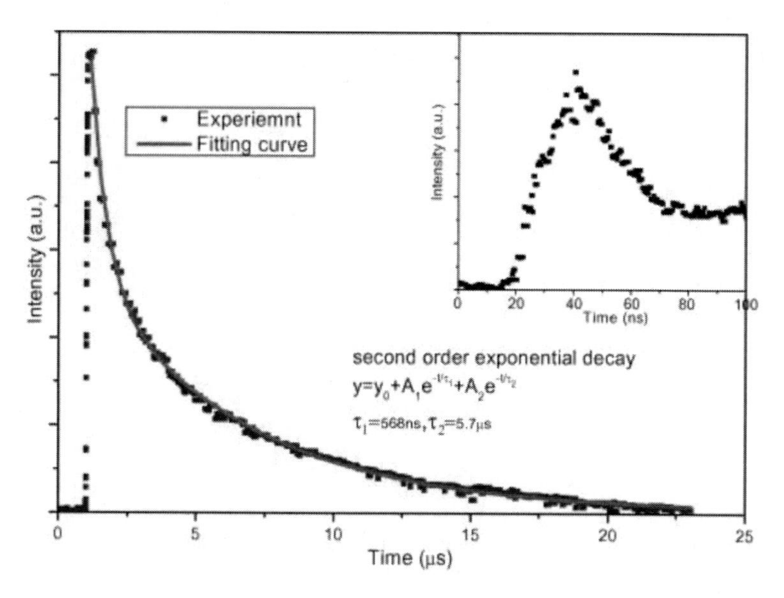

Figure 7. Temporal response of ZnO UV detectors excited by Nd–YAG laser pulses. The inset shows the enlarged impulse response. (Adapted with permission from Ref. [82]. Copyright © 2007 Elsevier B.V.).

Xu et al. fabricated photoconductors by depositing Al interdigitated electrodes (IDEs) on the c-plane sputtered ZnO thin films [83]. The device exhibited a peak responsivity of 18 A/W under 365 nm at 5 V bias. The photodetector showed a fast rise and fall time of 100 ns and 1.5 μs, respectively. The performance of the sputtered ZnO films was significantly improved to a peak responsivity of 2069 A/W and a fast rise time of 45 ns by post-annealing at 900°C in pure O_2 ambient for 1 h [100]. For PLD deposited ZnO thin films with Al IDEs, Zheng et al. measured an unprecedented responsivity of ~9000 A/W under 360 nm at 5 V bias, with ~ 90% quantum efficiency [84]. However, this device showed very slow rise and decay times of 5 min and 7 min, respectively. The high-quality epitaxial ZnO film grown by Mandalapu et al. showed a peak responsivity of ~0.4 A/W at 5 V bias under

374 nm using Al/Ti Ohmic contacts [91]. The presence of deep level defects in the epitaxial ZnO film contributed to the slow transient response.

Different surface functionalization methods have been utilized to improve the performance of the thin film based UV photoconductors. By oxygen plasma treatment on the ultra-thin ZnO epitaxial film, Liu et al. enhanced the peak responsivity significantly from 1 A/W to 10 A/W with a fast decay time below 50 μs [101]. The device also showed a high photoconductive gain of 1000. The performance enhancement originates from the suppressed oxygen vacancy and chemisorption effects due to the oxygen plasma treatment. Change et al. reported that the responsivity of ZnO thin film based UV PD can be improved with HCl treatment by a factor of 3 [85]. The HCl processed films showed a much rougher surface that increases the dark current of the devices. The measured peak responsivity reaches to 141 mA/W at 370 nm with 10 V bias. Unlike the unintentional doped *n*-type ZnO, Sun et al. prepared the Ga doped ZnO thin films using RF magnetron sputtering and fabricated the UV photoconductors [102]. The device showed a photoconductive gain of more than two orders of magnitude, and a peak responsivity of 2.6 A/W at 370 nm with 10 V bias. Moreover, the photoconductor exhibited a rapid rise and fall time of 10 ns and 960 ns, respectively.

By incorporating Mg contents with ZnO, the bandgap of the alloyed ZnMgO can be tuned from 3.3 eV to 7.8 eV. Yang et al. demonstrated visible-blind UV photoconductors based on epitaxially grown ZnMgO thin films [103]. The $Zn_{0.66}Mg_{0.34}O$ film was grown using PLD on c-plane sapphires with Cr/Au as ohmic IDEs. At 5 V bias and under an illumination of 308 nm, the fabricated detector showed a peak responsivity of 1200 A/W. The rise and fall times were 8 ns and 1.4 μs, respectively. By using RF magnetron sputtering technique, Jiang et al. fabricated ZnMgO thin film based PDs with a responsivity of ~ 1.3 A/W at 320 nm [96]. The device showed a high-speed response of 16 ns (rise time) and 250 ns (fall time). Table 2 summaries the thin film based UV photoconductors.

One-dimensional (1D) nanostructures, including nanowire (NW), nanorod (NR), microtube and nanobelt, have been widely explored as potential candidates for high-performance UV photodetectors [3]. Due to size effects, these nanostructures exhibit novel physical, chemical, and biological properties that are different from those of their bulk counterparts and especially sensitive to UV light. A 1D UV photoconductor with very high photoresponse has been demonstrated by Kind et al. [105]. The ZnO NW based photoconductor was very sensitive to 365 nm UV light with a photoconductive gain in the range of 10^4 to 10^6. The PD also showed a relative fast rise and fall response both within 1 s although the photoresponse was strongly dependent on the ambient gas conditions and the incident light power.

Table 2. Comparison of the thin film based TMO photoconductors

Materials-Ohmic contacts	Dark current-Bias voltage	Responsivity (A/W)	UV light	Rise time	Fall time	Ref
ZnO-Au	8 μA-1.5 V	0.006	350	-	~960 s	[81]
ZnO(Ga)-Al	6.7 μA/cm^2-10 V	2.6	370	10 ns	960 ns	[102]
ZnO-Al	38 μA-5 V	18	365	100 ns	1.5 μs	[83]
ZnO-Au	250 nA-3 V	30	360	20 ns	10 μs	[82]
ZnO-Al/Ti	40 mA-20 V	1.68	374	-	-	[91]
ZnO(N)-Al	450 nA-5 V	400	300	1 μs	1.5 μs	[104]
ZnO	400 pA-3 V	-	325	-	< 50 μs	[101]
ZnO-Al/ITO	640 μA-5 V	1616	365	71.2 ns	377 μs	[100]
$Zn_{0.6}Mg_{0.4}O$-Au	40 nA-3 V	1.3	320	16 ns	250 ns	[96]
$Zn_{0.66}Mg_{0.34}O$-Cr/Au	40 nA-5 V	1200	308	8 ns	1.4 μs	[103]

The 1D devices are also compatible with large scale integration processing, required for practical use as device elements and interconnects. CVD deposited ZnO NWs also demonstrated high photoconductive gain [106]. The as-grown ZnO NWs were 10 μm to 15 μm long with the diameter ranging from 150 nm to 300 nm. The deposited Ti/Au electrodes worked well as ohmic contacts on the ZnO NWs. Under the illumination of 390 nm UV, the ZnO NW based device showed an unprecedented high internal photoconductive gain of 10^8. The extremely high gain was attributed to surface oxygen defects associated with the ZnO NWs as illustrated in Figure 8. The temporal responses in air and under vacuum of the ZnO NWs revealed the presence of fast ~ 20 ns and slow ~ 10 s carrier relaxation processes. The lifetime of the electrons was extended in an oxygen deficient environment yielding diminished oxygen adsorption. Table 3 summaries the TMO based photoconductors with 1D nanostructures.

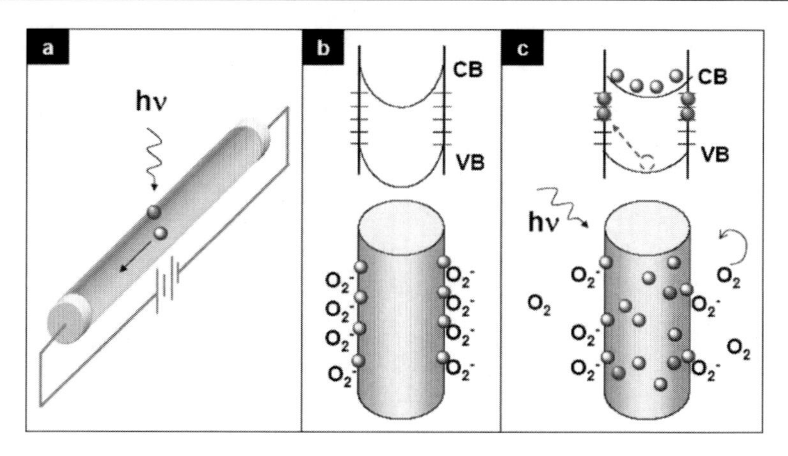

Figure 8. (a) Schematic of the photoconduction in NW photodetectors. Upon illumination with photon energy greater than the bandgap, electron-hole pairs are generated and holes are readily trapped at the surface. Under an applied electric field, the unpaired electrons are collected at the anode, which leads to the increase in conductivity. (b) and (c) Trapping and photoconduction mechanism in ZnO NWs. The top drawing in (b) shows the schematic of the energy band diagrams of a NW in dark, indicating band-bending and surface trap states. VB and CB are the valence and conduction band, respectively. The bottom drawing shows oxygen molecules adsorbed at the NW surface that capture the free electron present in the n-type semiconductor forming a low-conductivity depletion layer near the surface. (c) Under UV illumination, photogenerated holes migrate to the surface and are trapped, leaving behind unpaired electrons in the NW that contribute to the photocurrent. In ZnO NWs, the lifetime of the unpaired electrons is further increased by oxygen molecules desorption from the surface that binds holes. (Adapted with permission from Ref. [106]. Copyright © 2007 American Chemical Society).

Table 3. Comparison of the one-dimensional TMO photoconductors

Materials-nanostructure	Dark Current-Bias voltage	Photocurrent Gain	UV light (nm)	Rise time	Fall time	Ref
ZnO-NW	40 nA-5 V	$10\text{-}10^5$	350	1 s	1 s	[107]
ZnO-microtube	5 μA-5 V	~15-40	370	15 s	1850 s	[108]
ZnO-NW	6 nA-5V	$>10^8$	390	-	-	[106]
ZnO-NW	~1 pA-1 V	$10^4\text{-}10^6$	365	<1 s	<1 s	[105]
ZnO(Cu)-NW	10 pA-10 V	7000	365	-	-	[109]
ZnO-NR	1 nA-2 V	22	325	3.7 s	63.6 s	[110]
ZnO-NW	80 nA-1 V	25	254	66 s	115 s	[111]
ZnO-NW	0.04 nA-1 V	1500	365	0.6 s	6 s	[112]

b. Schottky Photodiode

A Schottky photodiode is a simple architecture, consisting of a metal-semiconductor junction. The schematic shows the energy band diagram between a metal and n-type semiconductor (Figure 9). At thermal equilibrium, band bending occurs at the interface and a Schottky barrier is established with a barrier height, Φ_B, that is the difference between the metal work function, Φ_M, (the energy difference between the metal Fermi level and the vacuum level) and the electron affinity, X, of the semiconductor (the difference between the semiconductor conduction band edge and the vacuum level). Schottky barrier diodes have several advantages including high speed, low dark current and high quantum efficiency. In addition, the devices can be operated in either photoconductive or photovoltaic (zero bias) mode although they normally operate in the former mode in applications requiring high speed.

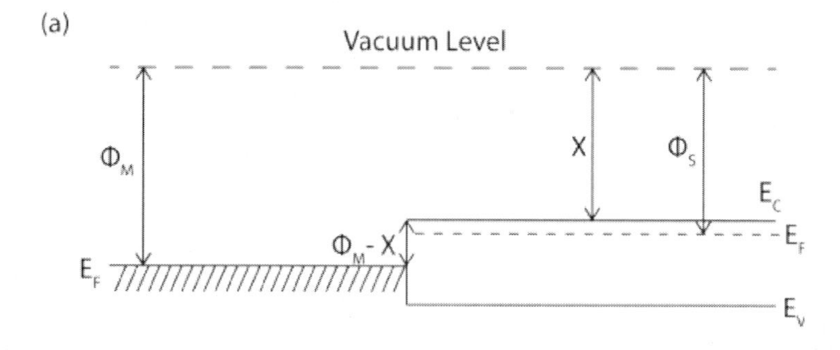

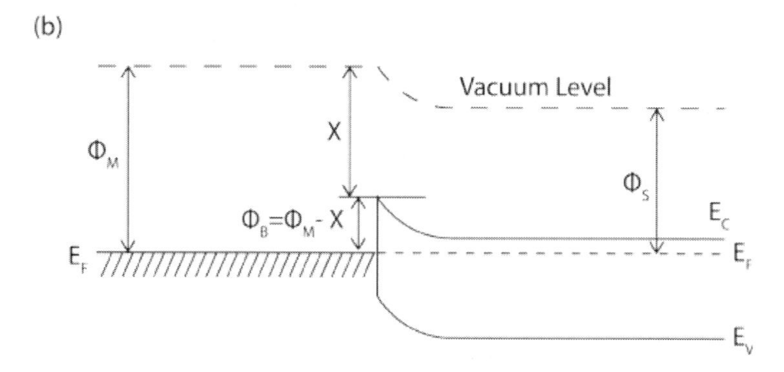

Figure 9. Energy band diagram of a metal and n-type semiconductor under (a) non-equilibrium and (b) thermal equilibrium conditions.

A UV Schottky barrier diode employing TMOs such as ZnO was reported three decades ago [116]. The architecture consisted of a glass substrate with Mn electrode as bottom contact, sputtered ZnO as active material and Au as top metal electrode. The diodes exhibited conventional I-V characteristics with rectifying behavior. However, the quantum efficiency was very low (of the order of 1%) due to high recombination rates in polycrystalline ZnO film. The device did exhibit relatively short rise time around 20 μs and decay time of 30 μs. ZnO photodiodes were also fabricated on GaN/Al_2O_3 substrates [115]. A reverse saturation current of $\sim 10^{-8}$ A in the dark, and a large photo current of $\sim 10^3$ A were achieved under ultraviolet light illumination with stable diode characteristics. The diodes had a large bandwidth of 195 nm, where the short-wavelength cutoff and the long-wavelength cutoff were 195 nm and 390 nm, respectively (Figure 10).

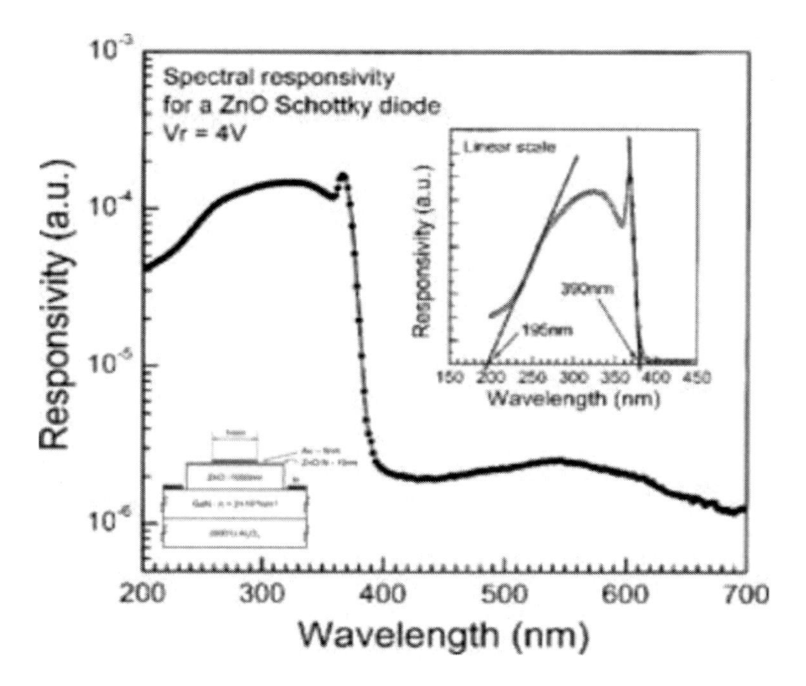

Figure 10. Spectral responsivity of ZnO Schottky barrier diodes as a function of wavelength using a 450W Xe lamp. Insets show the device structure with short- and long-wavelength cutoff regimes (Adapted with permission from Ref. [115]. Copyright © 2006 American Vacuum Society).

The responsivity was further improved when the photodiode was fabricated using single crystal (0001) ZnO by the hydrothermal growth method [117]. The device used a semitransparent Pt electrode as the Schottky contact and Al for the ohmic contact. The current-voltage characteristics and responsivity depended on the polarity of the surface (Figure 11). The responsivity was 0.185 A/W at a wavelength of 365 nm for Zn-polar surface and for O-polar, it was almost halved (0.09A/W). The results were attributed to the polarity dependences of surface chemical reactivity and the density of surface states on the ZnO. Pt metal was also used to form Schottky junctions with single ZnO nanowire grown by site-selective molecular-beam epitaxy [118]. The diodes exhibited excellent ideality factors of 1.1 at 25°C and reverse current of only 1.5×10^{-10} A at −10 V bias. The devices showed a strong photoresponse under 366 nm UV illumination with on-off current ratio of ~6. Instead of metal, a transparent polymer Schottky contact had also been utilized for high performance photodiode [119]. Organic poly(3,4-ethylenedioxythio phene) poly(styrenesulfonate) (PEDOT:PSS), which is a conducting polymer with low resistivity and high transparency, was spin coated on ZnO at room temperature in air. Quantum efficiency was approaching unity in the ultraviolet region and a visible rejection ratio of about 10^3 was achieved under zero-bias conditions. The normalized detectivity of the photodiode was estimated to be 3.6×10^{14} Jones or cm $Hz^{0.5}W^{-1}$ at 370 nm.

A dielectric passivation technique was implemented in Au/ZnO Schottky diodes to improve the rectification in which a high electric material $CaHfO_3$ was deposited on ZnO [120]. The leakage current of ZnO-based Schottky diodes were greatly suppressed. Ali et al. reported the performance of ZnO-based MSM Schottky barrier PDs processed at different temperature [121]. They found that the performance of the device improved with increasing post metal deposition annealing temperature up to 250°C, but it degraded while annealing above 250°C. Variation in electrical and photoresponse of these devices were attributed to interfacial reaction and phase transition during the annealing process.

A Schottky diode based on TiO_2 nanowire array was fabricated via a low-temperature hydrothermal method on fluorine-doped tin oxide (FTO)-coated glass substrate at relatively low temperatures (below 150°C) with Ag electrode. At -5 V bias, the dark current of the detector was less than 35 nA and a high responsivity of 3.1 A/W was achieved under 350 nm UV illumination. Other Schottky diodes made of Cr/TiO_2, Au//TiO_2 and Ni/TiO_2 were also reported [122].

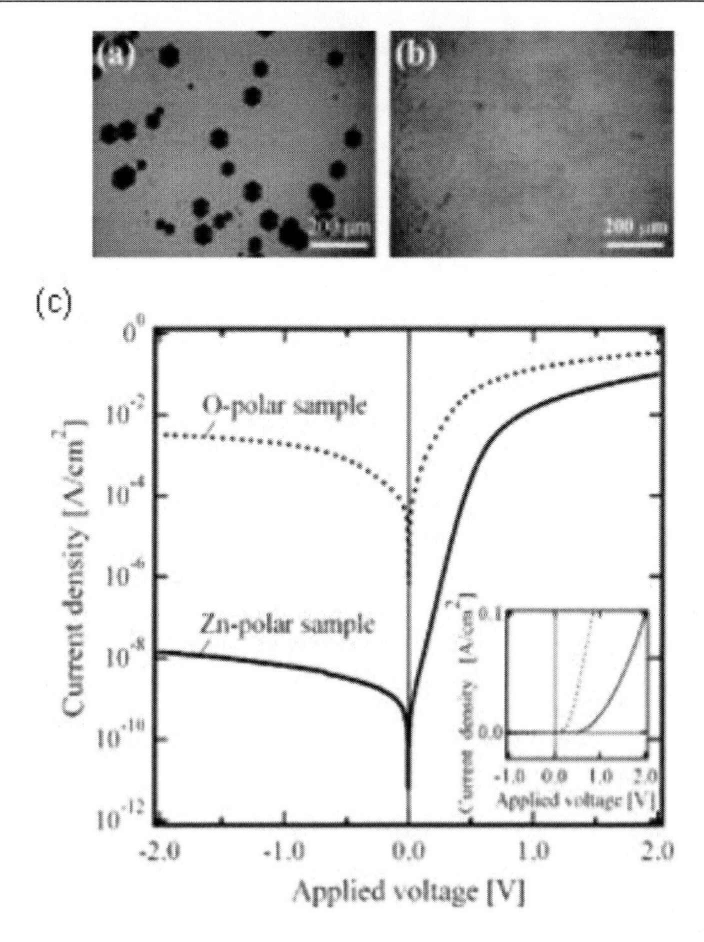

Figure 11. Optical micrographs of ZnO: (a) Zn-polar and (b) O-polar surfaces after etching in HCl (0.7 vol%) solution for 1 min at 60°C. (c) Current-voltage characteristics of the Schottky photodiode with a Pt Schottky semitransparent electrode. (Adapted with permission from Ref. [117]. Copyright © 2007 AIP Publishing LLC).

c. MSM

MSM PD is another class of devices which are comprised of two back-to-back Schottky diodes employing interdigitated electrodes on the top of the active light absorbing layer. Usually, a semiconducting material, selected for its bandgap, is epitaxially grown on a substrate and metal electrodes are patterned using a single lithography step. As the electrodes are on top of the active region, shadowing is unavoidable and the responsivity is low. Despite such drawbacks, these PDs have several advantages including (i) ease of fabrication and simplicity, (ii) process compatibility with transistors for integration in optoelectronic circuits and (iii) high speed operation due to the low capacitance per unit area.

Among various TMOs, ZnO thin film MSM photodiodes have been widely investigated and they have been grown using methods including RF magnetron sputtering [123, 124], pulse laser deposition (PLD) [125, 126], molecular beam epitaxy (MBE) [86, 127-129], laser assisted molecular beam deposition (LAMBD) [130], metal-organic chemical vapor deposition (MOCVD) [104, 131, 132], hybrid beam deposition [133], electrostatic spray deposition [134], sol-gel [135-137] and spray pyrolysis [138, 139]. The performance of the PDs depends on the types of metal contacts used in the devices. Large Schottky barrier height at the metal-semiconductor interface leads to high breakdown voltage and small leakage current that will improve the ratio between photo- and dark current, however quantum efficiency and responsivity will decrease. There have been large variations in the reported figures of merits of ZnO PDs. Oxygen plays a major role in photoconduction under UV illumination as explained earlier. When it gets absorbed on ZnO surface, a low-conductivity depletion region is created near the surface; oxygen desorption under UV illumination increases the electron concentration and this influences the barrier height between metal and semiconductor [140]. Thus, it can be expected that photoresponse would be dependent on the surface band bending and surface trap density in ZnO, which could account for the large variation of properties observed in the literature.

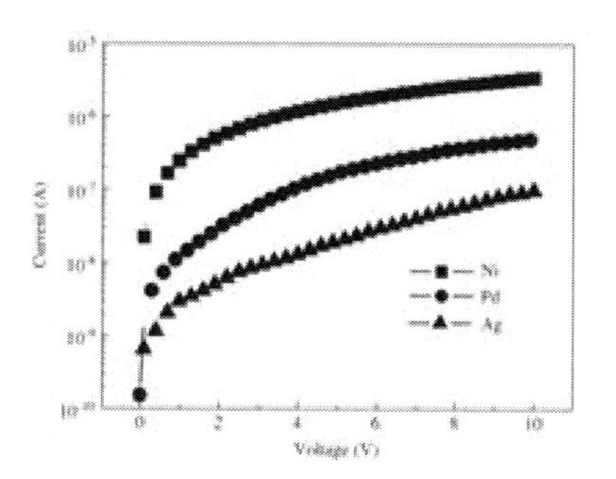

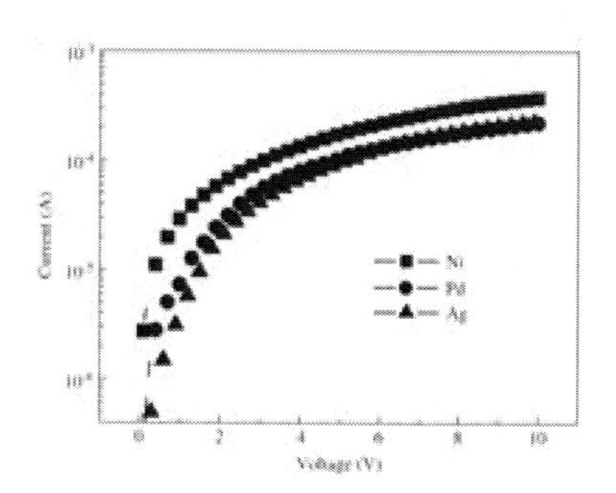

Figure 12. (left) Dark I–V characteristics of ZnO MSM UV sensors with different contact electrodes. (Right) Illuminated I–V characteristics of ZnO MSM UV sensors with different contact electrodes. (Adapted with permission from Ref. [129]. Copyright © 2006 Elsevier B.V.).

As metal work function is an important factor to determine the barrier height, various metals have been utilized to assess the performance of ZnO MSM PDs [127, 129, 132, 141]. Young et al. studied the effects of Ag, Pd and Ni contact electrodes on MBE-grown ZnO on sapphire substrates [129]. Figure 12 shows the dark and light current–voltage (I–V) characteristics of the fabricated sensors with Ag, Pd and Ni contact electrodes. The Schottky barrier height have been derived from the I–V curves using the equation [142]:

$$J = J_0\left[exp\left(\frac{qV}{nkT}\right) - 1\right]$$

where $J_0 = AT^2 exp\left(-\frac{\emptyset_B}{kT}\right)$ is the saturation current density, n is the ideality factor and k is the Boltzmann's constant, T is the absolute temperature, A is the effective Richardson coefficient, and $\emptyset_B$ is the barrier height. Using the above equation, the barrier heights for Ag/ZnO, Pd/ZnO and Ni/ZnO have been estimated to be 0.736, 0.701 and 0.613 eV, respectively. The dark current of the fabricated sensors at 1 V were 2.9×10^{-9}, 1.19×10^{-8} and 2.45×10^{-7} A, respectively. Ag/ZnO/Ag MSM UV sensors shows the lowest dark current because it has the highest Schottky barrier height. On the other hand, all the sensors show different level of photocurrent under UV illumination. Specially, sensors with Ni electrodes have the highest photocurrent because it has the lowest Schottky barrier height. Hence, quantum efficiency or responsivity of these devices varied accordingly. NEP which corresponds to the incident rms optical power required to produce a signal-to-noise ratio of one in 1-Hz bandwidth is a measure of the sensitivity of a photodetector or detector system and for Ag, Pd and Ni electrodes, NEPs were estimated to be 6.8×10^{-13}, 1.13×10^{-12} and 6.4×10^{-12} W, corresponding to the normalized detectivity of 1.04×10^{12}, 6.25×10^{11} and 1.1×10^{11} cm $Hz^{0.5}$ W^{-1}, respectively. Similarly, ZnO MSM photodetectors grown by LAMBD show three orders magnitude lower dark current for Au film than the same with Cr [130]. The responsivity was also reduced from 1.05 mA/W to 11.3 μA/W since Au has a higher work function, hence higher barrier height than Cr. For ZnO films epitaxially grown on sapphire substrates by MBE, Schottky barrier height for Ru/ZnO interface was estimated to be 0.76 eV and photocurrent to dark current contrast ratio of ~225 was achieved from the fabricated devices [127].

Although most of the MSM PDs have been fabricated on rigid substrates, Ji et al. recently demonstrated flexible PDs grown on poly(ethylene terephthalate) (PET) flexible substrates by RF sputtering [143]. A ZnO capping layer (stack structure: ZnO/Ag/ZnO/PET) was employed which showed excellent UV-to-visible rejection ratio with a responsivity of 3.80×10^{-2} A/W. Other than PET, ZnO thin film was also grown on flexible poly propylene carbonate plastic substrate using DC sputtering and the maximum responsivity of the Ni/ZnO/Ni MSM PD was found to be 1.59 A/W [144].

Alloying ZnO with MgO will allow the detection of UVB radiation [145-148]. Using PLD technique, $Mg_{0.34}Zn_{0.66}O$ thin films with a bandgap of 4.05 eV were epitaxially grown on c-plane sapphire substrates and planar geometry photoconductive type MSM PDs were fabricated with a high responsivity of 1200 A/W at 308 nm with an applied bias of 5V [103]. The response time was of the order of few ns. Similarly, $Mg_{0.47}Zn_{0.53}O$ was grown by RF magnetron co-sputtering and the PD showed the peak response at 290 nm with a cutoff wavelength at 312 nm [149]. They exhibited a very low dark current (~ 3 pA at 5 V), very high UV−visible rejection ratio and extremely fast transient response (of the order of ns). The reason for the short response time was related to the Schottky structure. Using MBE, cubic $Zn_{0.51}Mg_{0.49}O$ was deposited on the a-face sapphire substrate [145]. At 10 V, the PD device with a finger gap of 2 μm showed a low dark current of 3 pA with maximum responsivity of 5.188 A/W at 302 nm (Figure 13). The temporal photoresponse revealed a fast response speed (of the order of μs) with good reproducibility and stability.

In addition to ZnO and its alloys, TiO_2 is also a wide bandgap photoactive semiconductor which is sensitive to UV light. Xing et al. demonstrated highly sensitive fast-response UV photodetectors based on epitaxial TiO_2 films which were fabricated on $LaAlO_3$ (LAO) single crystal substrates by RF magnetron sputtering [150]. Due to the smaller lattice mismatch with TiO_2, LAO is one of the most suitable substrates for the epitaxial deposition of high-quality TiO_2 thin films. The fabricated TiO_2 PDs exhibited a highly sensitive photoresponse in the UV region with a maximum photocurrent responsivity of 3.63 A/W at 310 nm (Figure 14) and an ultrafast response time with a full-width at half-maximum of ~90 ns under a pulsed laser illumination. Xue et al. demonstrated a nanocrystalline TiO_2 film PD employing Au Schottky contacts, and the devices showed high responsivity of 199 A/W at 260 nm [151]. Similarly, Kong et al. fabricated MSM TiO_2 PD with responsivity of 899.6 A/W under UV light [152]. Although a very high responsivity in the UV region was achieved, the response time was of the order of tens of seconds.

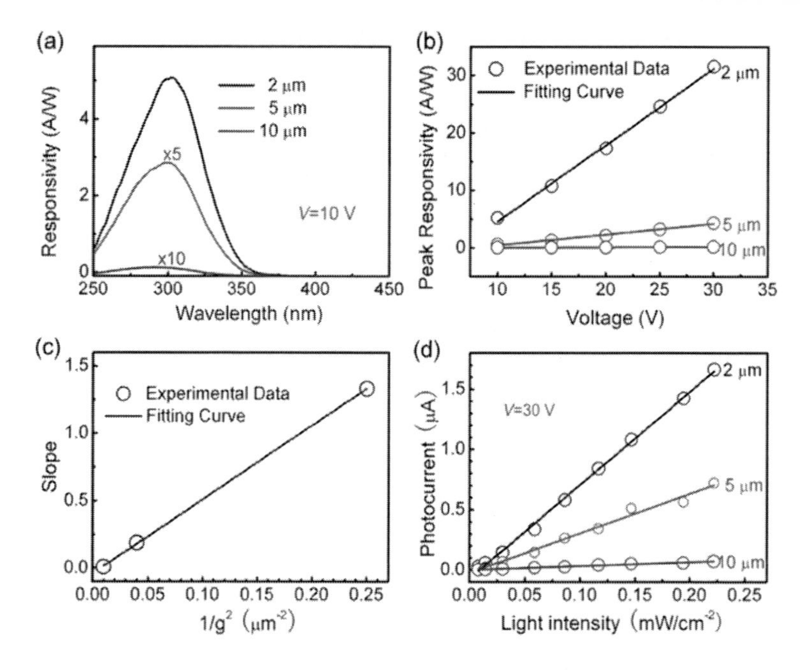

Figure 13. (a) Spectral response at 10 V bias, (b) responsivity as a function of bias, (c) slope of the plot in (b) as a function of $1/g^2$ (g is the gap between two electrodes), and (d) photocurrent as a function of light intensity at 30 V bias of the c-ZnMgO MSM photodetectors. (Adapted with permission from Ref. [145]. Copyright © 2015 American Chemical Society).

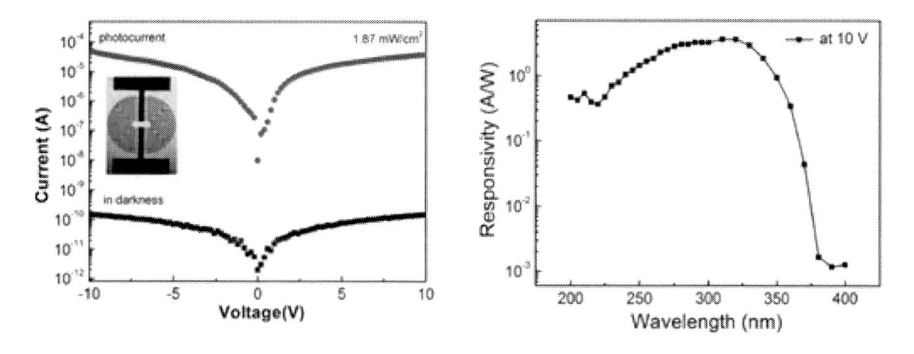

Figure 14. (left) Photo- and dark current characteristics of TiO_2 MSM UV sensors with different applied voltages. Inset shows the optical micrograph of the device. (Right) Spectral response of the TiO_2 detector with 10 μm finger width at 10 V bias. (Adapted with permission from Ref. [150]. Copyright © 2011 IOP Publishing Ltd.).

Various oxide nanostructures have also been widely studied as potential candidates for high-performance MSM UV photodetectors [3]. Nanostructured ZnO materials have received great attention due to their unique opto-electronic properties. Because of the high surface-to-volume ratio, and superior properties, ZnO has been widely used for nanometer-scale visible-blind UV-light detection with a high sensitivity and selectivity. As an example, a "*visible-blind*" solution-processed UV photodetector was realized on the basis of colloidal ZnO nanoparticles and the devices exhibited high UV photocurrent with a responsivity of 61 A/W at 370 nm with an average intensity of 1.06 mW/cm^2 [140]. Ji et al. fabricated ZnO nanorod (NR) arrays which were selectively grown in between the interdigitated electrodes by a chemical method [149]. The fabricated NR PDs showed much higher responsivity as compared to planar ZnO thin films. ZnO NW PD was fabricated by a simple method using two patterned zinc electrodes [153]. The self-catalytic growth takes place on patterned Zn electrodes at 700°C in an O_2/Ar gas flow of 20 sccm (standard cubic centimeter per minute at STP)/80 sccm, respectively, at atmospheric pressure for 3h. The device demonstrated fast response (< 0.4ms) under 370 nm UV illumination in air, which has been attributed to the adsorption, desorption, and diffusion of water molecules in the air onto the NW significantly influencing the photoresponse. Highly sensitive PDs based on PVA-coated ZnO nanoparticles and gold interdigitated electrodes patterned by optical lithography had been fabricated and the results indicated that PVA-coated ZnO nanoparticles had less surface defects [154]. The PDs exhibited photo-to-dark current ratios ranging between 3.85×10^6 to 1.34×10^8, depending on the gap and a peak responsivity of 731.42 A/W was achieved at 345 nm with a cutoff wavelength of 375 nm.

Niobium pentoxide (Nb_2O_5), is another important TMO with a bandgap of ~3.4 eV which is an ideal candidate for visible-blind UV PD. A high-performance PD made of an individual Nb_2O_5 nanobelt has been demonstrated and the results are shown in Figure 15 [155]. These quasi-aligned Nb_2O_5 nanobelts are typically 100–500 nm wide and 2–10 μm long and were synthesized using a hydrothermal treatment of a niobium foil in a potassium hydroxide solution followed by proton exchange and calcination treatment. The PD exhibited linear photocurrent characteristics, excellent light selectivity, and high EQE of 6070% with long-term stability at an applied voltage of 1.0 V. By improving the crystalline quality and engineering the defects of Nb_2O_5 nanobelt crystals, the performance was further improved. Tamang et al. also studied the photoresponse of isolated Nb_2O_5 nanowires with platinum electrodes using a focused laser beam [156]. Higher photocurrent was achieved with localized laser irradiation under reverse bias. They proposed that the localized thermal heating due to focused laser beam irradiation was responsible for the photocurrent at the NW–Pt interface due to the reduced Schottky barrier/width resulting from an increase in charge carriers and thermoelectric effects. Finally, we summarize the figures of merits of various MSM PD in Table 4.

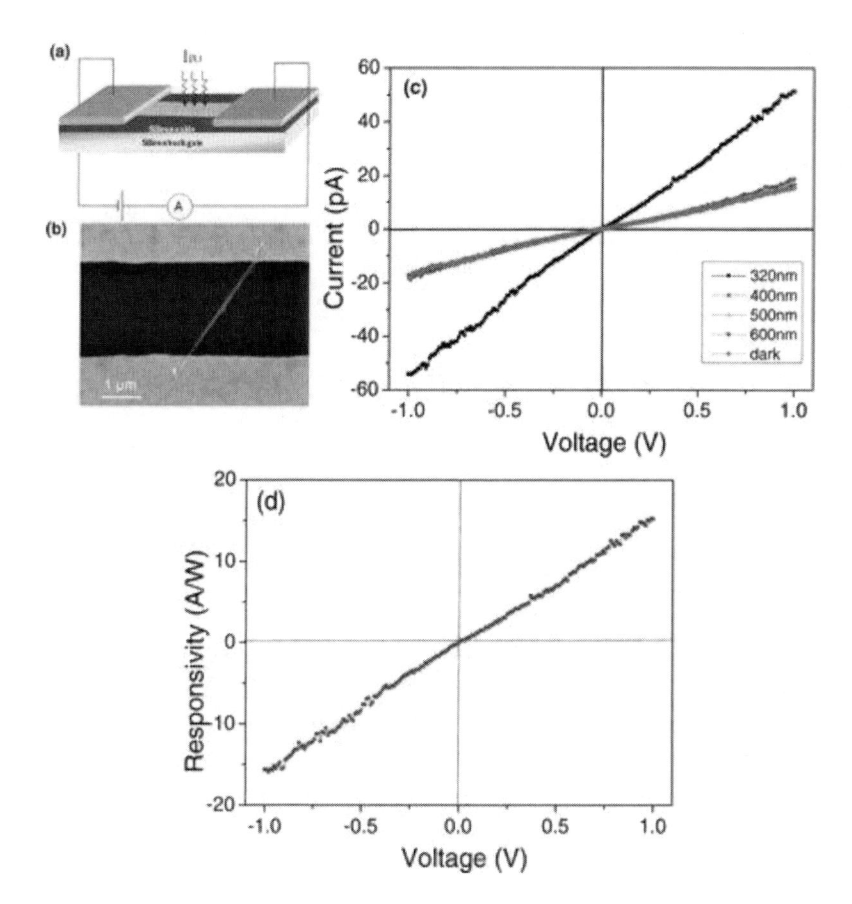

Figure 15. (a, b) Schematic diagram and SEM image of the individual Nb_2O_5-nanobelt device with interdigitated electrodes. (c) Current-voltage characteristics of an individual Nb_2O_5 nanobelt PD in dark and under illumination with different wavelength of light. (d) Responsivity versus applied-voltage characteristics under illumination of 320 nm light [155]. Copyright © 2011 Wiley-VCH Verlag GmbH and Co. KGaA. Reproduced with permission).

Table 4. Comparison of TMO based MSM PDs

Materials-Schottky contacts	Dark Current-Bias voltage	Responsivity (A/W)	UV light (nm)	Detectivity (Jones or cm $Hz^{0.5}$ W^{-1})	Rise time	Fall time	Ref.
ZnO-Ag	2.9 nA-1 V	0.066	370	1.04×10^{12}	-	-	[129]
ZnO-Pd	11.9 nA-1 V	0.051	370	6.25×10^{11}	-	-	
ZnO-Ni	245 nA-1 V	0.09	370	1.1×10^{11}	-	-	
ZnO-Cr	13.2μA-4.5 V	1.05×10^{-3}	-	-	-	-	[130]
ZnO-Au	19.7nA-4.5 V	11.3×10^{-6}	-	-	-	-	
ZnO-Ru	80 nA-1 V	-	-	-	-	13 ms	[127]
$Zn_{0.53}Mg_{0.47}O$-Au	3 pA-5 V	10.6×10^{-3}	290	-	10 ns	30 ns	[149]
$Zn_{0.8}Mg_{0.2}O$-Au	4.2 nA-5 V	0.022	350	3.1×10^{11}	10 ns	170 ns	[157]

d. *p-n* Junction

The *p–n* junction type PD, also known as photodiode, is the most common and well-studied class of semiconductor devices. A photodiode consists of an abrupt *p-n* junction with two ohmic contacts. Figure 16 shows the schematic diagram and the typical *I-V* characteristic of a *p-n* photodiode. When *n-* and *p-*type semiconductors are in contact, the mobile electrons from the *n* side diffuse into the adjacent *p* side and verse versa. A built-in electrical field is produced across the junction.

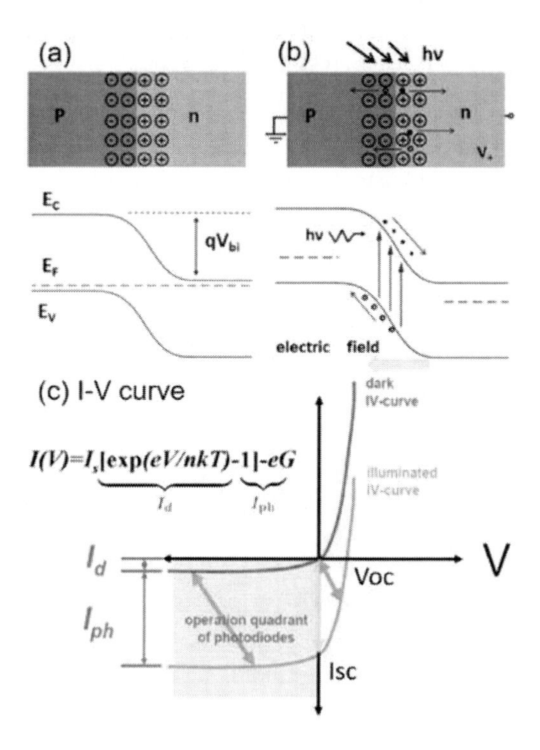

Figure 16. Diagram of a p-n junction under (a) equilibrium and (b) illumination conditions. (Adapted with permission from Ref. [6]. Copyright © 2014 IOP Publishing Ltd.) (c) *I-V* characteristic of *p-n* photodiode [146].

Under UV illumination, the generated electrons and holes are separated by the electric field and collected at *n* and *p* sides, respectively. A photodiode is usually operated under reverse bias but not extended to the avalanche region. If the photodiode is operated under zero current, the photogenerated electrons and holes are accumulated at two ends of the junction, leading to an open-circuit voltage as indicated in Figure 16c. The short-circuit current is measured under the zero external bias condition. The total current can be expressed as following equation [8, 146],

$$I_{tot}(V,\Phi) = I_s[\exp(qV/nkT) - 1] - I_{ph}(\Phi)$$

where $I_s[\exp(qV/nkT) - 1]$ is the dark current and $I_{ph}(\Phi)$ is the photocurrent. In the dark current term, q, V, n, k, T are the elementary charge, the applied voltage, the ideality factor, Boltzmann's constant, and the temperature in Kelvin, respectively. I_s is the reverse saturation current, which is expresses as

$$I_s = qA\left(\sqrt{\frac{D_p}{\tau_p}\frac{n_i^2}{N_D}} + \sqrt{\frac{D_n}{\tau_n}\frac{n_i^2}{N_A}}\right)$$

where q is the elementary charge, A is the cross-section of the device, $D_{p,n}$ are the diffusion coefficients of holes and electrons, respectively, $N_{D,A}$ are the densities of acceptors and donors, respectively, $\tau_{p,n}$ are the carrier life-times of holes and electrons, respectively and n_i is the intrinsic carrier concentration. The same equations govern the transportation of electrons in both Schottky and *p-n* junction PDs. However, in contrast to the Schottky type, the photo-effects in the *p-n* junction photodiodes are produced by the minority carriers. Therefore, the *p-n* junction type photodiode is generally more sensitive to the incident UV light than the Schottky device [6]. Three dominant factors, namely t_s, t_0, and t_{RC}, determine the response rate of a *p-n* photodiode [8] in which t_s is the drift time of the carrier across the depletion region, t_0 is the diffusion time of the carrier through the bulk to the electrodes and t_{RC} is the RC constant time due to the parasitic capacitance C and resistance R of the photodiode. Thereby, to enhance the response time of the photodiodes, it is desired to generate the photoelectron-hole pairs in the *p* side of the depletion region, owing to the high mobility of electrons [8]. We will now discuss various p-n homo- and heterojunction photodiodes using different TMOs.

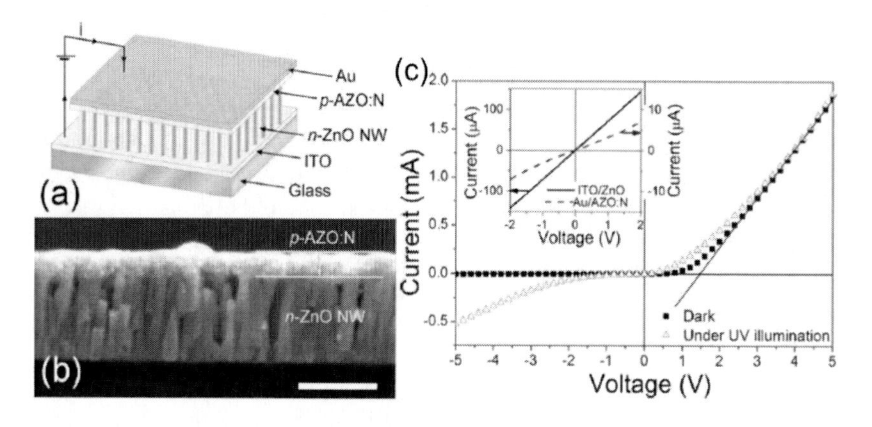

Figure 17. (a) Schematic diagram of the ZnO homojunction device. The circuit indicates that the device is connected with a positive bias. (b) An SEM image of the cross-section of the homojunction device. (c) I-V characteristics of the homojunction device measured in the dark and under UV illumination at 384 nm. The inset shows the I-V characteristics of the ITO/ZnO NWs contact and the Au/AZO:N film contact. (Adapted with permission from Ref. [158]. Copyright © 2010 AIP Publishing LLC).

d.1. p-n Homojunction

It is widely accepted that the unintentionally doped ZnO shows *n*-type semiconducting behavior which is due to the intrinsic oxygen vacancies in ZnO [159]. Thus, the fabrication of stable p-type ZnO remains a challenge and thus poses difficulty in making *p-n* homojunction photodiodes. However, there have been significant efforts to realize these homojunction devices. By depositing ZnO thin film on the GaAs substrate using RF magnetron sputtering and post-annealing the sample, Moon et al. achieved high quality *p*-ZnO film using As a dopant and fabricated homojunction photodiodes [160]. The device exhibited a notable photocurrent ~ 2 mA at 325 nm with negligible visible light response. Leung et al. reported a simple and low-cost fabrication methods of ZnO *p-n* homojunction [158]. The schematic illustration and SEM image of the photodiode are shown in Figures 17a and b, respectively. The *n*-ZnO NWs arrays was grown by hydrothermal method and *p*-type ZnO film was achieved using Al, N dopant by a sol-gel method. The solution was prepared with zin acetate dehydrate, ammonium acetate, and aluminum nitrate nonahydrate in propan-2-ol and diethanolamine. The fabricated junction showed a good rectification of 150 at 3V in the dark with a peak responsivity of 4 A/W at 384 nm. By doping ZnO with Li and N using the sol-gel method, a self-powered ZnO *p-n* homojunction photodiode was demonstrated by Shen et al. [161]. The photodiode showed a spectrum-selective response with the peak responsivity of 18.5 μA/W at 380 nm under zero bias. The narrow spectrum response was due to the filtering effect of the neutral region in the *p*-ZnO layer. The responsivity increased to 4 mA/W and 10 mA/W at an external reverse bias of 3 V and 10 V, respectively. To enhance the responsivity of the ZnO homojunction at zero bias, Sun et al. inserted an intrinsic ZnO (i-ZnO) layer in between *p*-ZnO and *n*-ZnO layers [162]. The addition of intrinsic ZnO layer broadened the depletion region with smaller parasitic capacitance. The films were grown by MBE method at high temperature (over 750°C) and dopants such as Li, N were utilized to achieve p-type ZnO. The maximum responsivity of such *p-i-n* device reached to 0.45 mA/W at 390 nm without any external bias.

The first ZnMgO *p-n* homojunction based photodiode was demonstrated by Liu et al. plasma-assisted MBE on sapphire substrate [98]. The *p*-type ZnMgO were also prepared by the same growth method [163]. Ni/Au and In were deposited and annealed at 300°C in vacuum to form ohmic contacts on *p*- and *n*-ZnMgO sides, respectively. By increasing the reserve bias from 0 V to 9 V, the responsivity at 325 nm increased linearly from 3.7 μA/W to 400 μA/W and the photodiode showed a high visible rejection ratio ($R_{325\ nm}/R_{400\ nm}$) of

100 at 0 V bias and 10000 at 6 V bias. The photodiode exhibited a fast temporal response with 10 ns rise and 150 ns fall time.

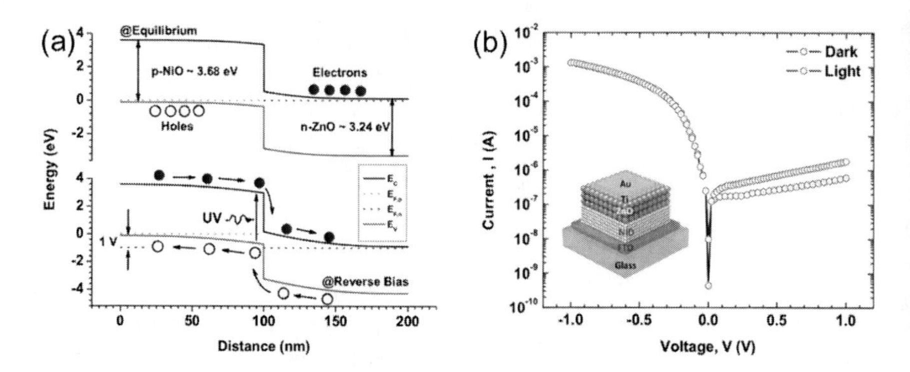

Figure 18. (a) Schematic band alignment of p-NiO in contact with n-ZnO, with (bottom) and without (top) UV illumination. (b) Diode characteristic of the device: both in dark and UV illumination at 325 nm with 6.5 μW power. The rectifying nature of the diode as well as the photoresponse is evident from the I-V data. The inset illustrates the schematic of the PD showing NiO/ZnO heterojunction along with FTO and Ti/Au metal contacts. (Adapted with permission from Ref. [164]. Copyright © 2015 Royal Society of Chemistry).

d.2. p-n Heterojunction

In order to circumvent the difficulty of forming the stable *p*-type ZnO, heterojunction architecture with *n*-ZnO and other *p*-type semiconductors, including wide-bandgap materials, Si and polymers, has attracted great attention recently. Being an intrinsic *p*-type semiconductor with a bandgap of 3.7 eV, NiO shows a matching band alignment with ZnO, as shown in Figure 18a. Thus, *p*-NiO has been exploited extensively to realize *p-n* heterojunction photodiode with *n*-ZnO [164-168]. Ohta et al. firstly demonstrated the photoresponse of a heterojunction made of *p*-NiO and *n*-ZnO [165]. The single crystal oxide films were grown on an Yttrium stabilized Zirconia substrate by PLD. Au and ITO were used as ohmic contacts. The diode showed a clear rectifying behavior with an ideality factor of 2 and a threshold voltage of 1 V. Although the fabricated device showed a negligible responsivity at 0 V bias, it was enhanced to 0.3 A/W at a reverse bias of 6 V. Using a low-cost solution process method, Debnath et al. fabricated UV photodiodes using NiO and ZnO [164]. The oxide films were deposited on a transparent glass/FTO substrate and Ti/Au ohmic contacts were engineered to be UV transparent, allowing the illumination of light from both sides. The schematic diagram of the diode is shown as an inset in Figure 18b.The device showed a low-dark current with a high detectivity of 6.3×10^{11} Jones at 365 nm under 1 V reverse bias. The responsivity of the photodiode was 0.28 A/W under the same conditions. By doping ZnO with Mg using the similar sol-gel method, Xie et al. demonstrated the tunable UV photoresponse with *p*-NiO and *n*-$Zn_{1-x}Mg_xO$ (x = 0 – 0.1) photodiodes [169]. The bandgaps of the $Zn_{1-x}Mg_xO$ films increased linearly from 3.24 eV to 3.49 eV with the increase of Mg contents. Since the incorporation of Mg into ZnO suppresses the oxygen vacancy and thus the carrier density, the dark current decreased with the increase of Mg content. Figure 19 shows the spectral response of the photodiodes made with NiO and ZnMgO. The increase of ZnMgO bandgap with higher Mg concentration lead to a blue-shift of the peak responsivity and detectivity. The devices showed maximum responsivity of 0.22 A/W and 0.4 A/W for x = 0.05 and 0.1, respectively. The detectivities of $Zn_{0.95}Mg_{0.05}O$ and $Zn_{0.9}Mg_{0.1}O$ based devices were 0.17×10^{12} Jones and 2.2×10^{12} Jones, respectively. Both the photodiodes showed excellent UV-to-visible rejection ratios. Hasan et al. reported low-power and low-temperature sputtered NiO/ZnO films on plastic substrate for UV PD application [167]. The prepared photodiode showed a self-powered behavior with an open-circuit voltage of 270 mV and short circuit current of 85 nA. The responsivity reached to 0.19 A/W at 370 nm under a reverse bias of 1.2 V.

In addition to the NiO, other *p*-type wide-bandgap materials such as SiC and GaN have been employed with ZnO for *p-n* junction. Alivov et al. reported a *n*-ZnO/*p*-SiC heterojunction photodiode grown by plasma-assisted MBE [170]. The ZnO layer was epitaxially grown on 6H-SiC substrate and annealed at 600°C. The fabricated device showed a good rectifying current-voltage characteristic with a dark current less than 2×10^{-4} A/cm^2 at -10 V bias. At -7.5 V, the peak responsivity reached at 0.045 A/W under 378 nm illumination. The ohmic contacts were made by Au/Al and Au/Ni on the ZnO and SiC sides, respectively. Bie et al. demonstrated ZnO/GaN nanoscale *p-n* junctions for visible-blind UV detection [171]. The photodiode made of *n*-ZnO NW and *p*-GaN film showed good rectification characteristics under dark condition with a threshold voltage of 3 V and a leakage current less than 1 pA. The device also showed self-powered behavior under 325 nm UV illumination. An open-circuit voltage of 2.7 V and short circuit current of 2 μA I_{sc} with a fill factor of 0.20 were observed under 325 nm.

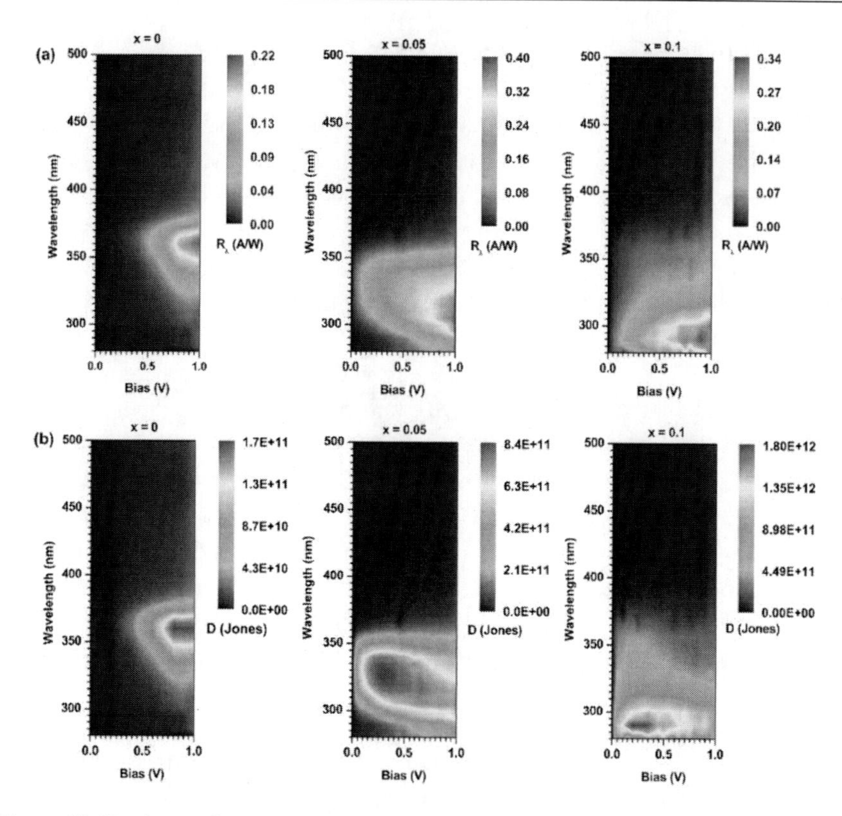

Figure 19. Device performance of NiO/$Zn_{1-x}Mg_xO$ heterojunctions with various Mg content, x as a function of applied reverse bias: (a) responsivity and (b) detectivity. (Adapted with permission from Ref. [169]. Copyright © 2015 American Chemical Society).

Owing to the advances in microfabrication technology, Si is also frequently used for the ZnO heterojunction photodiodes despite the large lattice mismatch. Jeong et al. reported the photo effect from the RF sputtered *n*-ZnO/*p*-Si diodes [173]. The Au-Al alloy (atomic ratio 3:1) was deposited on ZnO film to form ohmic contacts. Due to the small bandgap of Si (~1.1 eV), the diodes showed strong responsivities of 0.5 A/W and 0.3 A/W at 310 nm (UV) and 650 (visible) wavelength, respectively at 30 V reverse bias. The responsivity increased lineraly with the reverse bias. To eliminate the visible light response from Si, Zhang et al. inserted an intrinsic MgO layer in between the p-n junction and suppressed the visible light response from Si resulting in a visible-blind UV PD [172]. The high quality single crystalline 500 nm ZnO and 50 nm MgO films were grown by MBE on *p*-Si substrate and Ti/Au metal layers were used as ohmic contacts. The fabricated *p-i-n* photodiode showed a rectification ratio of 10^4 at 2 V and a dark current of 0.5 nA at -2 V. A sharp cutoff edge was observed at 378 nm, which corresponds to the near band edge absorption of ZnO. The addition of MgO layer introduced a potential barrier between Si/MgO and MgO/ZnO as illustrated in Figure 20. Therefore, the response to the visible light in the Si layer was rejected due to the high potential barrier at the junction. In contrast, the UV generated holes in the ZnO side can drift through the low potential barrier, which leads to the large collected photocurrent.

By employing a *p*-type polymer with ZnO, Lin et al. reported the nanostructure UV PD based on ZnO NR/polyfluorene hybrid [174]. The ZnO NRs were grown by electrodeposition on ITO substrate whereas polyfluorene which was synthesized using a sol-gel method was then spin-coated on ZnO. The fabricated device showed a clear rectifying *I-V* characteristics with a ratio of 122 at 1 V. Under 350 nm UV illumination, the photodiode showed a self-powered behavior with an open circuit voltage of 335 mV, a short circuit current density of 18 μA/cm^2 and a fill factor of 0.46. The peak responsivity 0.18 A/W was observed at 300 nm under 2 V reverse bias.

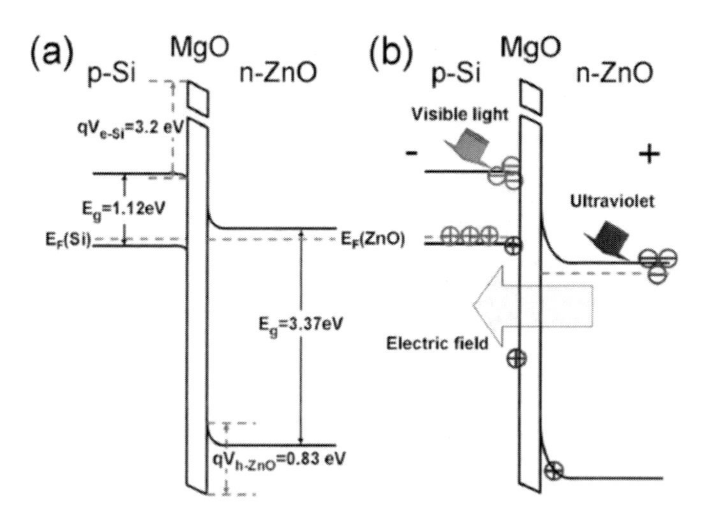

Figure 20. The energy-band diagrams of n-ZnO/insulator-MgO/ p-Si heterojunctions (a) under zero bias and in dark and (b) under reverse bias and in light illumination. The circles with a line and a cross inside stand for the photogenerated electrons and holes, while the brown and blue relate to the visible and UV excitation in depletion regions of p-Si and n-ZnO, respectively. (Adapted with permission from Ref. [172]).

Table 5. Summary of TMO based PDs with *p-n* junction structures

Materials	Bias (V)	Responsivity (A/W)	UV light (nm)	Turn on voltage (V)	Detectivity (Jones or cm $Hz^{0.5}$ W^{-1})	Rise time	Fall time	Ref.
Ni-NiO-ZnO-ITO	-1	0.01	380	~0.6	-	-	37	[177]
In-ZnO-NiO-Au	0	4.93×10^{-4}	370	1.7	-	10 μs	30.3 μs	[166]
FTO-ZnO-NiO-Ag	-3	2.27	365	2.3	-	-	30 s	[178]
ITO-NiO-ZnO-Al	-1	10.2	340	-	1×10^{12}	0.2 s	0.18 s	[179]
ITO-NiO-ZnO-Al	-5	21.8	310	-	1.6×10^{12}	-	-	[168]
FTO-NiO-ZnO-Ti/Au	-1	0.28	325	0.3	6.3×10^{11}	0.28 s	5.4 s	[164]
Pt-ZnO-GaN-Ni/Au	0	-	325	3	-	20 μs	219 μs	[171]
Ni/Au-*p*ZnO-*n*ZnO-In	-3	4×10^{-3}	380	3.3	-	-	-	[161]
Ni/Au-*p*ZnO-*n*ZnO-In	0	4.5×10^{-4}	390	4.2	-	-	260 ns	[162]
Au/Al-ZnO-Si-In	-30	0.5	310	-	-	-	-	[173]
Ag-Cu_2O-TiO_2-Ag	-5	-	365	0.9	-	<1 s	-	[175]
Ag-NiO-TiO_2-FTO	-	-	365	3	-	0.1 s	0.1 s	[176]

TiO_2 has also been used as UV PD due to wide-bandgap (3.2 eV for anatase and 3.0 eV for rutile). Tsai et al. reported *p*-Cu_2O/*n*-TiO_2 NW heterojunction as UV PD [175]. TiO_2 NWs were thermally grown on TiO_2/glass templates and DC sputtered Cu_2O film was infiltrated in between those NWs that formed a *p-n* heterojunction. The device showed a clear rectifying characteristic with a turn-on voltage of 0.9 V. Khun et al. demonstrated another fast and sensitive UV PD based on *p*-NiO/*n*-TiO_2 heterojunction [176]. The TiO_2 NRs and NiO film were hydrothermally grown on FTO glass substrate. The device exhibited very fast response under UV illumination. We now summarize various p-n junction PDs in Table 5.

CONCLUSION

The fundamental concepts of PDs and the recent progress in using various TMOs for UV PD applications have been concisely reviewed. TMOs and the related devices made out of them have been a subject of research over the decades. The efforts have been intensified recently for better understanding the physical properties of various metal oxide films/nanostructures for various optoelectronic device applications. The high quality TMO films and 1D nanostructures can now be easily grown using various physical and chemical approaches. With a proper selection of materials and detector scheme, the spectral response of a TMO based PD can also be tuned in the UVA, UVB and UVC regimes. The PDs composed of these wide-bandgap oxides offer compact size, fast photoresponse speed, high photoresponsivity, ruggedness, and intriguing application in optoelectronics. Some of the demonstrated devices have already shown advantages over the current commercial UV detectors. However, there are further challenges and opportunities to improve the figures of merits of such TMO based UV PDs. First of all, it is difficult to achieve *p*-type TMO, specially p-ZnO with the requisite reproducibility and quality. *p*-type TMOs are crucial for achieving high performance p-n homojunction based UV PDs as such configuration provides the best suitable band alignment, resulting in high quality p-n junction with improved performance. Although the realization of *p*-ZnO and *p*-ZnMgO has progressed significantly, it is still difficult to produce stable and reproducible ZnO with precisely controlled carrier concentration and mobility. On the other hand, the phase segregation and low crystal quality are the key issues in ZnMgO films. In addition to that, a cost-effective approach for producing high quality TMOs is necessary to commercialize TMO based UV PDs. The potential techniques include, but not limited to, sputtering, CVD, sol–gel and spray pyrolysis, which are capable for mass production and compatible with microfabrication process. There are plenty of opportunities to develop UV PDs with excellent figures of merits and novel concepts both in material epitaxy, device fabrication as well as new device architecture would be beneficial to expedite the progress further.

REFERENCES

[1] T. I. O. f. Standardization, "ISO 21348 definitions of solar irradiance spectral categories," vol. ISO 21348:2007(E), ed. Switzerland, 2007.

[2] L. Sang, M. Liao, and M. Sumiya, "A comprehensive review of semiconductor ultraviolet photodetectors: from thin film to one-dimensional nanostructures," *Sensors (Basel),* vol. 13, pp. 10482-518, 2013.

[3] L. Peng, L. Hu, and X. Fang, "Low-Dimensional Nanostructure Ultraviolet Photodetectors," *Advanced Materials,* vol. 25, pp. 5321-5328, 2013.

[4] F. Omnes, E. Monroy, E. Munoz, and J.-L. Reverchon, "Wide bandgap UV photodetectors: A short review of devices and applications," in *Gallium Nitride Materials and Devices II*. vol. 6473, H. Morkoc and C. W. Litton, Eds., ed, 2007.

[5] E. Monroy, F. Omnes, and F. Calle, "Wide-bandgap semiconductor ultraviolet photodetectors," *Semiconductor Science and Technology,* vol. 18, pp. R33-R51, 2003.

[6] Y. N. Hou, Z. X. Mei, and X. L. Du, "Semiconductor ultraviolet photodetectors based on ZnO and $Mg_xZn_{1-x}O$," *Journal of Physics D-Applied Physics,* vol. 47, p. 283001, 2014.

[7] Y. A. Goldberg, "Semiconductor near-ultraviolet photoelectronics," *Semiconductor Science and Technology,* vol. 14, pp. R41-R60, 1999.

[8] M. Razeghi and A. Rogalski, "Semiconductor ultraviolet detectors," *Journal of Applied Physics,* vol. 79, pp. 7433-7473, 1996.

[9] A. Tsukazaki, S. Akasaka, K. Nakahara, Y. Ohno, H. Ohno, D. Maryenko, et al., "Observation of the fractional quantum Hall effect in an oxide," *Nature Materials,* vol. 9, pp. 889-893, 2010.

[10] F. D. Auret, S. A. Goodman, M. Hayes, M. J. Legodi, H. A. van Laarhoven, and D. C. Look, "Electrical characterization of 1.8 MeV

proton-bombarded ZnO," *Applied Physics Letters,* vol. 79, pp. 3074-3076, 2001.

[11] S. Das, S. Chakrabarti, and S. Chaudhuri, "Optical transmission and photoluminescence studies of ZnO-MgO nanocomposite thin films," *Journal of Physics D-Applied Physics,* vol. 38, pp. 4021-4026, 2005.

[12] P. L. Richards, "Bolometers for infrared and millimeter waves," *Journal of Applied Physics,* vol. 76, pp. 1-24, 1994.

[13] P. Carcia, R. McLean, M. Reilly, and G. Nunes Jr, "Transparent ZnO thin-film transistor fabricated by rf magnetron sputtering," *Applied Physics Letters,* vol. 82, pp. 1117-1119, 2003.

[14] Ü. Özgür, Y. I. Alivov, C. Liu, A. Teke, M. A. Reshchikov, S. Doğan, et al., "A comprehensive review of ZnO materials and devices," *Journal of Applied Physics,* vol. 98, p. 041301, 2005.

[15] B. Khuri-Yakub, J. Smits, and T. Barbee, "Reactive magnetron sputtering of ZnO," *Journal of Applied Physics,* vol. 52, pp. 4772-4774, 1981.

[16] J. J. Chen, Y. Gao, F. Zeng, D. M. Li, and F. Pan, "Effect of sputtering oxygen partial pressures on structure and physical properties of high resistivity ZnO films," *Applied Surface Science,* vol. 223, pp. 318-329, 2004.

[17] S.-H. Jeong, B.-S. Kim, and B.-T. Lee, "Photoluminescence dependence of ZnO films grown on Si (100) by radio-frequency magnetron sputtering on the growth ambient," *Applied Physics Letters,* vol. 82, pp. 2625-2627, 2003.

[18] Y. M. Lu, W. S. Hwang, W. Y. Liu, and J. S. Yang, "Effect of RF power on optical and electrical properties of ZnO thin film by magnetron sputtering," *Materials Chemistry and Physics,* vol. 72, pp. 269-272, 2001.

[19] I.-S. Kim and B.-T. Lee, "Structural and optical properties of single-crystal ZnMgO thin films grown on sapphire and ZnO substrates by RF magnetron sputtering," *Journal of Crystal Growth,* vol. 311, pp. 3618-3621, 2009.

[20] Y. Kim, C. An, H. Cho, J. Kim, H. Lee, E. Jung, et al., "High-temperature growth and in-situ annealing of MgZnO thin films by RF sputtering," *Thin Solid Films,* vol. 516, pp. 5602-5606, 2008.

[21] S. Han, D. Shen, J. Zhang, Y. Zhao, D. Jiang, Z. Ju, et al., "Characteristics of cubic MgZnO thin films grown by radio frequency reaction magnetron co-sputtering," *Journal of Alloys and Compounds,* vol. 485, pp. 794-797, 2009.

[22] H. Sato, T. Minami, S. Takata, and T. Yamada, “Transparent conducting p-type NiO thin films prepared by magnetron sputtering,” *Thin solid films,* vol. 236, pp. 27-31, 1993.

[23] H. Ryu, G. Choi, W. Lee, and J. Park, “Preferred orientations of NiO thin films prepared by RF magnetron sputtering,” *Journal of materials science,* vol. 39, pp. 4375-4377, 2004.

[24] A. M. Reddy, A. S. Reddy, K.-S. Lee, and P. S. Reddy, “Growth and characterization of NiO thin films prepared by dc reactive magnetron sputtering,” *Solid State Sciences,* vol. 13, pp. 314-320, 2011.

[25] H.-L. Chen, Y.-M. Lu, and W.-S. Hwang, “Characterization of sputtered NiO thin films,” *Surface and Coatings Technology,* vol. 198, pp. 138-142, 2005.

[26] S. Schiller, G. Beister, W. Sieber, G. Schirmer, and E. Hacker, “Influence of deposition parameters on the optical and structurcal properties of TiO_2 films produced by reactive DC plasmatron sputtering,” *Thin Solid Films,* vol. 83, pp. 239-245, 1981.

[27] T. Xie, N. Sullivan, K. Steffens, B. Wen, G. Liu, R. Debnath, et al., “UV-assisted room-temperature chemiresistive NO_2 sensor based on TiO_2 thin film,” *Journal of Alloys and Compounds,* vol. 653, pp. 255-259, 2015.

[28] D. Wicaksana, A. Kobayashi, and A. Kinbara, “Process Effects on Structural-Properties of TiO_2 Thin-Films by Reactive Sputtering,” *Journal of Vacuum Science and Technology A-Vacuum Surfaces and Films,* vol. 10, pp. 1479-1482, 1992.

[29] P. Fons, K. Iwata, S. Niki, A. Yamada, and K. Matsubara, “Growth of high-quality epitaxial ZnO films on α-Al_2O_3,” *Journal of Crystal Growth,* vol. 201, pp. 627-632, 1999.

[30] N. Izyumskaya, V. Avrutin, W. Schoch, A. El-Shaer, F. Reuss, T. Gruber, et al., “Molecular beam epitaxy of high-quality ZnO using hydrogen peroxide as an oxidant,” *Journal of Crystal Growth,* vol. 269, pp. 356-361, 2004.

[31] Y. F. Chen, D. M. Bagnall, H. J. Koh, K. T. Park, K. Hiraga, Z. Q. Zhu, et al., “Plasma assisted molecular beam epitaxy of ZnO on c-plane sapphire: Growth and characterization,” *Journal of Applied Physics,* vol. 84, pp. 3912-3918, 1998.

[32] Y. Chen, D. Bagnall, Z. Zhu, T. Sekiuchi, K.-t. Park, K. Hiraga, et al., “Growth of ZnO single crystal thin films on c-plane (0 0 0 1) sapphire by plasma enhanced molecular beam epitaxy,” *Journal of Crystal Growth,* vol. 181, pp. 165-169, 1997.

[33] K. Ogata, K. Koike, T. Tanite, T. Komuro, F. Yan, S. Sasa, et al., "ZnO and ZnMgO growth on a-plane sapphire by molecular beam epitaxy," *Journal of Crystal Growth,* vol. 251, pp. 623-627, 2003.
[34] S. Choopun, R. Vispute, W. Noch, A. Balsamo, R. Sharma, T. Venkatesan, et al., "Oxygen pressure-tuned epitaxy and optoelectronic properties of laser-deposited ZnO films on sapphire," *Applied Physics Letters,* vol. 75, pp. 3947-3949, 1999.
[35] A. Ohtomo, M. Kawasaki, T. Koida, K. Masubuchi, H. Koinuma, Y. Sakurai, et al., "$Mg_xZn_{1-x}O$ as a II-VI widegap semiconductor alloy," *Applied Physics Letters,* vol. 72, pp. 2466-2468, 1998.
[36] S. Choopun, R. D. Vispute, W. Yang, R. P. Sharma, T. Venkatesan, and H. Shen, "Realization of band gap above 5.0 eV in metastable cubic-phase $Mg_xZn_{1-x}O$ alloy films," *Applied Physics Letters,* vol. 80, pp. 1529-1531, 2002.
[37] B. M. Ataev, A. M. Bagamadova, V. V. Mamedov, A. K. Omaev, and M. R. Rabadanov, "Highly conductive and transparent thin ZnO films prepared in situ in a low pressure system," *Journal of Crystal Growth,* vol. 198–199, Part 2, pp. 1222-1225, 1999.
[38] T. Naoyuki, K. Kazuhiko, N. Takato, M. Yoshimi, and Y. Hajime, "Growth of ZnO on Sapphire (0001) by the Vapor Phase Epitaxy Using a Chloride Source," *Japanese Journal of Applied Physics,* vol. 38, p. L454, 1999.
[39] K. Kaiya, K. Omichi, N. Takahashi, T. Nakamura, S. Okamoto, and H. Yamamoto, "Epitaxial growth of ZnO thin films exhibiting room-temperature ultraviolet emission by atmospheric pressure chemical vapor deposition," *Thin Solid Films,* vol. 409, pp. 116-119, 2002.
[40] F. T. J. Smith, "Metalorganic chemical vapor deposition of oriented ZnO films over large areas," *Applied Physics Letters,* vol. 43, pp. 1108-1110, 1983.
[41] W. Kern and R. Heim, "Chemical Vapor Deposition of Silicate Glasses for Use with Silicon Devices: I. Deposition Techniques," *Journal of The Electrochemical Society,* vol. 117, pp. 562-568, 1970.
[42] Y. Kashiwaba, K. Haga, H. Watanabe, B. P. Zhang, Y. Segawa, and K. Wakatsuki, "Structures and photoluminescence properties of ZnO films epitaxially grown by atmospheric pressure MOCVD," *Physica Status Solidi B-Basic Research,* vol. 229, pp. 921-924, 2002.
[43] T. Yasuda and Y. Segawa, "Zinc oxide thin films synthesized by metal organic chemical reactions," *Physica Status Solidi B-Basic Research,* vol. 241, pp. 676-679, 2004.

[44] C. R. Gorla, N. W. Emanetoglu, S. Liang, W. E. Mayo, Y. Lu, M. Wraback, et al., "Structural, optical, and surface acoustic wave properties of epitaxial ZnO films grown on (01$\bar{1}$2) sapphire by metalorganic chemical vapor deposition," *Journal of Applied Physics,* vol. 85, pp. 2595-2602, 1999.

[45] W. I. Park, G.-C. Yi, and H. M. Jang, "Metalorganic vapor-phase epitaxial growth and photoluminescent properties of $Zn_{1-x}Mg_xO(0\leq x\leq 0.49)$ thin films," *Applied Physics Letters,* vol. 79, p. 2022, 2001.

[46] W. Lee, M.-C. Jeong, and J.-M. Myoung, "Catalyst-free growth of ZnO nanowires by metal-organic chemical vapour deposition (MOCVD) and thermal evaporation," *Acta Materialia,* vol. 52, pp. 3949-3957, 2004.

[47] J. Zhong, S. Muthukumar, G. Saraf, H. Chen, Y. Chen, and Y. Lu, "ZnO nanotips grown on Si substrates by metal-organic chemical-vapor deposition," *Journal of Electronic Materials,* vol. 33, pp. 654-657, 2004.

[48] W. I. Park, D. H. Kim, S.-W. Jung, and G.-C. Yi, "Metalorganic vapor-phase epitaxial growth of vertically well-aligned ZnO nanorods," *Applied Physics Letters,* vol. 80, pp. 4232-4234, 2002.

[49] R. Kling, C. Kirchner, G. Th, F. Reuss, and A. Waag, "Analysis of ZnO and ZnMgO nanopillars grown by self-organization," *Nanotechnology,* vol. 15, p. 1043, 2004.

[50] W. I. Park, S. J. An, J. L. Yang, G.-C. Yi, S. Hong, T. Joo, et al., "Photoluminescent Properties of $ZnO/Zn_{0.8}Mg_{0.2}O$ Nanorod Single-Quantum-Well Structures," *The Journal of Physical Chemistry B,* vol. 108, pp. 15457-15460, 2004.

[51] P. Babelon, A. S. Dequiedt, H. Mostéfa-Sba, S. Bourgeois, P. Sibillot, and M. Sacilotti, "SEM and XPS studies of titanium dioxide thin films grown by MOCVD," *Thin Solid Films,* vol. 322, pp. 63-67, 1998.

[52] D.-J. Won, C.-H. Wang, H.-K. Jang, and D.-J. Choi, "Effects of thermally induced anatase-to-rutile phase transition in MOCVD-grown TiO_2 films on structural and optical properties," *Applied Physics A,* vol. 73, pp. 595-600, 2001.

[53] J. Yan, D. C. Gilmer, S. A. Campbell, W. L. Gladfelter, and P. G. Schmid, "Structural and electrical characterization of TiO_2 grown from titanium tetrakis-isopropoxide (TTIP) and TTIP/H_2O ambients," *Journal of Vacuum Science and Technology B,* vol. 14, pp. 1706-1711, 1996.

[54] G. K. Boschloo, A. Goossens, and J. Schoonman, "Photoelectrochemical Study of Thin Anatase TiO_2 Films Prepared by Metallorganic Chemical

Vapor Deposition," *Journal of The Electrochemical Society,* vol. 144, pp. 1311-1317, 1997.

[55] J. Livage, M. Henry, and C. Sanchez, "Sol-gel chemistry of transition metal oxides," *Progress in Solid State Chemistry,* vol. 18, pp. 259-341, 1988.

[56] A. E. Gash, T. M. Tillotson, J. H. Satcher Jr, L. W. Hrubesh, and R. L. Simpson, "New sol–gel synthetic route to transition and main-group metal oxide aerogels using inorganic salt precursors," *Journal of Non-Crystalline Solids,* vol. 285, pp. 22-28, 2001.

[57] N.-L. Wu, S.-Y. Wang, and I. A. Rusakova, "Inhibition of Crystallite Growth in the Sol-Gel Synthesis of Nanocrystalline Metal Oxides," *Science,* vol. 285, pp. 1375-1377, 1999.

[58] J. Rockenberger, E. C. Scher, and A. P. Alivisatos, "A New Nonhydrolytic Single-Precursor Approach to Surfactant-Capped Nanocrystals of Transition Metal Oxides," *Journal of the American Chemical Society,* vol. 121, pp. 11595-11596, 1999.

[59] K. K. Banger, Y. Yamashita, K. Mori, R. L. Peterson, T. Leedham, J. Rickard, et al., "Low-temperature, high-performance solution-processed metal oxide thin-film transistors formed by a 'sol-gel on chip' process," *Nature Materials,* vol. 10, pp. 45-50, 2011.

[60] D. M. Antonelli and J. Y. Ying, "Synthesis of Hexagonally Packed Mesoporous TiO_2 by a Modified Sol–Gel Method," *Angewandte Chemie International Edition in English,* vol. 34, pp. 2014-2017, 1995.

[61] L. Spanhel and M. A. Anderson, "Semiconductor clusters in the sol-gel process: quantized aggregation, gelation, and crystal growth in concentrated zinc oxide colloids," *Journal of the American Chemical Society,* vol. 113, pp. 2826-2833, 1991.

[62] M. Ohyama, H. Kouzuka, and T. Yoko, "Sol-gel preparation of ZnO films with extremely preferred orientation along (002) plane from zinc acetate solution," *Thin Solid Films,* vol. 306, pp. 78-85, 1997.

[63] D. Bao, H. Gu, and A. Kuang, "Sol-gel-derived c-axis oriented ZnO thin films," *Thin Solid Films,* vol. 312, pp. 37-39, 1998.

[64] J.-H. Lee and B.-O. Park, "Transparent conducting ZnO:Al, In and Sn thin films deposited by the sol–gel method," *Thin Solid Films,* vol. 426, pp. 94-99, 2003.

[65] K.-F. Lin, H.-M. Cheng, H.-C. Hsu, L.-J. Lin, and W.-F. Hsieh, "Band gap variation of size-controlled ZnO quantum dots synthesized by sol–gel method," *Chemical Physics Letters,* vol. 409, pp. 208-211, 2005.

[66] D. Bera, L. Qian, S. Sabui, S. Santra, and P. H. Holloway, "Photoluminescence of ZnO quantum dots produced by a sol–gel process," *Optical Materials,* vol. 30, pp. 1233-1239, 2008.

[67] L. Mädler, W. J. Stark, and S. E. Pratsinis, "Rapid synthesis of stable ZnO quantum dots," *Journal of Applied Physics,* vol. 92, pp. 6537-6540, 2002.

[68] Y.-S. Fu, X.-W. Du, S. A. Kulinich, J.-S. Qiu, W.-J. Qin, R. Li, et al., "Stable Aqueous Dispersion of ZnO Quantum Dots with Strong Blue Emission via Simple Solution Route," *Journal of the American Chemical Society,* vol. 129, pp. 16029-16033, 2007.

[69] L. Zhang, L. Yin, C. Wang, N. lun, Y. Qi, and D. Xiang, "Origin of Visible Photoluminescence of ZnO Quantum Dots: Defect-Dependent and Size-Dependent," *The Journal of Physical Chemistry C,* vol. 114, pp. 9651-9658, 2010/06/03 2010.

[70] D. A. Schwartz, N. S. Norberg, Q. P. Nguyen, J. M. Parker, and D. R. Gamelin, "Magnetic Quantum Dots: Synthesis, Spectroscopy, and Magnetism of Co^{2+}- and Ni^{2+}-Doped ZnO Nanocrystals," *Journal of the American Chemical Society,* vol. 125, pp. 13205-13218, 2003.

[71] J. Yu, X. Zhao, and Q. Zhao, "Effect of surface structure on photocatalytic activity of TiO_2 thin films prepared by sol-gel method," *Thin Solid Films,* vol. 379, pp. 7-14, 2000.

[72] J. Yu, X. Zhao, and Q. Zhao, "Photocatalytic activity of nanometer TiO_2 thin films prepared by the sol–gel method," *Materials Chemistry and Physics,* vol. 69, pp. 25-29, 2001.

[73] D. J. Kim, S. H. Hahn, S. H. Oh, and E. J. Kim, "Influence of calcination temperature on structural and optical properties of TiO2 thin films prepared by sol–gel dip coating," *Materials Letters,* vol. 57, pp. 355-360, 2002.

[74] Z. Wang, U. Helmersson, and P.-O. Käll, "Optical properties of anatase TiO_2 thin films prepared by aqueous sol–gel process at low temperature," *Thin Solid Films,* vol. 405, pp. 50-54, 2002.

[75] T. Watanabe, S. Fukayama, M. Miyauchi, A. Fujishima, and K. Hashimoto, "Photocatalytic Activity and Photo-Induced Wettability Conversion of TiO_2 Thin Film Prepared by Sol-Gel Process on a Soda-Lime Glass," *Journal of Sol-Gel Science and Technology,* vol. 19, pp. 71-76, 2000.

[76] Y. Djaoued, S. Badilescu, P. V. Ashrit, D. Bersani, P. P. Lottici, and R. Brüning, "Low Temperature Sol-Gel Preparation of Nanocrystalline

TiO_2 Thin Films," *Journal of Sol-Gel Science and Technology,* vol. 24, pp. 247-254, 2002.

[77] D. Avnir, V. R. Kaufman, and R. Reisfeld, "Organic fluorescent dyes trapped in silica and silica-titania thin films by the sol-gel method. Photophysical, film and cage properties," *Journal of Non-Crystalline Solids,* vol. 74, pp. 395-406, 1985.

[78] L. Hu, T. Yoko, H. Kozuka, and S. Sakka, "Effects of solvent on properties of sol—gel-derived TiO_2 coating films," *Thin Solid Films,* vol. 219, pp. 18-23, 1992.

[79] I. Ichinose, H. Senzu, and T. Kunitake, "A Surface Sol−Gel Process of TiO_2 and Other Metal Oxide Films with Molecular Precision," *Chemistry of Materials,* vol. 9, pp. 1296-1298, 1997.

[80] K. Jacobi, G. Zwicker, and A. Gutmann, "Work function, electron affinity and band bending of zinc oxide surfaces," *Surface science,* vol. 141, pp. 109-125, 1984.

[81] D. Basak, G. Amin, B. Mallik, G. K. Paul, and S. K. Sen, "Photoconductive UV detectors on sol–gel-synthesized ZnO films," *Journal of Crystal Growth,* vol. 256, pp. 73-77, 2003.

[82] K. W. Liu, J. G. Ma, J. Y. Zhang, Y. M. Lu, D. Y. Jiang, B. H. Li, et al., "Ultraviolet photoconductive detector with high visible rejection and fast photoresponse based on ZnO thin film," *Solid-State Electronics,* vol. 51, pp. 757-761, 2007.

[83] Q. A. Xu, J. W. Zhang, K. R. Ju, X. D. Yang, and X. Hou, "ZnO thin film photoconductive ultraviolet detector with fast photoresponse," *Journal of Crystal Growth,* vol. 289, pp. 44-47, 2006.

[84] X. G. Zheng, Q. S. Li, J. P. Zhao, D. Chen, B. Zhao, Y. J. Yang, et al., "Photoconductive ultraviolet detectors based on ZnO films," *Applied Surface Science,* vol. 253, pp. 2264-2267, 2006.

[85] S. Chang, R. W. Chuang, S.-J. Chang, C. Lu, Y. Chiou, and S. Hsieh, "Surface HCl treatment in ZnO photoconductive sensors," *Thin Solid Films,* vol. 517, pp. 5050-5053, 2009.

[86] S. P. Chang, S. J. Chang, Y. Z. Chiou, C. Y. Lu, T. K. Lin, Y. C. Lin, et al., "ZnO photoconductive sensors epitaxially grown on sapphire substrates," *Sensors and Actuators A: Physical,* vol. 140, pp. 60-64, 2007.

[87] Z. Bi, X. Yang, J. Zhang, X. Bian, D. Wang, X. Zhang, et al., "A back-illuminated vertical-structure ultraviolet photodetector based on an RF-sputtered ZnO film," *Journal of electronic materials,* vol. 38, pp. 609-612, 2009.

[88] H. Sheng, N. Emanetoglu, S. Muthukumar, B. Yakshinskiy, S. Feng, and Y. Lu, "Ta/Au ohmic contacts to n-type ZnO," *Journal of Electronic Materials,* vol. 32, pp. 935-938, 2003.

[89] J.-M. Lee, K.-K. Kim, S.-J. Park, and W.-K. Choi, "Low-resistance and nonalloyed ohmic contacts to plasma treated ZnO," *Applied Physics Letters,* vol. 78, pp. 3842-3844, 2001.

[90] H.-K. Kim, S.-H. Han, T.-Y. Seonga, and W.-K. Choi, "Low-resistance Ti/Au ohmic contacts to Al-doped ZnO layers," *Applied Physics Letters,* vol. 77, 2000.

[91] L. J. Mandalapu, F. X. Xiu, Z. Yang, and J. L. Liu, "Ultraviolet photoconductive detectors based on Ga-doped ZnO films grown by molecular-beam epitaxy," *Solid-State Electronics,* vol. 51, pp. 1014-1017, 2007.

[92] H.-K. Kim, K.-K. Kim, S.-J. Park, T.-Y. Seongb, and I. Adesida, "Formation of low resistance nonalloyed Al/Pt ohmic contacts on n-type ZnO epitaxial layer," *Journal of applied physics,* vol. 94, 2003.

[93] H.-K. Kim, K.-K. Kim, S.-J. Park, T.-Y. Seong, and Y. S. Yoon, "Thermally stable and low resistance Ru ohmic contacts to n-ZnO," *Japanese journal of applied physics,* vol. 41, p. L546, 2002.

[94] K. Ip, K. H. Baik, Y. W. Heo, D. P. Norton, S. J. Pearton, J. R. LaRoche, et al., "Annealing temperature dependence of contact resistance and stablity for Ti/Al/Pt/Au ohmic contacts to bulk n-ZnO," *Journal of Vacuum Science and Technology B,* vol. 21, pp. 2378-2381, 2003.

[95] T. Akane, K. Sugioka, and K. Midorikawa, "Nonalloy Ohmic contact fabrication in a hydrothermally grown n-ZnO (0001) substrate by KrF excimer laser irradiation," *Journal of Vacuum Science and Technology B,* vol. 18, pp. 1406-1408, 2000.

[96] D. Y. Jiang, J. Y. Zhang, K. W. Liu, Y. M. Zhao, C. X. Cong, Y. M. Lu, et al., "A high-speed photoconductive UV detector based on an Mg0.4Zn0.6O thin film," *Semiconductor Science and Technology,* vol. 22, pp. 687-690, 2007.

[97] H. Sheng, N. Emanetoglu, S. Muthukumar, S. Feng, and Y. Lu, "Nonalloyed Al ohmic contacts to MgxZn1−xO," *Journal of electronic materials,* vol. 31, pp. 811-814, 2002.

[98] K. W. Liu, D. Z. Shen, C. X. Shan, J. Y. Zhang, B. Yao, D. X. Zhao, et al., "$Zn_{0.76}Mg_{0.24}O$ homojunction photodiode for ultraviolet detection," *Applied Physics Letters,* vol. 91, p. 201106, 2007.

[99] Y. Takahashi, M. Kanamori, A. Kondoh, H. Minoura, and Y. Ohya, "Photoconductivity of ultrathin zinc oxide films," *Japanese Journal of Applied Physics,* vol. 33, p. 6611, 1994.

[100] Z. Bi, J. Zhang, X. Bian, D. Wang, X. a. Zhang, W. Zhang, et al., "A high-performance ultraviolet photoconductive detector based on a ZnO film grown by RF sputtering," *Journal of Electronic Materials,* vol. 37, pp. 760-763, 2008.

[101] M. Liu and H. K. Kim, "Ultraviolet detection with ultrathin ZnO epitaxial films treated with oxygen plasma," *Applied Physics Letters,* vol. 84, p. 173, 2004.

[102] J. Sun, F.-J. Liu, H.-Q. Huang, J.-W. Zhao, Z.-F. Hu, X.-Q. Zhang, et al., "Fast response ultraviolet photoconductive detectors based on Ga-doped ZnO films grown by radio-frequency magnetron sputtering," *Applied Surface Science,* vol. 257, pp. 921-924, 2010.

[103] W. Yang, R. D. Vispute, S. Choopun, R. P. Sharma, T. Venkatesan, and H. Shen, "Ultraviolet photoconductive detector based on epitaxial $Mg_{0.34}Zn_{0.66}O$ thin films," *Applied Physics Letters,* vol. 78, pp. 2787-2789, 2001.

[104] Y. Liu, C. R. Gorla, S. Liang, N. Emanetoglu, Y. Lu, H. Shen, et al., "Ultraviolet detectors based on epitaxial ZnO films grown by MOCVD," *Journal of Electronic Materials,* vol. 29, pp. 69-74, 2000.

[105] H. Kind, H. Yan, B. Messer, M. Law, and P. Yang, "Nanowire ultraviolet photodetectors and optical switches," *Advanced Materials,* vol. 14, p. 158, 2002.

[106] C. Soci, A. Zhang, B. Xiang, S. A. Dayeh, D. Aplin, J. Park, et al., "ZnO nanowire UV photodetectors with high internal gain," *Nano Letters,* vol. 7, pp. 1003-1009, 2007.

[107] Y. Li, F. Della Valle, M. Simonnet, I. Yamada, and J.-J. Delaunay, "High-performance UV detector made of ultra-long ZnO bridging nanowires," *Nanotechnology,* vol. 20, p. 045501, 2008.

[108] J. Cheng, Y. Zhang, and R. Guo, "ZnO microtube ultraviolet detectors," *Journal of Crystal Growth,* vol. 310, pp. 57-61, 2008.

[109] N. Kouklin, "Cu-Doped ZnO Nanowires for Efficient and Multispectral Photodetection Applications," *Advanced Materials,* vol. 20, pp. 2190-2194, 2008.

[110] S. E. Ahn, J. S. Lee, H. Kim, S. Kim, B. H. Kang, K. H. Kim, et al., "Photoresponse of sol-gel-synthesized ZnO nanorods," *Applied Physics Letters,* vol. 84, pp. 5022-5024, 2004.

[111] Q. Li, Q. Wan, Y. Liang, and T. Wang, "Electronic transport through individual ZnO nanowires," *Applied Physics Letters,* vol. 84, pp. 4556-4558, 2004.
[112] J. Zhou, Y. Gu, Y. Hu, W. Mai, P.-H. Yeh, G. Bao, et al., "Gigantic enhancement in response and reset time of ZnO UV nanosensor by utilizing Schottky contact and surface functionalization," *Applied Physics Letters,* vol. 94, p. 191103, 2009.
[113] J. He, P. Chang, C. Chen, and K. Tsai, "Electrical and optoelectronic characterization of a ZnO nanowire contacted by focused-ion-beam-deposited Pt," *Nanotechnology,* vol. 20, p. 135701, 2009.
[114] C. S. Lao, M.-C. Park, Q. Kuang, Y. Deng, A. K. Sood, D. L. Polla, et al., "Giant enhancement in UV response of ZnO nanobelts by polymer surface-functionalization," *Journal of the American Chemical Society,* vol. 129, pp. 12096-12097, 2007.
[115] D. C. Oh, T. Suzuki, T. Hanada, T. Yao, H. Makino, and H. J. Ko, "Photoresponsivity of ZnO Schottky barrier diodes," *Journal of Vacuum Science and Technology B,* vol. 24, pp. 1595-1598, 2006.
[116] H. Fabricius, T. Skettrup, and P. Bisgaard, "Ultraviolet detectors in thin sputtered ZnO films," *Applied Optics,* vol. 25, pp. 2764-2767, 1986.
[117] H. Endo, M. Sugibuchi, K. Takahashi, S. Goto, S. Sugimura, K. Hane, et al., "Schottky ultraviolet photodiode using a ZnO hydrothermally grown single crystal substrate," *Applied Physics Letters,* vol. 90, p. 121906, 2007.
[118] Y. W. Heo, L. C. Tien, D. P. Norton, S. J. Pearton, B. S. Kang, F. Ren, et al., "Pt/ZnO nanowire Schottky diodes," *Applied Physics Letters,* vol. 85, pp. 3107-3109, 2004.
[119] M. Nakano, T. Makino, A. Tsukazaki, K. Ueno, A. Ohtomo, T. Fukumura, et al., "Transparent polymer Schottky contact for a high performance visible-blind ultraviolet photodiode based on ZnO," *Applied Physics Letters,* vol. 93, p. 123309, 2008.
[120] H. von Wenckstern, S. Müller, G. Biehne, H. Hochmuth, M. Lorenz, and M. Grundmann, "Dielectric Passivation of ZnO-Based Schottky Diodes," *Journal of Electronic Materials,* vol. 39, pp. 559-562, 2009.
[121] G. M. Ali and P. Chakrabarti, "Effect of thermal treatment on the performance of ZnO based metal-insulator-semiconductor ultraviolet photodetectors," *Applied Physics Letters,* vol. 97, p. 031116, 2010.
[122] X. Hailin, C. Weiyou, L. Caixia, K. Xiangzi, Q. Pengfei, L. Ziran, et al., "Fabrication of TiO_2 Schottky barrier diodes by RF magnetron

sputtering," in *Nano/Micro Engineered and Molecular Systems, 2008. NEMS 2008. 3rd IEEE International Conference on*, 2008, pp. 108-111.

[123] F. S. Mahmood, R. D. Gould, A. K. Hassan, and H. M. Salih, "D.C. properties of ZnO thin films prepared by r.f. magnetron sputtering," *Thin Solid Films,* vol. 270, pp. 376-379, 1995.

[124] X. L. Zhang, K. S. Hui, and K. N. Hui, "High photo-responsivity ZnO UV detectors fabricated by RF reactive sputtering," *Materials Research Bulletin,* vol. 48, pp. 305-309, 2013.

[125] D. A. Zhilin, N. V. Lyanguzov, L. A. Nikolaev, V. I. Pushkariov, and E. M. Kaidashev, "Photoelectric properties of MSM structure based on ZnO nanorods, received by thermal evaporation and carbothermal syntesis," *Journal of Physics: Conference Series,* vol. 541, p. 012038, 2014.

[126] X. G. Zheng, Q. S. Li, W. Hu, D. Chen, N. Zhang, M. J. Shi, et al., "Photoconductive properties of ZnO thin films grown by pulsed laser deposition," *Journal of Luminescence,* vol. 122–123, pp. 198-201, 2007.

[127] T. K. Lin, S. J. Chang, Y. K. Su, B. R. Huang, M. Fujita, and Y. Horikoshi, "ZnO MSM photodetectors with Ru contact electrodes," *Journal of Crystal Growth,* vol. 281, pp. 513-517, 2005.

[128] S. J. Young, L. W. Ji, T. H. Fang, S. J. Chang, Y. K. Su, and X. L. Du, "ZnO ultraviolet photodiodes with Pd contact electrodes," *Acta Materialia,* vol. 55, pp. 329-333, 2007.

[129] S. J. Young, L. W. Ji, S. J. Chang, and Y. K. Su, "ZnO metal–semiconductor–metal ultraviolet sensors with various contact electrodes," *Journal of Crystal Growth,* vol. 293, pp. 43-47, 2006.

[130] M. Li, W. Anderson, N. Chokshi, R. L. DeLeon, and G. Tompa, "Laser annealing of laser assisted molecular beam deposited ZnO thin films with application to metal-semiconductor-metal photodetectors," *Journal of Applied Physics,* vol. 100, p. 053106, 2006.

[131] H. Shen, M. Wraback, C. R. Gorla, S. Liang, N. Emanetoglu, Y. Liu, et al., "High-Gain, High-Speed ZnO MSM Ultraviolet Photodetectors," *MRS Online Proceedings Library Archive,* vol. 595, p. F99W11.16, 1999.

[132] S. Liang, H. Sheng, Y. Liu, Z. Huo, Y. Lu, and H. Shen, "ZnO Schottky ultraviolet photodetectors," *Journal of Crystal Growth,* vol. 225, pp. 110-113, 2001.

[133] L. Li, Y. Ryu, H. W. White, and P. Yu, "Optical properties of metal-semiconductor-metal ZnO UV photodetectors," *Proc. SPIE 7603, Oxide-based Materials and Devices,* vol. 7603, pp. 76031A-76031A-9, 2010.

[134] K.-S. Hwang, J.-H. Jeong, Y.-S. Jeon, K.-O. Jeon, and B.-H. Kim, "Electrostatic spray deposited ZnO thin films," *Ceramics International,* vol. 33, pp. 505-507, 2007.

[135] K. J. Chen, F. Y. Hung, S. J. Chang, and S. J. Young, "Optoelectronic characteristics of UV photodetector based on ZnO nanowire thin films," *Journal of Alloys and Compounds,* vol. 479, pp. 674-677, 2009.

[136] Z.-Q. Xu, H. Deng, J. Xie, Y. Li, and X.-T. Zu, "Ultraviolet photoconductive detector based on Al doped ZnO films prepared by sol–gel method," *Applied Surface Science,* vol. 253, pp. 476-479, 2006.

[137] Z. Xu, H. Deng, J. Xie, Y. Li, Y. Li, X. Zu, et al., "Photoconductive UV Detectors Based on ZnO Films Prepared by Sol-Gel Method," *Journal of Sol-Gel Science and Technology,* vol. 36, pp. 223-226, 2005.

[138] S. I. Inamdar and K. Y. Rajpure, "High-performance metal–semiconductor–metal UV photodetector based on spray deposited ZnO thin films," *Journal of Alloys and Compounds,* vol. 595, pp. 55-59, 2014.

[139] S. I. Inamdar, V. V. Ganbavle, and K. Y. Rajpure, "ZnO based visible–blind UV photodetector by spray pyrolysis," *Superlattices and Microstructures,* vol. 76, pp. 253-263, 2014.

[140] Y. Jin, J. Wang, B. Sun, J. C. Blakesley, and N. C. Greenham, "Solution-Processed Ultraviolet Photodetectors Based on Colloidal ZnO Nanoparticles," *Nano Letters,* vol. 8, pp. 1649-1653, 2008.

[141] D. Jiang, J. Zhang, Y. Lu, K. Liu, D. Zhao, Z. Zhang, et al., "Ultraviolet Schottky detector based on epitaxial ZnO thin film," *Solid-State Electronics,* vol. 52, pp. 679-682, 2008.

[142] S. M. Sze and K. K. Ng, *Physics of Semiconductor Devices*: John Wiley and Sons, 2006.

[143] J. Liang-Wen, W. Cheng-Zhi, L. Chih-Ming, M. Teen-Hang, L. Kin-Tak, P. Shi-Ming, et al., "Characteristic Improvements of ZnO-Based Metal–Semiconductor–Metal Photodetector on Flexible Substrate with ZnO Cap Layer," *Japanese Journal of Applied Physics,* vol. 49, p. 052201, 2010.

[144] N. N. Jandow, F. K. Yam, S. M. Thahab, H. Abu Hassan, and K. Ibrahim, "Characteristics of ZnO MSM UV photodetector with Ni contact electrodes on poly propylene carbonate (PPC) plastic substrate," *Current Applied Physics,* vol. 10, pp. 1452-1455, 2010.

[145] M.-M. Fan, K.-W. Liu, X. Chen, Z.-Z. Zhang, B.-H. Li, H.-F. Zhao, et al., "Realization of cubic ZnMgO photodetectors for UVB applications," *Journal of Materials Chemistry C,* vol. 3, pp. 313-317, 2015.

[146] K. Liu, M. Sakurai, and M. Aono, "ZnO-Based Ultraviolet Photodetectors," *Sensors,* vol. 10, p. 8604, 2010.

[147] Z. Zhang, H. von Wenckstern, and M. Grundmann, "Energy-selective multichannel ultraviolet photodiodes based on (Mg,Zn)O," *Applied Physics Letters,* vol. 103, p. 171111, 2013.

[148] G. Shukla, "$Zn_{1-x}Mg_{xO}$ Homojunction-Based Ultraviolet Photodetector," *IEEE Photonics Technology Letters,* vol. 21, pp. 887-889, 2009.

[149] Y. Zhao, J. Zhang, D. Jiang, C. Shan, Z. Zhang, B. Yao, et al., "Ultraviolet Photodetector Based on a MgZnO Film Grown by Radio-Frequency Magnetron Sputtering," *ACS Applied Materials and Interfaces,* vol. 1, pp. 2428-2430, 2009.

[150] X. Jie, W. Huiyun, G. Er-Jia, and Y. Fang, "Highly sensitive fast-response UV photodetectors based on epitaxial TiO_2 films," *Journal of Physics D: Applied Physics,* vol. 44, p. 375104, 2011.

[151] H. Xue, X. Kong, Z. Liu, C. Liu, J. Zhou, W. Chen, et al., "TiO_2 based metal-semiconductor-metal ultraviolet photodetectors," *Applied Physics Letters,* vol. 90, p. 201118, 2007.

[152] X. Kong, C. Liu, W. Dong, X. Zhang, C. Tao, L. Shen, et al., "Metal-semiconductor-metal TiO_2 ultraviolet detectors with Ni electrodes," *Applied Physics Letters,* vol. 94, p. 123502, 2009.

[153] J. B. K. Law and J. T. L. Thong, "Simple fabrication of a ZnO nanowire photodetector with a fast photoresponse time," *Applied Physics Letters,* vol. 88, p. 133114, 2006.

[154] L. Qin, C. Shing, and S. Sawyer, "Metal-Semiconductor-Metal Ultraviolet Photodetectors Based on Zinc-Oxide Colloidal Nanoparticles," *IEEE Electron Device Letters,* vol. 32, pp. 51-53, 2011.

[155] X. Fang, L. Hu, K. Huo, B. Gao, L. Zhao, M. Liao, et al., "New Ultraviolet Photodetector Based on Individual Nb2O5 Nanobelts," *Advanced Functional Materials,* vol. 21, pp. 3907-3915, 2011.

[156] T. Rajesh, V. Binni, G. M. Subodh, T. Eng Soon, and S. Chorng Haur, "Probing the photoresponse of individual Nb_2O_5 nanowires with global and localized laser beam irradiation," *Nanotechnology,* vol. 22, p. 115202, 2011.

[157] K. W. Liu, J. Y. Zhang, J. G. Ma, D. Y. Jiang, Y. M. Lu, B. Yao, et al., "$Zn_{0.8}Mg_{0.2}O$-based metal–semiconductor–metal photodiodes on quartz for visible-blind ultraviolet detection," *Journal of Physics D: Applied Physics,* vol. 40, p. 2765, 2007.

[158] Y. H. Leung, Z. B. He, L. B. Luo, C. H. A. Tsang, N. B. Wong, W. J. Zhang, et al., "ZnO nanowires array p-n homojunction and its

application as a visible-blind ultraviolet photodetector," *Applied Physics Letters,* vol. 96, p. 053102, 2010.

[159] H. J. Kim and J. H. Lee, "Highly sensitive and selective gas sensors using p-type oxide semiconductors: Overview," *Sensors and Actuators B-Chemical,* vol. 192, pp. 607-627, 2014.

[160] T.-H. Moon, M.-C. Jeong, W. Lee, and J.-M. Myoung, "The fabrication and characterization of ZnO UV detector," *Applied Surface Science,* vol. 240, pp. 280-285, 2005.

[161] H. Shen, C. X. Shan, B. H. Li, B. Xuan, and D. Z. Shen, "Reliable self-powered highly spectrum-selective ZnO ultraviolet photodetectors," *Applied Physics Letters,* vol. 103, p. 232112, 2013.

[162] F. Sun, C.-X. Shan, S.-P. Wang, B.-H. Li, Z.-Z. Zhang, C.-L. Yang, et al., "Ultraviolet photodetectors fabricated from ZnO p–i–n homojunction structures," *Materials Chemistry and Physics,* vol. 129, pp. 27-29, 2011.

[163] Z. Wei, B. Yao, Z. Zhang, Y. Lu, D. Shen, B. Li, et al., "Formation of p-type MgZnO by nitrogen doping," *Applied physics letters,* vol. 89, p. 2104, 2006.

[164] R. Debnath, T. Xie, B. Wen, W. Li, J.-Y. Ha, N. Sullivan, et al., "A solution-processed high-efficiency p-NiO/n-ZnO heterojunction photodetector," *RSC Advances,* vol. 5, pp. 14646-14652, 2015.

[165] H. Ohta, M. Hirano, K. Nakahara, H. Maruta, T. Tanabe, M. Kamiya, et al., "Fabrication and photoresponse of a pn-heterojunction diode composed of transparent oxide semiconductors, p-NiO and n-ZnO," *Applied Physics Letters,* vol. 83, pp. 1029-1031, 2003.

[166] P. N. Ni, C. X. Shan, S. P. Wang, X. Y. Liu, and D. Z. Shen, "Self-powered spectrum-selective photodetectors fabricated from n-ZnO/p-NiO core-shell nanowire arrays," *Journal of Materials Chemistry C,* vol. 1, pp. 4445-4449, 2013.

[167] M. R. Hasan, T. Xie, S. C. Barron, G. Liu, N. V. Nguyen, A. Motayed, et al., "Self-powered p-NiO/n-ZnO heterojunction ultraviolet photodetectors fabricated on plastic substrates," *APL Materials,* vol. 3, p. 106101, 2015.

[168] N. Park, K. Sun, Z. Sun, Y. Jing, and D. Wang, "High efficiency NiO/ZnO heterojunction UV photodiode by sol-gel processing," *Journal of Materials Chemistry C,* vol. 1, pp. 7333-7338, 2013.

[169] T. Xie, G. Liu, B. Wen, J. Y. Ha, N. V. Nguyen, A. Motayed, et al., "Tunable Ultraviolet Photoresponse in Solution-Processed p–n Junction Photodiodes Based on Transition-Metal Oxides," *ACS Applied Materials and Interfaces,* vol. 7, pp. 9660-9667, 2015.

[170] Y. I. Alivov, Ü. Özgür, S. Doğan, D. Johnstone, V. Avrutin, N. Onojima, et al., "Photoresponse of n-ZnO/ p-SiC heterojunction diodes grown by plasma-assisted molecular-beam epitaxy," *Applied Physics Letters,* vol. 86, 2005.

[171] Y. Q. Bie, Z. M. Liao, H. Z. Zhang, G. R. Li, Y. Ye, Y. B. Zhou, et al., "Self-powered, ultrafast, visible-blind UV detection and optical logical operation based on ZnO/GaN nanoscale p-n junctions," *Advanced Materials,* vol. 23, pp. 649-53, 2011.

[172] T. C. Zhang, Y. Guo, Z. X. Mei, C. Z. Gu, and X. L. Du, "Visible-blind ultraviolet photodetector based on double heterojunction of n-ZnO/insulator-MgO/p-Si," *Applied Physics Letters,* vol. 94, p. 113508, 2009.

[173] I. S. Jeong, J. H. Kim, and S. Im, "Ultraviolet-enhanced photodiode employing n-ZnO/p-Si structure," *Applied Physics Letters,* vol. 83, p. 2946, 2003.

[174] Y.-Y. Lin, C.-W. Chen, W.-C. Yen, W.-F. Su, C.-H. Ku, and J.-J. Wu, "Near-ultraviolet photodetector based on hybrid polymer/zinc oxide nanorods by low-temperature solution processes," *Applied Physics Letters,* vol. 92, p. 233301, 2008.

[175] T.-Y. Tsai, S.-J. Chang, T.-J. Hsueh, H.-T. Hsueh, W.-Y. Weng, C.-L. Hsu, et al., "p-Cu2O-shell/n-TiO2-nanowire-core heterostucture photodiodes," *Nanoscale research letters,* vol. 6, pp. 1-7, 2011.

[176] K. Khun, Z. H. Ibupoto, and M. Willander, "Development of fast and sensitive ultraviolet photodetector using p-type NiO/n-type TiO_2 heterostructures," *Physica Status Solidi (a),* vol. 210, pp. 2720-2724, 2013.

[177] Y. Vygranenko, K. Wang, and A. Nathan, "Low leakage p-NiO/i-ZnO/n-ITO heterostructure ultraviolet sensor," *Applied Physics Letters,* vol. 89, p. 172105, 2006.

[178] M. A. Abbasi, Z. H. Ibupoto, A. Khan, O. Nur, and M. Willander, "Fabrication of UV photo-detector based on coral reef like p-NiO/n-ZnO nanocomposite structures," *Materials Letters,* vol. 108, pp. 149-152, 2013.

[179] D. Y. Kim, J. Ryu, J. Manders, J. Lee, and F. So, "Air-Stable, Solution-Processed Oxide p–n Heterojunction Ultraviolet Photodetector," *ACS Applied Materials and Interfaces,* vol. 6, pp. 1370-1374, 2014.

BIOGRAPHICAL SKETCHES

Ting Xie

Dr. Ting Xie is currently a Postdoctoral Associate in the Department of Electrical and Computer Engineering at University of Maryland, College Park. He received his Ph.D. degree in Electrical Engineering from University of Maryland, College Park in 2016 and graduated from Zhejiang University with a B.E. degree in 2010. Since 2013 he has been working as a guest researcher at National Institute of Standards and Technology. His research endeavors range from nanoscale electronics devices including spintronic, gas sensors and UV detectors, to analytical surface science technologies.

Selected Publications

1. T. Xie and R. D. Gomez, "Maskless formation of magnetic nanoparticles by annealing continuous CoFeB/Cu bilayer thin films," IEEE Transactions on Magnetics, 51(11), 1-4, 2015.
2. T. Xie, G. Liu, B. Wen, J.Y. Ha, N.V. Nguyen, A. Motayed and R. Debnath, "Tunable ultraviolet photoresponse in solution-processed p-n junction photodiodes based on transition metal oxides," ACS Applied Materials and Interfaces, 7(18), 9600-9667, 2015.
3. T. Xie, N. Sullivan, K. Steffens, B. Wen, G. Liu, R. Debnath, A. Davydov, R. D. Gomez and A. Motayed, "UV-assisted NO_2 sensing at room-temperature using rf-sputtered TiO_2 thin-film," Journal of Alloys and Compounds, 653, 255-259, 2015.
4. T. Xie, R. Hasan, B. Qiu, E. Arinze, N. Nguyen, A. Motayed, S. Thon and R. Debnath, "High-performing visible-blind photodetectors based on SnO_2/CuO nanoheterojunctions," Applied Physics Letters,107(24), 241108, 2015.
5. T. Xie, D. Romero and R. D. Gomez, "Surface compositions of atomic layer deposited $Zn_{1-x}Mg_xO$ thin films studied using Auger electron spectroscopy," Journal of Vacuum Science and Technology A, 33(5), 05E110, 2015.
6. T. Xie, R. Hasan, S. C. Barron, G. Liu, N. Nguyen, A. Motayed, M. Rao and R. Debnath, "Self-powered heterojunction ultraviolet photodetectors fabricated on plastic substrates," APL Materials, 3(10), 106101, 2015.

7. R. Debnath, T. Xie, B. Wen, W. Li, J.Y. Ha, N. Sullivan, N. Nguyen and A. Motayed, "A solution-processed high-efficiency p-NiO/n-ZnO heterojunction photodetector," RSC Advances, 5(19), 14646-14652, 2015.
8. G. Liu, B. Wen T. Xie, A. Castillo, J.Y. Ha, N. Sullivan, R. Debnath, A. Davydov, M. Peckerar and A. Motayed, "Top-Down Approach for Fabrication of Horizontally-Aligned Gallium Nitride Nanowire Arrays," Microelectronic Engineering, 142, 58-63, 2015.

Albert Davydov

Albert Davydov is a Staff Member of the National Institute for Standards and Technology. He received his Ph.D. in Chemistry from Moscow State University (Russia) in 1989. He was an Assistant Professor of Chemistry at Moscow State University (1989-1993), an Invited Researcher at the University of Sheffield, UK (1992-1993), a Research Scientist at the University of Florida (1993-1997), and a NIST Research Associate at the University of Maryland (1997-2005).

He joined NIST fulltime in 2005 and is now active in the area of semiconductor thin films, nanowires and 2D materials and devices. He is presently a Leader of the Functional Nanostructured Materials Group and a Project Leader on "Low-dimensional semiconductors for sensors, optoelectronics and energy applications" at the Materials Science and Engineering Division at Material Metrology Laboratory at NIST.

Dr. Davydov has more than 30 years of experience and over 100 publications related to growth of bulk crystals, deposition of thin films, and the fabrication, characterization, and processing of a wide range of nanostructured electronic and optical materials. He is also involved in thermodynamic modeling of phase diagrams for metal-semiconductor systems.

He is a Head of the Semiconductor Task Group for the International Centre for Diffraction Data, Co-chair of the Reference Materials Task Group at ASTM Subcommittee on Compound Semiconductors, leader of the review panel for the NSF-NRI program on "Nanoelectronics for 2020 and Beyond," and an Associate Editor of the Journal of Mining and Metallurgy.

Selected Publications

1. D. Ruzmetov, K. Zhang, G. Stan, B. Kalanyan, G.R. Bhimanapati, S.M. Eichfeld, R.A. Burke, P.B. Shah, T.P. O'Regan, F.J. Crowne, A.G. Birdwell, J.A. Robinson, A.V. Davydov, and T.G. Ivanov, "Vertical

2D/3D semiconductor heterostructures based on epitaxial Molybdenum Disulfide and Gallium Nitride," ACS Nano, 2016, v. 10, 3580.

2. M. Müller, G. Schmidt, S. Metzner, P. Veit, F. Bertram, S. Krylyuk, R. Debnath, J-Y. Ha, B. Wen, P. Blanchard, A. Motayed, M.R. King, A.V. Davydov, and J. Christen, "Structural and optical nanoscale analysis of GaN core–shell microrod arrays fabricated by combined top-down and bottom-up process on Si(111)," Jap. J. Appl. Phys., 2016, v. 55, 05FF02.
3. D. Sharma, A. Motayed, P.B. Shah, M. Amani, M. Georgieva, A.G. Birdwell, M. Dubey, Q. Li, and A.V. Davydov, "Transfer characteristics and low-frequency noise in single- and multi-layer MoS_2 field-effect transistors," Appl. Phys. Lett., 2015, v. 107, 162102.
4. Arunima K. Singh, A.K., Hennig, R.G., Davydov, A.V., and Tavazza, F., Al2O3 as a suitable substrate and a dielectric layer for n-layer MoS_2, Appl. Phys. Lett., 2015, v. 107, 053106.
5. Liu, G., Wen, B., Xie, T., Castillo, A., Ha, J-Y., Sullivanc, N., Debnath, R., Davydov, A.V., Peckerar, M., Motayed, A., "Top–down fabrication of horizontally-aligned gallium nitride nanowire arrays for sensor development," Microelectronic Engineering, 2015, v. 142, 58.
6. E. H. Williams, J-Y. Ha, M. Juba, B. Bishop, S. Krylyuk, A. Motayed, M.V. Rao, J.A. Schreifels, A.V. Davydov, "Real-time electrical detection of the formation and destruction of lipid bilayers on silicon nanowire devices, Sensing and Bio-Sensing Research," 2015, v. 4, 103.
7. T. Xie, N. Sullivan, K. Steffens, B. Wen, G. Liu, R, Debnath, A.V. Davydov, R. Gomez, and A. Motayed, "UV-assisted room-temperature chemiresistive NO_2 sensor based on TiO_2 thin film," J. Alloys and Compounds, 2015, v. 653, 255.
8. P. Zhang, P. Liu, S. Siontas, A. Zaslavsky, D. Pacifici, J-Y. Ha, S. Krylyuk, and A. V. Davydov, "Dense nanoimprinted silicon nanowire arrays with passivated axial p-i-n junctions for photovoltaic applications," J. Appl. Phys., 2015, v. 117, 125104.
9. S. Berweger, J.C. Weber, J. John, J.M. Velazquez, A. Pieterick, N.A. Sanford, A.V. Davydov, B. Brunschwig, N.S. Lewis, T.M. Wallis, and P. Kabos, "Microwave Near-Field Imaging of Two-Dimensional Semiconductors," Nano Letters, 2015, v. 15, 1122.
10. S. Krylyuk, R. Debnath, H.P. Yoon, M.R. King, J-Y Ha, B. Wen, A. Motayed, and A.V. Davydov, "Faceting control in core-shell GaN micropillars using selective epitaxy," APL-Materials, 2014, v. 2, 106104.
11. D. Sharma, M. Amani, A. Motayed, P.B. Shah, A.G. Birdwell, S. Najmaei, P.M. Ajayan, J. Lou, M. Dubey, Q. Li and A.V. Davydov,

"Electrical transport and low-frequency noise in chemical vapor deposited single-layer MoS_2 devices," Nanotechnology, 2014, v. 25, 155702.
12. D. Zhang, H. Baek, J.-H. Ha, T. Zhang, J. Wyrick, A.V. Davydov, Y. Kuk, and J.A. Stroscio, "Quasiparticle scattering from topological crystalline insulator SnTe (001) surface states," Phys. Rev. B, 2014, v. 89, 245445.
13. R. Debnath, J-Y. Ha, B. Wen, D. Paramanik, A. Motayed, M.R. King, and A.V. Davydov, "Top-down fabrication of large-area GaN micro- and nanopillars," J. Vac. Sci. Technol. B, 2014, v. 32, p. 021204.
14. V.P. Oleshko, T. Lam, D. Ruzmetov, P. Haney, H.J. Lezec, A.V. Davydov, S. Krylyuk, J. Cumings, A.A. Talin, "Miniature all-solid-state heterostructure nanowire Li-ion batteries as a tool for engineering and structural diagnostics of nanoscale electrochemical processes," Nanoscale, 2014, v. 6, 11756.
15. E.H. Williams, A.V. Davydov, V.P. Oleshko, K.L. Steffens, I. Levin, N.J. Lin, K.A. Bertness, A.K. Manocchi, J.A. Schreifels, and M.V. Rao, "Solution-based functionalization of gallium nitride nanowires for protein sensor development," Surface Science, 2014, v. 627, 23.
16. H. Yima, W.Y. Kong, S-J. Yoon, S. Nahm, H-W. Jang, Y-E. Sung, J-Y. Ha, A.V. Davydov, J-W. Choi, "3-dimensional hemisphere-structured $LiSn_{0.0125}Mn_{1.975}O_4$ thin-film cathodes, Electrochem. Communications, 2014, v. 43, 36.
17. D. Sharma, A. Motayed, S. Krylyuk, Q. Li, and A.V. Davydov, "Detection of Deep-Levels in Doped Silicon Nanowires Using Low-Frequency Noise Spectroscopy," IEEE Transactions on Electron Devices, 2013, v. 60, p 4206.
18. E.H. Williams, JA Schreifels, MV Rao, AV Davydov, VP Oleshko, NJ Lin, KL Steffens, S Krylyuk, KA Bertness, AK Manocchi, Y Koshka, "Selective streptavidin bioconjugation on silicon and silicon carbide nanowires for biosensor applications," J. Mater. Res., 2013, v. 28, 68.
19. Z. Ma, D McDowell, E. Panaitescu, A.V. Davydov, M. Upmanyu and L. Menon, "Vapor–liquid–solid growth of serrated GaN nanowires: shape selection driven by kinetic frustration," J. Mater. Chem. C, 2013, v. 1, 7294.
20. J Shi, Z.D. Li, A. Kvit, S. Krylyuk, A.V. Davydov, X.D. Wang, "Electron Microscopy Observation of TiO_2 Nanocrystal Evolution in High-Temperature Atomic Layer Deposition," Nano Letters, 2013, v. 13, 5727.
21. G.S. Aluri, A. Motayed, A.V. Davydov, V. P. Oleshko, K.A. Bertness, and M.V. Rao, "Nitro-Aromatic Explosive Sensing Using GaN Nanowire-

Titania Nanocluster Hybrids," IEEE Sensors Journal, 2013, v. 13, 1883.
22. M.A. Real, E.A. Lass, F-H. Liu, T. Shen, G.R. Jones, J.A. Soons, D.B. Newell, A.V. Davydov, and R.E. Elmquist, "Graphene Epitaxial Growth on SiC(0001) for Resistance Standards," IEEE Transact. on Instrument. and Measurement, 2013, v. 62, 1454.
23. R. Venkatesh K.G. Thirumalai, B. Krishnan, A.V. Davydov, J.N. Merrett, Y. Koshka, "SiC nanowire vapor–liquid–solid growth using vapor-phase catalyst delivery," J. Mater. Res., 2013, v. 28, 50.

Ratan Debnath

Ratan Debnath is the Director of Research at N5 Sensors Inc. - a startup company based in Maryland commercializing high-performance chemical and gas sensors for consumer and industrial applications. He received his Doctor of Engineering from Technical University, Aachen, Germany in 2009, PhD from University of Dundee, UK in 2005 and Bachelor's degree from Indian Institute of Technology, Roorkee (IITR), India in 1999. Since 2011, he has also been working as a guest researcher in Materials Science and Engineering Division at National Institute for Standards and Technology (NIST), Gaithersburg, Maryland. From 2011-2013, he was appointed as a Visiting Assistant Research Scientist in the Department of Materials Science and Engineering at University of Maryland, College Park. He is a recipient of e8/MITACS Postdoctoral Fellowship at University of Toronto, Canada (2008-2011), DAAD-Helmholtz Fellowship at Research Center Jülich, Germany (2005-2008), Bangabandhu Fellowship at University of Dundee (2001-2002) and ICCR scholarship at IITR (1995-1999).

Dr. Debnath has more than 15 years of experience and over 40 publications related to the growth of various thin film and nanostructure semiconducting materials (chalcogenides, nitrides, metal oxides etc.) as well characterization, fabrication and processing of these materials for various electronic and optoelectronic device applications. Foremost, he has been at the forefront of various SBIR projects funded by State of Maryland, EPA, DHS, NSF, NASA etc. and successfully completed them.

Selected Publications

1. R Debnath, CM Hangarter, D Josell, "3D Geometries: Enabling Optimization Toward the Inherent Limits of Thin-Film Photovoltaics" Semiconductor Materials for Solar Photovoltaic Cells, v. 218 of the series

Springer Series in Materials Science pp 1-24, 2016.
2. M. Müller, G. Schmidt, S. Metzner, P. Veit, F. Bertram, S. Krylyuk, R. Debnath, J-Y. Ha, B. Wen, P. Blanchard, A. Motayed, M.R. King, A.V. Davydov, and J. Christen, "Structural and optical nanoscale analysis of GaN core–shell microrod arrays fabricated by combined top-down and bottom-up process on Si(111)," Jap. J. Appl. Phys., 2016, v. 55, 05FF02.
3. T. Xie, G. Liu, B. Wen, J.Y. Ha, N.V. Nguyen, A. Motayed and R. Debnath, "Tunable ultraviolet photoresponse in solution-processed p-n junction photodiodes based on transition metal oxides," ACS Applied Materials and Interfaces, 7(18), 9600-9667, 2015.
4. T. Xie, N. Sullivan, K. Steffens, B. Wen, G. Liu, R. Debnath, A. Davydov, R. D. Gomez and A. Motayed, "UV-assisted NO_2 sensing at room-temperature using rf-sputtered TiO_2 thin-film," Journal of Alloys and Compounds, 653, 255-259, 2015.
5. T. Xie, R. Hasan, B. Qiu, E. Arinze, N. Nguyen, A. Motayed, S. Thon and R. Debnath, "High-performing visible-blind photodetectors based on SnO_2/CuO nanoheterojunctions," Applied Physics Letters,107(24), 241108, 2015.
6. T. Xie, R. Hasan, S. C. Barron, G. Liu, N. Nguyen, A. Motayed, M. Rao and R. Debnath, "Self-powered heterojunction ultraviolet photodetectors fabricated on plastic substrates," APL Materials, 3(10), 106101, 2015.
7. R. Debnath, T. Xie, B. Wen, W. Li, J.Y. Ha, N. Sullivan, N. Nguyen and A. Motayed, "A solution-processed high-efficiency p-NiO/n-ZnO heterojunction photodetector," RSC Advances, 5(19), 14646-14652, 2015.
9. G. Liu, B. Wen T. Xie, A. Castillo, J.Y. Ha, N. Sullivan, R. Debnath, A. Davydov, M. Peckerar and A. Motayed, "Top-Down Approach for Fabrication of Horizontally-Aligned Gallium Nitride Nanowire Arrays," Microelectronic Engineering, 142, 58-63, 2015.
10. S. Krylyuk, R. Debnath, Heayoung P. Yoon, Matthew R King, Jong-Yoon Ha, Baomei Wen, Abhishek Motayed and Albert V. Davydov, "Faceting Control in Core-Shell GaN Micropillars Using Selective Epitaxy," APL Materials, 2 (2014) 106104.
11. 3. R. Debnath, Jong-Yoon Ha, Baomei Wen, Yihua Liu, Dipak Paramanik, Abhishek Motayed, Matthew King, Albert V. Davydov and Alec Talin, "Fabrication of high quality GaN nanopillar arrays by dry and wet chemical etching," J. Vac. Sci. Technol. B, 32 (2014) 021204.

12. Carlos M Hangarter, Ratan Debnath, Jong Y Ha, Mehmet A Sahiner, Christopher J Reehil, William A Manners, Daniel Josell, "Photocurrent mapping of 3D CdSe/CdTe windowless solar cells," ACS Applied Materials and Interfaces, 5(18), 9120-9127, 2013.
13. D U Kim, C M Hangarter, R Debnath, J Y Ha, C R Beauchamp, M D Widstrom, J E Guyer, N Nguyen, B Y Yoo, D Josell, "Backcontact CdSe/CdTe windowless solar cells," Solar energy materials and solar cells, 109, 246-253, 2013.

In: Advances in Materials Science Research
Editor: Maryann C. Wythers
ISBN: 978-1-53610-059-4

Chapter 4

APPLICATIONS OF INFRARED THERMOGRAPHY TO THE ANALYSIS OF STEEL WELDED JUNCTIONS

M. Rodríguez-Martín[1,2], S. Lagüela[1,3], D. González-Aguilera[1] and P. Rodríguez-Gonzálvez[1]

[1]TIDOP Research Group, University of Salamanca, Avila, Spain
[2]Technological Department, Faculty of Science and Art, Catholic University of Avila, C/Canteros, Avila, Spain
[3]Applied Geotechnologies Research Group, University of Vigo, Vigo, Spain

ABSTRACT

The thermographic methodology based on the analysis of the response of the materials after an external stimulation can be used as a tool for the detection, characterization and evaluation of surface cracks in steel welds. These types of cracks are usually difficult to detect and their dangerousness could be a problem for the integrity of welded elements in machines and structures. Several studies have demonstrated the usefulness of the thermographic technique for the assessment of composites and metallic materials. However, the study of cracks in welds is a topic which has only been addressed in recent times. The thermographic technique, in combination with different processing algorithms allows the efficient detection of surface cracks in welds. If the procedure is complemented with the geometric calibration of the infrared camera, the accurate measurement of the surface dimensions of the cracks can be performed. What is more, the generation of the depth-prediction model of the cracks is possible through the correlation of the thermographic data with the three-dimensional data obtained using the macro-photogrammetry technique.

1. INTRODUCTION

Active thermography is a highly efficient technique based on the thermal excitation of materials in combination with their thermographic monitoring. This technique is mainly applied in order to obtain quantitative thermal data for the evaluation of the physical characteristics of materials in different fields of engineering and materials science.

The application of active infrared thermography (IRT) for the detection and characterization of defects and imperfections in materials began in 1960–1970, when the optics and sensors in the infrared band started to be applied in engineering fields with high requirements, such as aerospace and nuclear engineering. However, the limitations of the infrared systems and sensors in those years held back its application as NDT (Non-Destructive Technique) [1].

Currently, there are different active thermography modalities depending on the method used for material stimulation. The stimulation energy can be applied continuously or in a pulsed manner. The physical nature and the modality [2] of the stimulation process makes possible the existence of different modalities of active thermography. Thus, active quantitative thermography has been implemented using surface heating by pulsed laser [3] and flashlamps [4]; but also non-optical heat sources [5] such as mechanical vibration [6 7 8], acoustic wave excitation [9 10 11], magnetic stimulation [12] and microwaves [13]. Regarding its application as NDT method, active thermography enables the detection and characterization of flaws and pathologies in different materials [14], also in addition to allowing the computation of thermophysical parameters of the objects such as thermal diffusivity and thermal effusivity [15 16].

Current applications of active thermography in the different modalities include the following fields for assessment of materials [17]:

1. Civil engineering: In recent times, active thermography has been used for the analysis of different constructive and structural elements. Meola et al. (2007) [18] developed an experimentation with coats of

plaster over a brick and marble support, which contained little bubbles in order to simulate a masonry structure. In this experiment, authors concluded that the detection of defects is better during the cooling phase (after the thermal stimulation of the materials). Other studies have addressed the detection of defects in concrete [19] and composites materials [20]. In the latter, the evaluation of the contrast in the frequency domain allowed the analysis of the depth of the defect with low uncertainty.

2. Naval and aerospace engineering: In this field, the evaluation of materials based on active thermography has been widely developed since composite materials are very suitable for active thermographic studies due to their composition of different materials with different responses to external thermal stimuli. Recently, the pulsed phase optics thermography has been used for the three-dimensional evaluation of carbon fiber reinforced polymer (CFRP). For this aim, a processing method called TSR (Thermographic Signal Reconstruction) is applied in order to detect delamination. Polymers Reinforced with Fiberglass are common materials in aeronautic and naval structures. They present many of the advantages of CFRP in addition to higher levity and lower cost. However, they frequently present different defects such as delamination caused by small impacts. For this reason, the evaluation of this type of materials is important. In particular, the working methodology called lock-in thermography is used to analyse pathologies in composite materials. The principle of this technique is the introduction of a periodically modulated heat signal on an object with the consequent periodic thermographic monitoring of the surface of the material in order to analyse the thermal response in the phase to which the heat provided relates. This has been applied in order to detect different pathologies in FRP such as impacts, delamination, cracks or fiber breaking [21].
3. Metallurgical engineering: In the field of metallic materials, active thermography technique has been applied in fatigue tests. Galietti et al. (2010) [22] proposed a method in which the evolution of the phase change material is analyzed during fatigue stress through the acquisition of thermographic images. Recently, fatigue stress was also investigated using vibrotermography with an inversion algorithm in order to characterize vertical cracks in metals [6]. Welded joints require high requirements in safety, and are mainly evaluated using NDT. The first welding test based on the use of infrared

thermography dates back to the late nineties. It was developed and patented by Adams and Crisman in 1989, who proposed a method and a device for evaluating the thermal welding. In Meola et al. (2004) [23] the active thermography technique was compared with other conventional non-destructive testing. They concluded that lock-in thermography is a non-contact technique that allows easy, fast and in situ analysis. This technique is able to detect welding defects, recognize materials of different characteristics and extract information of the material and the changes caused by welding. Specifically, welds in pipelines constitute an attractive field for the application of the procedure, given the difficulty of access to their internal surface. To this end, Choi et al. (2011) [11] conducted an investigation for the detection of internal defects in pipe welds using the technique of active thermography assisted by ultrasounds. Nowadays, current research deals with simplified procedures that allow the detection and in-situ characterization of surface cracks in welded unions through active thermography, using step-shaped heat pulses without implementing more complex procedures [24].

In the last years, most of the research efforts in thermography have been focused on the integration and automation of procedures for defect detection and control of industrial processes. In this regard, the research conducted in 2001 by Sascha et al., [25] in which two cameras were integrated coaxially within the optical path of a laser processing must be taken into consideration. The first camera used an external light by a laser diode while the second camera operated in the near infrared spectrum. It was shown that the control system is an efficient solution for thermographic analysis of aluminium welding process in real time.

Within the metallurgical field, active thermography can be applied in non-destructive testing of welding, to allow the early detection of defects in different types of materials and for quality control in predictive maintenance and prediction of faults in machines, structures and even for troubleshooting in electrical equipment [26].

The main advantage of active thermography applied to non-destructive testing in welding (in comparison to other techniques) is the large surface area covered and the high rate of data acquisition and thus resolution of the temperature evolution [27], in addition to the possibility of providing information as complete as the combination of radiography or gammagraphy with ultrasound test. The evolution of the research so far has been towards the application of the technique of infrared thermography as NDT only in laboratories equipped with the sophisticated heating mechanisms aforementioned, which are difficult to use in external environments. For this reason, the European standard does not contemplate the IRT technique as another NDT method. However, recent studies have discussed the possibility of applying the heat source in the installation place. In 2014, Keo, et al., [13] have tested the application of heat with a microwave source directly to the specimen. This is a high efficiency thermal source, portable and which allows heating in a short time, but it is also an expensive resource, which requires a rigorous security study and presents difficulties to be used in Europe due to the existing safety standards for microwave heating systems [28].

In addition to this state of art about the thermographic technique applied to active testing of materials, this chapter focuses on the study of the technique applied to the detection and characterization of cracks with surface incidents in welds. Within the defectology of the welded materials, the study of the cracking process presents much importance to ensure the safety of machines, constructions and vehicles. The cracking process is extremely dangerous because it may result in the failure of the structural jointed element and thus, in the complete collapse of the structure with drastic implications. Therefore, the assessment, measurement and characterization of the type of crack is important towards the prediction of the direction of propagation and the probable type of failure. The different types of cracks in welds are defined by the international standards [29, 30]. The complex surfaces of the welds make the visual detection of little cracks a very complex task, creating consequently the need of using complementary NDTs for the welding inspector. Among these techniques are ultrasounds, penetrant liquids [31] and magnetic particles [32]. However, these methods require contact and, in some cases, the preparation of the surface of the material. Furthermore, their application is not possible in locations with difficult access and thus these methods do not allow the accurate measurement of cracks. For this reason, active thermography is a novel remote testing method, useful in welding inspection, which allows the fast and accurate detection, characterization and measurement of cracks in welds.

Summarizing, the possibility of performing thermographic non-destructive testing in welding is important due to its adequacy for the detection of different pathologies, flaws and imperfections. When compared to other inspection techniques, IRT can be used to improve the quality evaluation of welds.

In the next sections, the thermographic procedures for surface crack detection and characterization will be explained. Subsequently, the thermal rectification image process to extract bi-dimensional measures of surface cracks will be presented and, finally, new developments based on the thermographic approach for the three-dimensional measurement of cracks will be described.

2. Superficial Crack Detection and Characterization

In this section, different methods for the detection of cracks in welding developed by the authors will be explained. They all involve a simplified surface heating, with the aim at reducing the complexity of the heating process and make it applicable for in-situ inspections.

2.1. Temperature-Based Crack Detection and Characterization

The detection of both surface cracks and internal cracks with effect on the surface of steel welds can be done over a thermal image using previous heating (between 40ºC and 70ºC) (Figure 1). After this, the thermal contrast in the image is increased in order to obtain a clear visualization of the crack (Figure 2). Cracks are detected from a study of temperatures applied to each pixel of the thermal image. The difference of emissivity values and directions of infrared radiation in steel between defect and non-defect zone enable crack detection.

Two typologies of cracks in steel welded unions can be detected using this procedure according to [24]: toe cracks and longitudinal cracks. The first are internal cracks, which are propagated to the surface of the material, where they are detected. The second are cracks that arise at the surface of the material and are propagated in parallel to the direction of the weld edge.

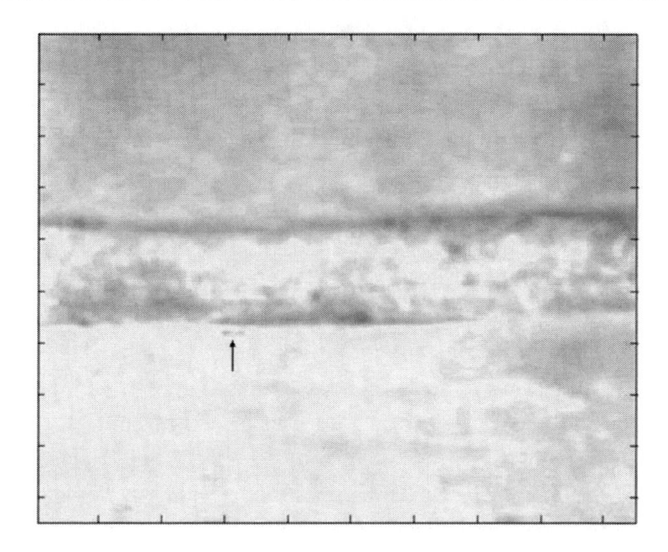

Figure 1. Thermographic image of a Steel joint (low carbon steel welded with Tungsten Inert Gas (TIG) technique) acquired after the transmission of the heating. A crack in the transition zone is detected in the infrared image (acquired with an IR Camera NEC TH9260 with 640 x 480 UFPA (Uncooled Focal Plane Array)).

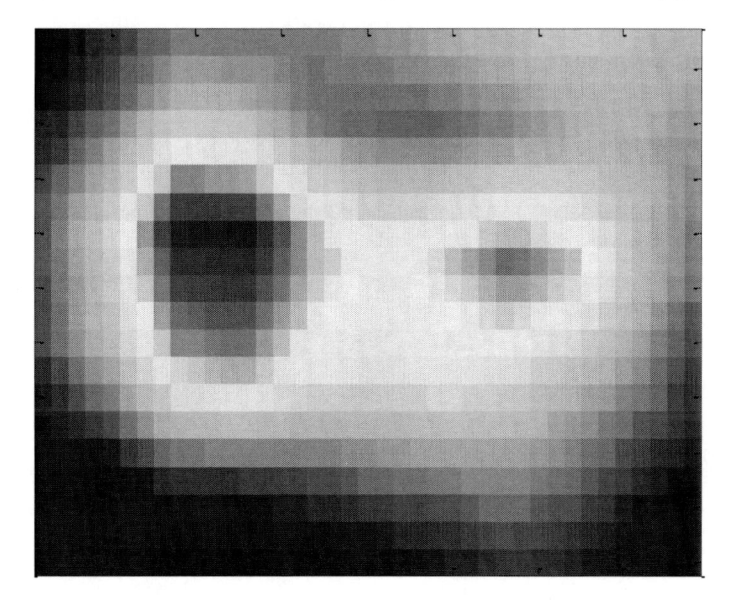

Figure 2. Pixel map for the crack extracted from the thermographic image shown in Figure 1, where the temperature distribution allows its characterization. In this case, a quasi-symmetrical structure with two surface incisions is detected. Its shape is compatible with a toe crack.

Heating can be applied in two different configurations regarding the crack and the thermographic camera: directly to the crack, next to the camera (reflection mode, since the radiation measured by the camera is the reflected fraction) or indirectly to the back face of the weld, while the camera is focused to the front face (transmission mode, given that the radiation measured by the camera is the fraction that travels through the welding). Each heating modality is more adequate for one type of crack. In particular, [24] demonstrated that in the case of the toe crack, greater contrast exists in the cooling rates after back heating (transmission mode) than after front heating (reflection mode), results that are reversed in the case of the longitudinal crack. This difference is due to the geometry of the cracks: the toe crack is an internal crack, so if back heating is applied, the thermal energy has to be transmitted through the thickness of the plaque and the temperature maps in the monitored surface are more homogenous. In this way, the definition of the toe crack is higher. In contrast, for the longitudinal crack, the visualization in the thermal image is better when direct heating is applied over the crack because of its location at the surface.

The analysis of the temperature distribution in the crack enables the characterization of the typology of crack through the morphology shown in the thermal image. In the case of the longitudinal crack, the temperature distribution in the defect zone is more homogenous because the crack is at the surface of the material; however, in the case of the toe crack, the temperature distribution in the defect zone is not homogenous due to the presence of zones with different depths (Figure 3). With these criteria, it is easy to characterize the type of crack for the two typologies studied.

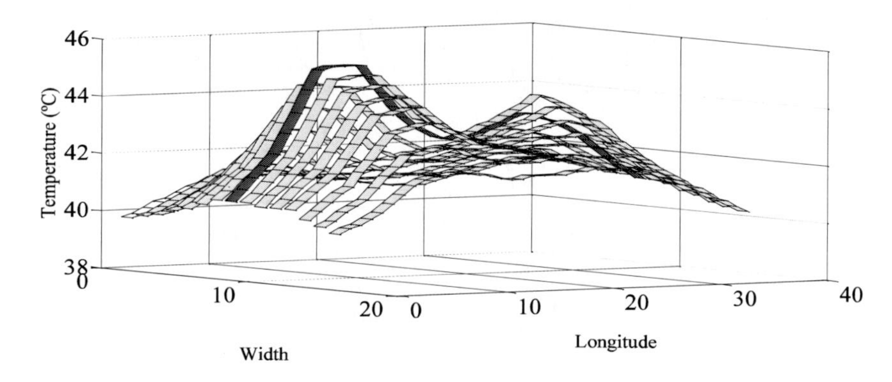

Figure 3. Temperature distribution in a toe crack. Green line shows the section with higher temperature variation.

2.2. Isotherm Processing for Crack Detection

Isotherm processing has been implemented in order to improve the previous study of the temperature distribution at the surface of the metal. Isotherms are lines that indicate the zones at the same level of temperature, and are plotted over the thermal image. In the example study [33], isotherms are generated through the implementation of an algorithm based on contour lines (Matlab ®) applied to the thermographic images. First, the number of temperature levels is determined by the inspector, which as a consequence determines the number of contour levels to draw. The contour algorithm interprets the input thermal matrix from the thermographic image as a regularly spaced grid, where each cell is linked to its nearest neighbors. The algorithm analyzes this temperature matrix and compares the values of each block of four neighboring cells in the matrix to the values of contour level established for the processing. If a contour level is placed within a cell, the algorithm implements a linear interpolation to locate the points at which the contour crosses the edges of the cell. The algorithm automatically links point to point to draw a segment for each contour line, which, in this case, represents the different isotherms.

In this way, the algorithm implements a correlation between similar temperature values, making a finite number of groups of pixels (zones) according to the number of levels established. Each group is plotted with a different color and the isotherms are the lines that define each group level. With this procedure, the contour of the defects is detected by thermal contrast with the surface without defect, since defects are embodied within closed isotherms, allowing for the easy characterization of the typology of crack based on the observation of the different thermal zones enclosed by the isotherms. This way, the processing based on the generation of isotherms has the capability of detecting cracks.

The quality of results depends mainly on the number of contour levels established (Figure 4). When the number of levels is increased, the difference of temperature between isotherms is reduced and the thermal representation is smoother. However, a large number of lines slow the computation. For cracks around 3 mm length, between 5 and 20 levels are sufficient for characterization [33], as shown in Figure 4.

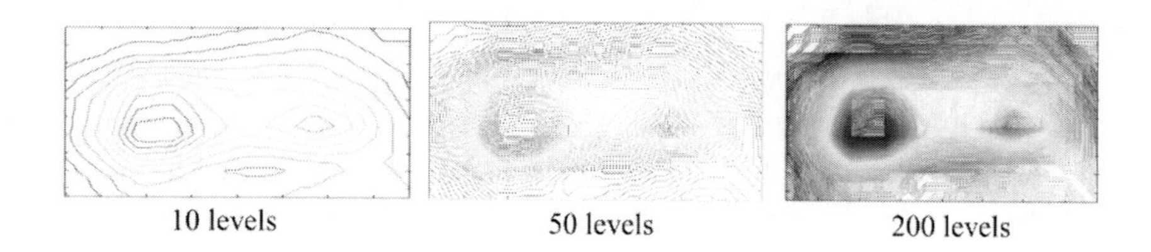

Figure 4. Different plots for the toe crack of Figure 1 varying the number of levels for the algorithm of contour lines.

2.3. Cooling Rate-Based Crack Detection

Another effective approach to defect surface cracks in welding is the study of the cooling rate of steel after the application of continuous heating. Unlike the methods previously explained, this approach is based on the study of a sequence of images. The objective of the study is to enable the detection of cracks. From the point of view of physics, different physical parameters are involved in the cooling. The dissipation of the heat depends (inter alia) on the geometry of the target under study, as considered by the Newton Cooling Law [24]. For these reasons, a cooling study can provide more information that materializes in a better detection of cracks.

Using the Pixelwise algorithm for Time-Derivative of Temperature (PATDT) presented in [34], a cooling rate matrix could be extracted. Intrinsically, the cooling rate is calculated by the algorithm based on the experimental results from the thermographic study. The algorithm automatically fits the different values of temperature and time, and studies the decreasing tendency of the temperature for each pixel during the cooling period. Then, a new matrix is formed where each pixel position is occupied by the integral average of cooling rate values.

Applying the contour line processing raised in the subsection 2.2, it is possible to extract the different cooling zones within the crack (Figure 5), allowing the detection and characterization of the cracks in a similar way as when absolute temperatures are used (Figure 4). In a toe crack, the cooling-rate is lower in the deeper zones.

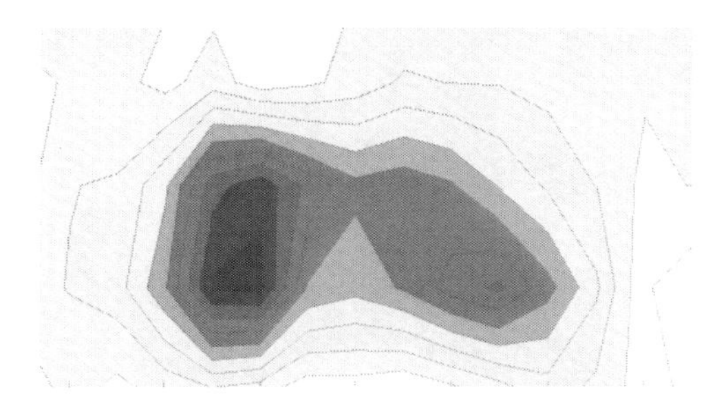

Figure 5. Different zones of cooling rate for the crack of Figure 1 (10 levels are used). The zones with slower cooling are shown in red. The contrast of the crack zone with respect to the non-crack zone is remarkably appreciable.

3. Measurement of Surface Cracks

In the last section, different thermographic approaches have been exposed for the detection and characterization of cracks. In this section, these thermographic methods are adapted in order to create new measurement tools for surface cracks in welding.

The size of the crack is important mainly due to two reasons: on one hand, knowing the size of the crack is necessary to evaluate the riskiness of the defect and to take a decision about the total or partial repair of the weld. On the other hand, the position and inclination of the crack is useful to predict if the fracture typology can provoke the collapse of the welded structure. This information is of great value for the welding inspector, which usually utilizes manual gadgets of low accuracy for this aim.

In order to use a thermographic camera as a tool for the measurement of surface cracks in welds, first, the camera has to be geometrically calibrated prior data acquisition. Different calibration fields have been designed in the last years, based on different thermal parameters: Figure 6 shows a passive calibration field based on differences of emissivity between the background (high emissivity value) and the targets (low emissivity value) [35]. Emissivity differences make elements to be measured as if they were at different temperatures when they are in thermal equilibrium. As an alternative, Figure 7 shows an active calibration field based on temperature differences between the background and the targets, which consist on light bulbs [36]. For each position of the focus of the IR camera used for data acquisition, the main parameters required for image rectification and 3D image orientation are focal length, coordinates of the principal point and the parameters of lens distortion.

These internal camera parameters are computed after the acquisition processing explained in [35 36].

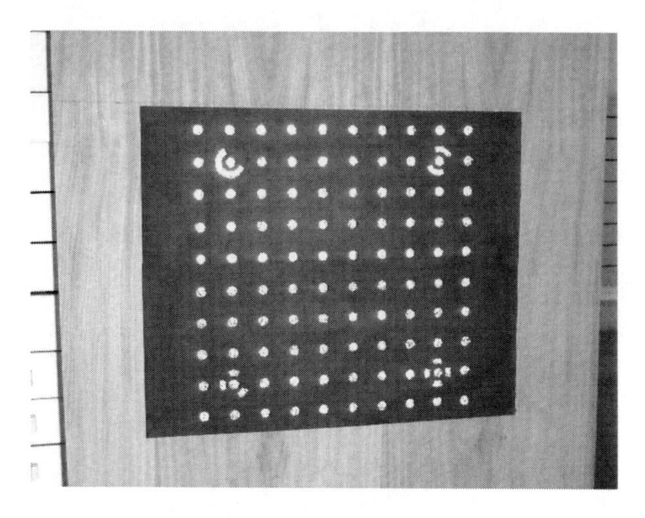

Figure 6. Passive calibration field of thermographic cameras based on the emissivity difference between the background and the targets.

Figure 7. Active calibration field for thermographic cameras consisting on a wooden plank as background with light bulbs as targets. Targets are detected in the background due to their heating after turned on.

The geometrical calibration of the thermographic camera allows the computation of more accurate geometric data from the cracks than that obtained without the knowledge of the internal camera parameters. This effect is especially important for the rectification processing, which is based on the computation from individual images and does not allow for error compensation between images.

3.1. Two-Dimensional Measurements and Characterization of the Crack

The application of an image rectification processing to the thermal images results on the generation of an image with 2D metrics where planar measurements of the cracks can be performed directly [33].

The first step of the rectification process for thermal images consists on the extraction of the temperature matrix of each thermographic image: each matrix cell contains the temperature value of the corresponding part of the plaque contained in the image pixel. Temperature values are corrected on an emissivity basis, using as reference the temperature values that have been measured at the beginning of the test (for example, with a contact thermometer).

The emissivity value can be calculated using Stefan-Boltzman's law, and is applied as correction to the matrix in order to obtain the real temperature values. Equation 1 shows the derivation of the emissivity (ε) value based on Stefan-Boltzman's law, assuming the relation between the black body, or real temperature (T_{BB}^4) measured with a contact thermometer, and the temperature measured (T_{RB}^4) with the IR camera:

$$\varepsilon = \frac{T_{RB}^4}{T_{BB}^4} \qquad (1)$$

Once the temperature values are corrected, the matrix is subjected to a rectification algorithm (Figure 8), which is based on a planar projective transformation [33]. This transformation requires the knowledge of the real coordinates of at least 4 points present in the image, in order to estimate the eight coefficients of the transformation. These coefficients enclose rotation, scale, translation and perspective.

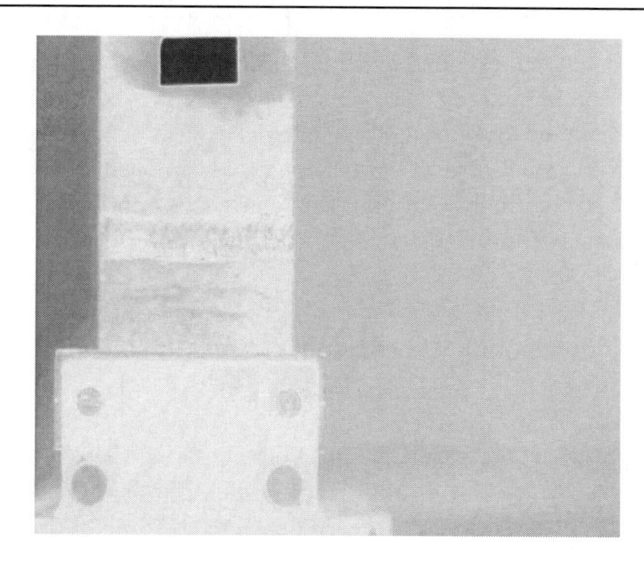

Figure 8. Rectified thermal image of a steel butt weld. In the borders of the image, a cut area is shown; it appears due to the geometrical transformation indicated in this section that could provoke loss of information in the extremes of the images. For this reason, placing the target in the centre is recommendable.

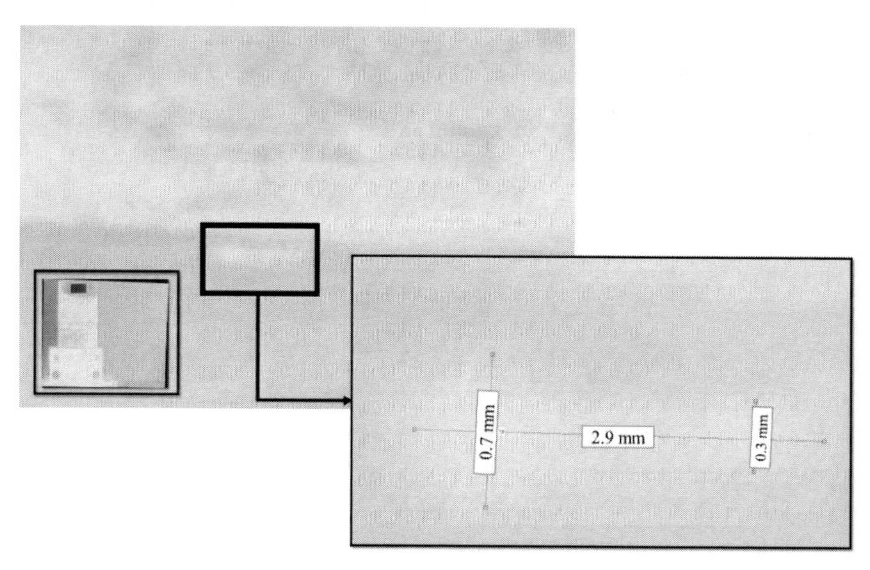

Figure 9. Example of measurements in the last rectified thermal image following the procedure established in [33]. The top display shows a rectified image, which zoomed allows the detection of the crack and the measurement of its length and width.

Once the thermal images are rectified, [33] demonstrated that dimension measurements can be directly performed on them (Figure 9). The measurements of width and length of the crack can be directly made on the image (already knowing the object sample distance for each pixel) or assisted by contour lines, which could be easily generated from rectified thermal images following the procedure for generation of isotherms exposed in subsection 2.2. Using isotherms, the extension of the crack can be delimited and the measurements can be made more accurately. Another parameter to be measured is the angle of the edge of the crack with respect to the direction of the welding. It is measured on the rectified images, and gives important information about the direction of propagation of the crack.

3.2. Three-Dimensional Information: Depth Prediction Model

Several studies [34 - 36] have shown that the combination of infrared thermography with close-range macro-photogrammetry allows the establishment of depth prediction models through the correlation of the geometrical values generated through photogrammetry and the thermal values obtained by active thermography. These prediction models can be designed from two different approaches: (i) using temperature values or (ii) using values of cooling rate. These two prediction models will be exposed and compared.

The generation of the photogrammetric 3D-model of the crack is implemented following the method based on photogrammetry and computer vision established in [37]. Summarizing the process: the first step is the acquisition of single photographic images with a commercial Digital Single Lens Reflex (DSLR) camera following a semi-spherical trajectory centred on the object and maintaining a fixed focal length and constant distance to the object. Once images are acquired, two processing steps are applied: first, the automatic determination of the spatial and angular position of each image, regardless the order of acquisition and without needing initial approximations or camera calibration. Second, the automatic computation of a dense 3D point cloud with submillimetric resolution, in which each pixel of the image renders a three-dimensional point of the model of the weld (Figure 10). This procedure allows the generation of a high accuracy 3D model of the surface crack. In this model, the geometrical depth for each zone of the crack can be extracted.

Figure 10. Three-dimensional model of the toe crack obtained using the macro-photogrammetric technique established in [37].

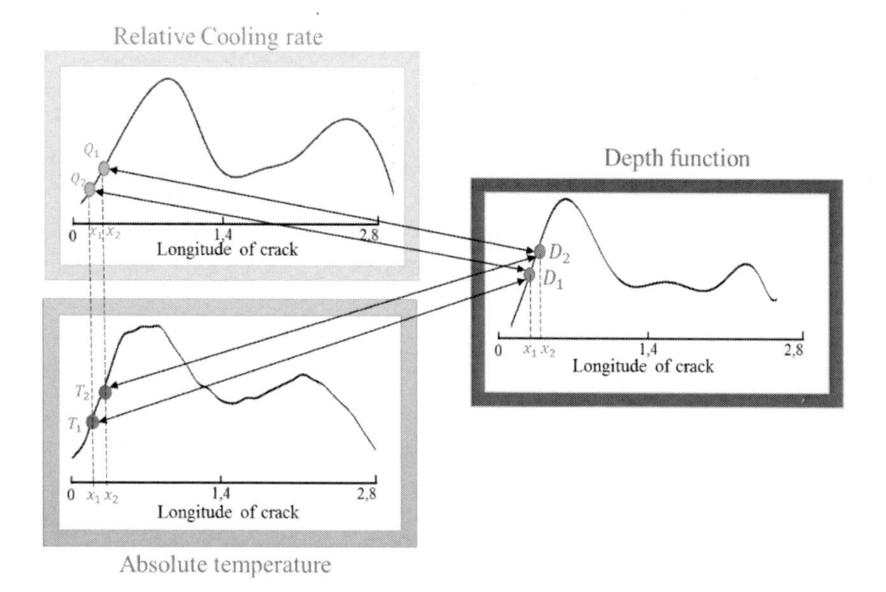

Figure 11. Different works published [34, 36] demonstrated that the cooling rate values (top) (Q_1, D_1), (Q_2, D_2), (Q_3, D_3), ..., (Q_n, D_n) and temperature values (bottom) (T_1, D_1), (T_2, D_2), (T_3, D_3), ..., (T_n, D_n) can be respectively correlated with the depth values extracted from the 3D macro-photogrammetric model (right) through a longitudinal section of the crack $(x_1, x_2, x_3 \dots x_n)$.

Based on the knowledge of the depth for different zones of the crack, the authors can establish two different correlations between depth and temperature values [36] or values of relative cooling rate [34] in the thermal image (Figure 11). Relative cooling rate is computed in the study mentioned as the difference between the cooling rate for each pixel in the cracking zone and the average of cooling rates in the non-cracking zone. The temperature distribution and cooling rate distribution are obtained through a longitudinal section of the crack. For the same longitudinal section, a depth map is obtained from the macro-photogrammetric procedure and results of temperature and cooling rate are independently correlated with the real depth data (Figure 11).

Trying to reinforce this analysis, the authors [34 36] performed two different studies. In the study based on absolute temperature, the knowledge of the values of temperature for each pixel in the thermal image allowed the computation of the depth of the crack, as a function of the temperature values captured by thermography. The error in the depth estimation is around 9%. In the study based on cooling rate analysis, the depth values for different zones of the crack lead to the establishment of a correlation between cooling rate and depth in this pixel region. The conclusion was that it is possible to establish a correlation between the depth of the surface crack and the relative cooling rate measured with thermography. This method allows the measurement of the depth of the crack with an error between 13% and 14%. The error is higher than in the first method for the selected case studies, but [38] demonstrated that values of cooling rate present a better correlation with the values of depth than the temperature values, because the correlation cooling rate-depth presents higher statistical adequacy than the correlation temperature-depth. This is because the cooling study involves many physical parameters, including the geometry of the crack. Areas more rugged into the crack present a better dissipation of the heat, and the enclosed zones of the crack are deeper and present a lower cooling rate because the heat radiated from the material provokes the formation of hot bags inside the crack. This allows for a good characterization of the depth of the cracks using the method based on the study of the cooling rate.

Conclusion

Active thermography presents great importance for the testing and assessment of materials. It has been applied in different fields of engineering such as mechanical, naval and oceanic, civil and aeronautics. Within the material testing field, the technique is highly developed for the study of composite materials. However, it has been currently studied for its application in the metallurgical field, particularly for the study of welded structures because welding inspection is a key field, which requires extensive development because of its complexity and the importance to ensure the safety of the structures, the machines and the vehicles.

As recent studies shown, the active thermographic technique has great potential for the detection, geometric measurement and characterization of cracks in welds. It can become a tool of enormous importance in the quality inspection of welds.

Using the thermal image, surface cracks that are difficult to identify can be detected in steel welds. In order to improve the detection process, an isotherm processing can be applied to the thermal image in order to delimit the contour of the crack. The number of thermal levels to establish depends on the morphology of the crack and the resolution desired. With the information given by the thermal image and by the processing mentioned, it is also possible to characterize the crack based on the study of the different thermal zones. Using a PATDT algorithm, a map of cooling rate can be generated, which helps on the characterization of the crack; especially for cracks with hot air enclosed such as toe crack.

The bi-dimensional measurement of cracks is an important criterion to assess the quality of welds. Length, width and inclination of the surface plane can be obtained from the thermal image if a rectification algorithm is applied. In this way, a ground truth is introduced considering the real pixel size allowing to extract 2D metric information from the image.

The three-dimensional measurement of surface cracks is a field currently under development. The results of recent studies shown that the depth of surface cracks can be predicted based on thermographic approaches. It can be implemented between the correlation of the three-dimensional depth data from a macro-photogrammetric process and the thermal data from active thermography. There are mainly two approaches for this aim. The depth prediction model based on the cooling rate study and the depth prediction model based on absolute temperatures. Currently, this last approach represents a hot topic for the scientific community. With the current advance of the new thermographic techniques applied to welding inspection, it is expected that the technique will become a non-destructive testing method at the same level of the ones traditionally used (X-rays, ultrasounds, magnetic particles, eddy currents and penetrant liquids).

REFERENCES

[1] Ishchuk N, Karpov G, Fesenko A. The detection of hidden subsurface objects in the infrared wavelength band by identifying their thermal properties. *Measurement Techniques* 2009. Vol. 52-4 pp. 388-393. DOI: 10.1007/s11018-009-9282-7.

[2] Vavilov V “Thermal NDT: historical milestones, state of-the-art and trends”. Quantitative InfraRed Thermography Journal 2014 Vol. 11(1), pp. 66-83

[3] L. Teng, D.P. Almond, D. Andrew, S. Rees. Crack imaging by scanning pulsed laser spot thermography. *NDT&E International* 2011. Vol 44. pp. 216–225.

[4] P. Broberg, Surface crack detection in welds using thermography. *NDT&E International* 2013. Vol 57. pp. 69–73.

[5] N. Tsopelas, N.J. Siakavellas. Experimental evaluation of electromagneticthermal non-destructive inspection by eddy current thermography in square aluminum plates. *NDT&E Interntional* 2011. Vol. 44. pp. 609–620.

[6] Castelo A, Mendioroz A, Celorrio R, et al. “Vertical cracks characterization and resolution from lock-in vibrothermography”. *The 12 International Conference on Quantitative Infrared thermography.* (Burdeos, 7-11 de Julio de 2014).

[7] L.D. Favro, X. Han, Z. Ouyang, G. Sun, H. Sui, R.L. Thomas, Infrared imaging of defects heated by a sonic pulse, *Rev. Sci. Instrum* 2000. Vol 71. pp. 2418–2421.

[8] F. Mabrouki, M. Thomas, M. Genest, A. Fahr, Numerical modeling of vibrothermography based on plastic deformation, *NDT&E International* 2010. Vol 43. pp. 476–483.

[9] K.A. Tsoi, N. Rajic, Interference fit fastener characterisation using sonic thermography, in: *5th Australasian Congress on Applied Mechanics,* Brisbane, 2000.

[10] Mendioroz A. Castelo A. Celorrio R et al. Characterization and spatial resolution of cracks using lock-in vibrothermography. NDT&E International 2014. Vol.66 pp. 8–15 DOI 10.1016/j.ndteint.2014.04.004.

[11] Choi M, Park H, Park J. et al. Internal Defect Detection of Dissimilar Weld Pipe Using Ultrasonic Infrared Thermography. 5th Pan American Conference for NDT. Cancun: 2011.

[12] Oswald-Tranta, B, Sorger, M, Localizing surface cracks with inductive thermographical inspection: From measurement to image processing, Quantitative InfraRed Thermography Journal 2011 Vol. 8 pp.149-164

[13] S.A. Keo, F. Brachelet, F. Breaban, D. Defer, Steel detection in reinforced concrete wall by microwave infrared thermography, NDT&E International 2014. Vol. 62. pp. 172–177.

[14] Q. Tang, J. Liu, Y. Wang, H. Liu, Subsurface interfacial defects of metal materials testing using ultrasound infrared lock-in thermography, Procedia Eng 2011. Vol. 16. pp. 499–505.

[15] P. Bison, S. Marinetti, A. Mazzoldi, E. Grinzato, C. Bressan, Cross-comparison of thermal diffusivity measurements by thermal methods, *Infrared Phys. Technol* 2002. Vol. 43. pp. 127–132.

[16] N. Tao, Z. Zeng, L. Feng, Y. Li, C. Zhang, Thermal effusivity measurement of subsurface heterogeneous materials using pulsed thermography, *Infrared Laser* Eng 2011. Vol. 40 (11) pp. 2098–2103.

[17] M. Rodríguez-Martín, S. Laguela, D Gonzalez-Aguilera, L Diez-Vilarino. Active Thermography: Theoretical Approach, Development and Applications for Engineering and Industry. DYNA Ingeniería e Industria 2015. Vol 90 (6) pp. 568-572. http://dx.doi.org/10.6036/7596.

[18] Meola C. A new approach for estimation of defects detection with infrared thermography. *Materials Letters* 2007 Vol. (61) pp. 747–750. DOI:10.1016/j.matlet.2006.04.120.

[19] Maierhofer CH, Brink A, Rollig M, et al. (2005). Quantitative numerical analysis of transient IR-experiments on buildings. *Infrared Physics and Technology*. 2004 Vol (46), pp. 173–180. DOI:10.1016/j.infrared.2004.03.022.

[20] Montanini R, Freni F. Non-destructive evaluation of thick glass fiber-reinforced composites by means of optically excited lock-in thermography. *Composites Part A: Applied Science and Manufacturing* 2012. Vol (43). p. 2075–2082. doi:10.1016/j.compositesa.2012.06.004.

[21] Meola C, Giovanni M. Infrared thermography to evaluate impact damage in glass/epoxy with manufacturing defects. *International Journal of Impact Engineering* 2014. Vol. 67 pp. 1-11. DOI:10.1016/j.ijimpeng.2013.12.010.

[22] Galietti, U, Palumbo D. Application of thermal methods for characterization of steel welded joints: X International Conference on Quantitative InfraRed Thermography. (Québec 27-30 de Julio de 2010).

[23] Meola C, Squillace A, Memola F, et al. Analysis of stainless steel welded joints: a comparison between destructive and non-destructive

techniques. Journal of Materials Processing Technology 2004. Vol. (155). pp.1893–1899. DOI: 10.1016/j.jmatprotec.2004.04.303.

[24] M. Rodríguez-Martín, S. Lagüela, D. Gonzalez-Aguilera, P. Arias. Cooling analysis of welded materials for crack detection using infrared thermography. *Infrared Physic and Technology* 2014, Vol. 67, pp. 547–54.

[25] Sascha F, Michael U, Rolser R. Coaxial control of Aluminum and Steel Laser Brazing Processes. *Physics Procedia* 2011 Vol. (12) pp. 752–760 DOI:10.1016/j.phpro.2011.03.094.

[26] A.S. Nazmul, S. Taib, A semi-automatic approach for thermographic inspection of electrical installations within buildings, *Energy Build* 2012. Vol. 55. pp.585–591.

[27] D. Balageas, Termografía Infrarroja: una técnica multifacética para la Evaluación No Destructiva (END), IV Conferencia Panamericana de END, Buenos Aires, 2007.

[28] Safety in electroheat installations – Part 6: specifications for safety in industrial microwave heating equipment, European Committee for Standardization. EN 60519-6: 2011, 2011.

[29] Welding and allied processes – classification of geometric imperfections in metallic materials – Part 1: fusion welding, European Committee for Standardization. UNE-EN-ISO 6520-1:2007, 2007.

[30] Welding, Fusion-welded joints in steel, nickel, titanium and their alloys (beam welding excluded). Quality levels for imperfections (ISO 5817:2003 corrected version.2005, including Technical Corrigendum), European Committee for Standardization. UNE-EN-ISO-5817: 2009, 2009.

[31] European Committee for Standardization. 2015. *"Non-destructive testing of welds – Penetrant testing - Acceptance levels"*. EN-ISO 23277:2015.

[32] European Committee for Standardization. 2015. *"Non-destructive testing of welds – Magnetic particle testing - Acceptance levels"*. EN-ISO 23278:2015.

[33] Rodríguez-Martín, S. Lagüela, D. González-Aguilera, J. Martínez. *Thermographic test for the geometric characterization of cracks in welding using IR image rectification.* Automation in Construction 2016. Volume 61. pp. 58–65.

[34] M. Rodríguez-Martín, S. Lagüela, D. González-Aguilera, and P. Rodríguez-Gonzálvez. Crack-Depth Prediction in Steel Based on

Cooling Rate, *Advances in Materials Science and Engineering* 2016, Vol. 2016, pp. 1-9 doi:10.1155/2016/1016482.

[35] Lagüela S, González-Jorge H, Armesto J, Herráez J. High performance grid for the metric calibration of thermographic cameras. *Measurement Science and Technology* 2012 Vol. 23 doi:10.1088/0957-0233/23/1/015402.

[36] M. Rodríguez-Martín, S. Lagüela, D. González-Aguilera, Martinez J. Prediction of depth model for cracks in steel using infrared thermography. *Infrared Physics and Technology* 2015. Vol 71. pp. 492–500.

[37] M. Rodríguez-Martín, S. Lagüela, D. Gonzalez-Aguilera, P, Rodríguez-Gonzálvez. Procedure for quality inspection of welds based on macro-photogrammetric 3D reconstruction. *Optics and Laser Technology* 2015. Vol 73. pp. 54-62.

[38] M. Rodríguez-Martín, S. Lagüela, D. Gonzalez-Aguilera, P, Rodríguez-Gonzálvez. Cooling rate VS temperature to establish thermographic prediction model in surface cracks. Quantitative Thermography International Conference (QIRT) 2016. Gandsk (Poland).

In: Advances in Materials Science Research
Editor: Maryann C. Wythers
ISBN: 978-1-53610-059-4

Chapter 5

Review of Recent Studies on Suspension Plasma Sprayed ZrO_2 Coatings

***Paweł Sokołowski, MSc*[1,2]**
***and Lech Pawłowski, PhD, DSc, HoF*[1,*]**
[1]UMR CNRS 7315, University of Limoges, Limoges, France
[2]Faculty of Mechanical Engineering, Wrocław University of Technology, Wrocław, Poland

Abstract

One of the most frequent topic of thermal spraying research field in recent years concerns the use of liquid feedstock instead of dry powder. In particular, the application of suspensions allowed introducing nanometric or submicrometric particles into flames and jets and enabled obtaining finely grained coatings having modified microstructure and interesting properties. This development may lead to create a fair competition with regard to vapor deposition methods in many industrial applications.

This chapter discusses Suspension Plasma Spraying (SPS) as a method that uses liquid feedstock. The present review should help to understand which are the advantages and drawbacks of this method. The suspension preparation and its influence on the deposition process and on

[*] Corresponding author: Email: lech.pawlowski@unilim.fr.

the coating microstructure is discussed. Then, the suspension injection into flame or jet using nozzles or atomizers is presented. Moreover, the different types of plasma torches which can be used in the SPS processes having axial and radial injectors are discussed. The different microstructures of SPS coatings are presented. The particular attention is paid to the grain size and their orientation as well as to the phases' composition. The build-up mechanisms of various types of coatings morphologies called *columnar* and *two-zones-microstructure* is discussed. Finally, two important applications of SPS zirconia coatings, namely Thermal Barrier Coatings (TBC) and Solid Oxide Fuel Cells (SOFC), are discussed in more detailed way.

Keywords: suspension plasma spraying, zirconia coatings, thermal barrier coatings, solid oxide fuel cells

THE ROLE OF POWDERS IN FORMULATION OF SUSPENSIONS USED TO SPRAY

The Use of Liquid Precursors to Obtain Nanostructures

Before starting the discussion about the use of liquid feedstock in thermal spraying, it is useful to define what is *nanotechnology*. K. Eric Drexler introduced the term *nano-scale* term as early as in 1981 [1]. Later on, he defined *nanotechnology* as a way of controlling atoms and molecules for creation of useful materials [2]. Since then, a growing interest in nanotechnology in many fields of materials science has been observed. This resulted from the fact, that small sized grains in material structure lead to reach many useful properties. The properties are often more interesting than that of the same material having large grains. In fact, the nanomaterials may be characterized by the better hardness, durability, plasticity, creep resistance, chemical resistance and reactivity, electrical conductivity,…than coarse-grain-sized materials [3-6].

Surface engineering followed this development. The manufacturing of coatings having a finely-grained structure has been the subject of intensive researches for two last decades. By now, the coatings with nanometer grains have been mainly obtained by vapor deposition methods including PVD or CVD [7-8]. On one hand, these technologies enabled to control the coating's growth on atomic level. But their low deposition rate and long deposition time together with the need of low vacuum to succesful processing are their drawbacks. Furthermore, the cost of CVD/PVD equipment is generally higher comparing to that of thermal spray methods. Finally, the applications of PVD and CVD methods are limited to thin films deposition.

Thermal spraying processes, especially conventional powder plasma spraying, are a cheaper alternative to vapor deposition ones. But, there are a lot of difficulties associated with feeding, transport of fine powder particles and of their injection in a plasma jet. The main problem is associated with low momentum of small particles making it difficult to penetrate the plasma jet [9 - 10]. The first idea was to agglomerate the nano-powders to prepare powder feedstock [11-14]. The selection of process parameters leading to obtain the useful nanostructured coatings using agglomerated nanometer powders was however very tedious.

Another idea came from Gitzhofer et al. [15]. They proposed that instead of using a dry powder, a suspension of finely-grained powder in a solvent can be used. In that way, a droplet of suspension including fine powder may get enough momentum to be injected into the plasma jet. The transport of liquid feedstock from feeder to injector is also easier than for fine and dry powder. This idea grew up and led to the development of new thermal spraying method - Suspension Plasma Spraying (SPS). This method will be discussed in details in this chapter.

On the other hand, another thermal spray process, namely Solution Precursor Plasma Spraying (SPPS), was developed [16- 17]. This method uses liquid precursors as a feedstock material. The precursors include solutes (some salts, nitrates or acetates) with the addition of solvents as water or ethanol. Furthermore, the solutions are, in fact, the molecular-level mixtures with desired stoichiometry of all compounds [18]. The general idea behind the SPPS method is to form solid particles while the liquid precursor get in contact with hot plasma jet. Afterwards, these particles can melt and build fine-grained and homogenous structures after impact with a substrate [19].

It should be mentioned that many attempts to deposit nanometer coatings based on discussed feedstock (Figure 1) have been carried out using other, than plasma spray, thermal spray methods. R.S. Lima et al. [20] deposited nanostructured WC-Co coatings by cold gas spray method using agglomerated and sintered nano-powders feedstock. High-Velocity Oxy-Fuel (HVOF) method was used by e.g., Berghaus et al. [21] to spray zirconia nano-suspensions in order to produce thin, dense and finely grained electrolyte for Solid Oxide Fuel Cell's. Finally, Saremi and Valefi [22] applied flame spray technology to deposit finely grained YSZ coatings for TBC applications by using solution precursor.

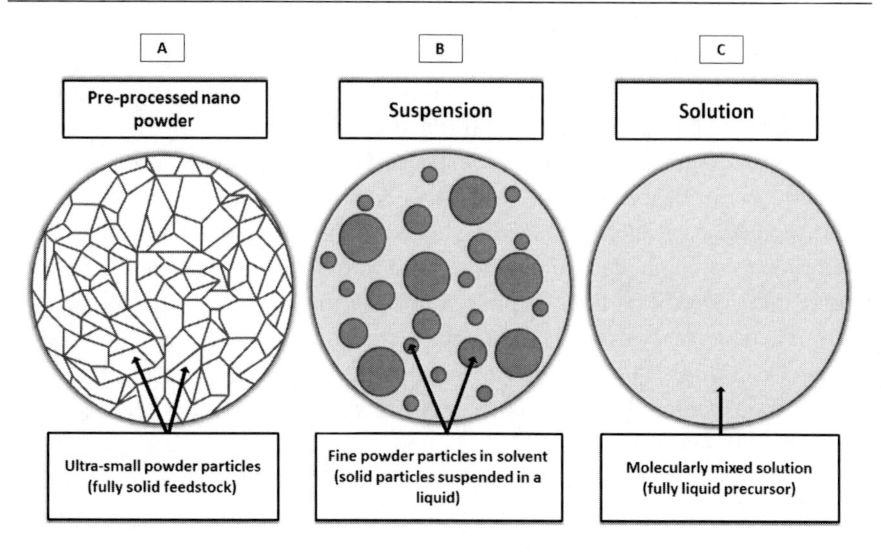

Figure 1. The possible feedstocks for thermal spraying processes enabling to obtain nano-structured coatings: (a) pre-processed nano-powder particle, (b) suspension droplet and (c) solution droplet.

Suspension Preparation

The knowledge related to suspension preparation belongs to colloid chemistry and its goal is to prepare a stable suspension. This means that the particles' sedimentation should be limited and well dispersed and that the rheological properties such as viscosity should be acceptable. The chemical agents, such as flocculating agents and viscosity enhancers are useful to reach these goals. As the suspension preparation influences strongly the spray process and coating microstructure, each component of suspension will be discussed separately.

Solid Powder Preparation

The quality of thermally sprayed coating is influenced by the quality of powder [23]. There are two principal ways to obtain finely grained powder useful for formulation of suspension: (i) direct manufacturing of nano- or submicrometer-sized powders; or (ii) milling of coarse powders. Generally speaking, one can say that the choice is between two approaches called: (i) bottom-up; and (ii) top-down. These approaches are used to classify the ways of fine powder preparation (see Table 1). These preparation method can be categorized following the phase in which the powders are synthesized as gas, liquid or solid one [24-25].

In the case of "bottom-up" approach a continuous progress can be observed and there are many methods which allow producing the finely grained powders. The use of liquid precursors in powder manufacturing process is a method in which powders precipitate from a solution [26]. Calcination of precipitated precursors is used e.g., by MEL Chemicals Inc. to manufacture zirconia nano-powders. Carpio et al. [27] were using such a suspension with such powder to spray finely-grained coatings. But the bottom-up approach can be also very complex. This is the case of e.g., sub-micrometer zirconia powder manufactured by Tosoh Corp. (Figure 2a). The following stages are used in powder manufacturing: hydrolysis of chemical precursors, drying, calcination, milling and finally spray drying [28]. This powder was already sprayed by SPS for Solid Oxide Fuel Cells manufacturing [29] as well as for Thermal Barrier Coatings [30]. On the other hand, the "top-down" approach [26] assumes a mechanical milling i.e., breaking of coarse, micrometer sized, particles to obtain the smaller ones (Figure 2b). As a base powders commercially manufactured for other thermal spraying processes can be used [9]. The production process depends on the kind of the material. For SPS process two ball-milling methods are useful, namely "low energy" and "high energy" milling [31]. Both of them are efficient and low-cost processes and enable the production of fine powders at low temperatures and can be applied to all classes advanced materials (including ductile, brittle and very hard compounds) [32]. The application of ball milling method in SPS method is frequently cited in literature. Kaβner et al. used ball-milling to crush alumina powders [33], Chen et al. ball-milled and mixed composite Al-ZrO_2 for SPS [34], Łatka et al. applied this method to reduce size of various types of zirconia powders [35], Jaworski et al. for hydroxyapatite [36], and, Berghaus et al. applied this method to break-down the agglomerates of cermet WC-Co [37].

Besides mechanical milling there is a possibility to apply mechanical alloying. This process needs higher temperature and higher kinetic energy than milling to achieve high atomic mobility [38]. Mechanically alloyed powders such as YSZ-NiCrAlY, YSZ-Ni, Ni, Ni-Al, etc. were successfully deposited by various thermal spraying methods [39-41]. This type of powder manufacturing becomes an interesting option for suspension preparation.

Table 1. The basic classification of bottom-up and top-down methods (inspired by [25])

Phase of powder synthesis	Processing	Approach
Gas	Physical vapor deposition	Bottom-up
	Chemical vapor deposition	
	Aerosol processing	
Liquid	Sol–gel process	Bottom-up
	Wet chemical synthesis	
Solid	Mechanical alloying Mechanical milling[1] Mechanochemical synthesis Sonication	Top-down
Combined	Vapor–liquid–solid	Bottom-up

Figure 2. Powder particles that can be used for suspension formulation: (a) as-produced submicrometer zirconia powder and (b) mechanically milled submicrometer zirconia powder (attrition milling).

Fraction of Solid

The content of solid phase in suspension is also an important parameter. The amount of powder influences the rheological properties of suspension [42]. Suspension viscosity is an important property affecting the formation of droplets after injection into plasma jet [43]. The size of droplet and the distribution of powder inside it affect, in turn, the phenomena related of heat and momentum transfer from plasma jet [44]. The amount of the powder particles in the suspension droplet influences the microstructure of coatings, i.e., porosity and surface topography [30, 41]. Furthermore, the suspension concentration can be crucial when taking into the account the transport of liquid feedstock from feeder to plasma torch (in the industrial practice the

[1] The process is most popular to prepare fine powders for Suspension Plasma Spraying.

feeder is placed usually outside of the spray booth, so the total length of the installation can exceed several meters). Finally, high concentration of the suspension is of interest for industrial productivity.

Solvent

As discussed before, the fine powder particles are shielded by the solvent fluid around them and are a part of suspension droplet (see Figure 1b). Water, alcohol or their mixtures are used in suspensions formulation. As alcohol the ethanol is mainly used but other kinds of alcohols as methanol, methylethylketone …[44] were also applied.

Two most important properties of liquid feedstock are influenced by the type of solvent namely surface tension and its thermodynamic properties concerning evaporation. The effects of surface tension on the spray process were discussed by Rampon et al. [45]. The authors showed that aqueous suspensions have significantly higher surface tension than alcohol based suspensions. They demonstrated that surface tension preserves liquid against the change of surface area and controls the break-up of suspension droplets. Afterwards, the droplet size affects its velocity in the plasma jet as well as its residence time in the plasma. Finally, it was proved that solvent has the strongest influence on the surface tension this property cannot be modified by addition of any chemical agent. Similar observations with regard to influence of solvent on the SPS process were made by Toma et al. [46]. The authors observed more efficient fragmentation of suspension droplets based on ethanol than that based on water. This was explained by the lower surface tension of ethanol.

The phenomena related to the surface tension and evaporation were reported by Chen et al. [47] for Solution Precursor Plasma Spraying (SPPS). They observed that droplets having high surface tension and high boiling point solvent did not fully evaporate in the plasma jet. At the same time the solution droplets formed by the solvent having low surface tension and low boiling point could evaporate rapidly and could have had a complete heat treatment in plasma jet. The droplet vaporization was investigated by Kaβner et al. [48]. The authors did calculations of the vaporization time of droplets using a model which considered latent heat of evaporation. The aqueous and alcohol based suspensions of droplets having different size were considered. The conclusion was that the water based droplets need nearly three-times longer time to evaporate in the plasma jet. In fact, the vaporization enthalpy of water equal to 2.3×10^6 Jkg^{-1} is almost three times greater than that of ethanol equal to 0.8×10^6 Jkg^{-1} [49].

It should be also mentioned that the type of solvent used to formulate suspension directly affects the velocity that is required to penetrate the plasma jet by the suspension droplets as showed by Pawłowski [50]. Due to the higher density of water when comparing to ethanol the aqueous suspension needs lower velocity to penetrate plasma jet. This is particularly important for radial injection modes that are mostly used in SPS practice due to the design of plasma torches.

Another property of the suspension which is influenced by the solvent type is its viscosity. Water and alcohol have different viscosity. This parameter, however, can be controlled by addition of chemical agents, called plasticizers. Their role will be discussed later.

The application of so called "green suspensions" is shown in the literature. Such suspension includes water as a reagent and a solvent. An example of such suspensions is nano-titania powder and deionized water [51]. The elimination of organic solvents and chemical agents is useful environment but needs a special powder production method (continuous hydrothermal flow synthesis).

The security risk, namely inflammability should be considered if alcohol based suspensions are used. Finaly, the economic reasons render useful replacement of alcohol by water.

Chemical Agents

Homogeneous and very stable liquid slurry is indispensable in order to provide reproducible Suspension Plasma Spraying process. The stability is especially important when very fine powders are used for suspension formulation. The nanoparticles agglomerate easily, because they have much great specific area [25].

The rheology of suspension depends on powder physical properties (particle size, particle shape, density, etc.) and on the interactions between individual powder particles [52]. Wang et al. [53] discussed the interactions between the solid particles in suspension and their separation (so called flocculation), which were attributed to the effects of electrostatic repulsion and the action of Van der Waals forces. More detailed information about the forces influencing the behavior of colloidal suspension was given by Ring [54]. The different forces affecting the movement of particles in a solvent i.e., electrostratic, Van der Waals attraction force, viscous, and gravitational forces were analysed. Larson suggested that the particle size has a significant influence on suspension stability [55]. The different approaches should be applied when formulating suspensions, which base on powders having various particle sizes. For coarse powders the gravitational forces are dominant, whereas for fine (sub-micrometer sized and smaller) the forces resulting in Brownian motion dominate.

To maintain the stability of suspensions of fine grained powders the use of repulsion between particles is useful. This can be done by modyfying the zeta potential (ζ). It represents the potential difference between a layer of fluid attached to a particle and its slipping plane. The suspension is stable for absolute value of zeta potential greater than 30mV. The ζ – potential can be changed by the careful addition of flocculating agents (called also dispersant agents) to assure the long range electrostatic repulsion.

Another important factor affecting the sedimentation speed of finely grained suspensions is viscosity, η. The sedimentation speed is determined by the Stokes equation:

$$V = \frac{2\Delta\rho g a^2}{9\eta} \tag{1}$$

According to this formula the decrease of sedimentation speed can be achieved by: (i) decreasing of solid size, a, (ii) ensuring that the difference of densities of dispersed and continuous phases, $\Delta\rho$, is small; (iii) increasing the value of viscosity, η. In fact, the viscosity is an important variable. It can be changed by the addition of viscosity enhancers called thickeners or plasticizers. It is important to recall that the viscosity increases when the ζ – potential increases, so the control of both parameters is necessary. These phenomena were discussed in details by Waldbillig and Kesler [56], who examined the influence of different types of dispersant on suspension rheology. They found out that the viscosity can be minimized when the optimum quantity of dispersant is added to the suspension. Otherwise the suspension viscosity increases strongly, what influences negatively the spray efficiency and may result in clogging of injector during spraying. The viscosity of suspension can be controlled by the amount of plasticizer [45].

On the other hand, if the suspension is prepared using sub-micrometer particles or if the particle size distribution is not monomodal, then the gravitational forces overcome the forces resulting in Brownian motion. Consequently, another approach is needed. It consists of controlling the isoelectric point of suspension leading to minimize the electrostatic forces. Finally, the sedimentation rate decreases when high viscosity suspension is used.

Many practical examples focused on selection of flocculating agents can be find in the literature for zirconia suspensions [57-59]. The example of well and poorly stabilized zirconia suspension is shown in Figure 3.

The suspension preparation process can be divided into following steps:

- powder preparation and its characterization;
- suspension formulation and homogenization and rheological properties measurements followed by stabilization test;
- suspension re-dispersion prior to spray process.

Another possibility is the use of commercial suspensions, what becomes are more and more popular nowadays. Such suspensions are delivered with appropriate solvent and chemical agents. The user should only well re-disperse the mixture prior to spraying. However, the high cost is the main reason which limits the use of commercial suspensions. To take an example, the price of one kilogram of 20 wt.% loaded suspension corresponds currently to the cost of 1 kg of the powder used to formulate the suspension.

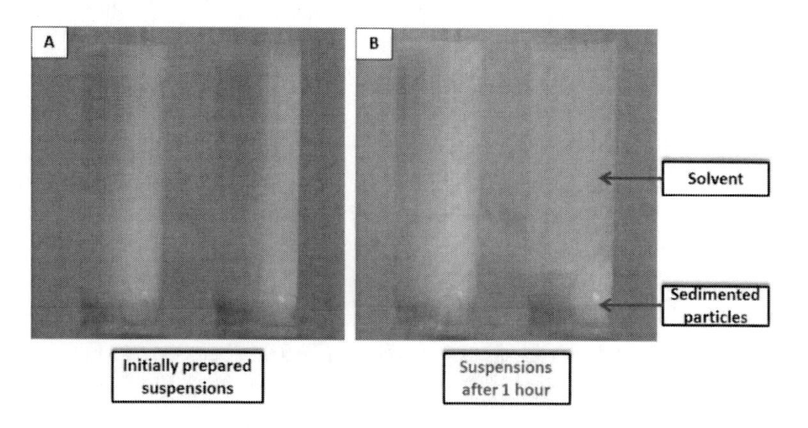

Figure 3. Two kinds of zirconia suspension: (a) well-stabilized suspension, (b) poorly-stabilized suspension.

Injection of Suspension

The liquid feedstock is transported from containers to the injector by action of compressed air (pressurized system) or by peristaltic pump. The transport system should provide a constant flowrate of liquid. Afterwards, the injection of slurry to the plasma occurs. The dynamic pressure of a liquid stream should be greater than that of the plasma jet to reach the injection [50]. Suspension injection is one of the key factors in Suspension Plasma Spraying. Consequently, the different types of injection will be discussed in details.

Continuous Stream Injection vs Atomization

Mechanical injection is probably the most frequently used in SPS process (Figure 4a). It can be realized radially or axially with regard to the torch axis. In this injection type the feedstock is delivered through a nozzle having fixed diameter and shape. Another characteristic appearance for mechanical injection is that the atomization of the slurry takes place inside the plasma jet. The physics of this atomization is out of the scope of this chapter and is presented elsewhere (see e.g., the work of Pawłowski [60]).

Another possibility is the injection associated with atomization of the liquid before its contact with plasma jet (Figure 4b). Such injection can be divided in two steps: (i) initially the slurry is introduced to a nozzle; (ii) action of an external energy results in a fragmented of slurry into droplets. In the practice, the simple atomizers are used in which the fragmentation of droplets results from action of gas flow. In this case gases having high specific mass are used. More information can be found in the refence [60]. The droplet formulation mechanism is complex and depends on the design of atomizer, external energy and on the properties of liquid slurry. Furthermore, the possible use of atomizers is limited rather to external and radial injection modes. Based on the literature different practical information of choosing proper injection system can be found. The comparison of the described injection methods is summarized in Table 2.

The discussion of the influence of suspension injection mode on the coatings microstructure can be also found in the literature. A direct comparison was done e.g., by Kozerski et al. [67-68]. The authors found out that the use of atomizer resulted in uniform, finely grained microstructure. The splats had a spherical shape and were not strongly deformed. The particles were molten and often solidified before they reached the substrate. When the mechanical injection mode was used the coating had more complex microstructure. Well molten lamellas as well as sintered particles with loosely bound grains were observed (so called *two-zone microstructure*). Authors observed also that the use of mechanical injection mode caused stronger and more irregular deformation of grains in the coating structure.

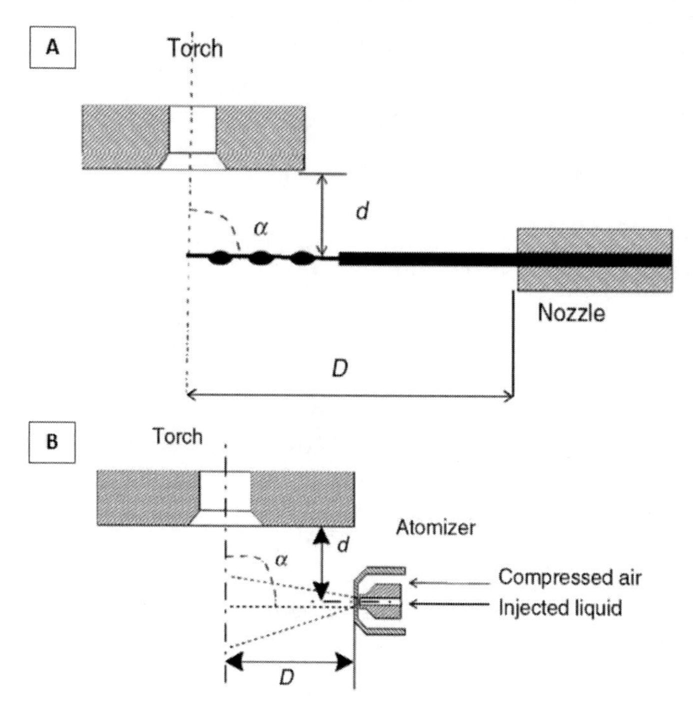

Figure 4. The sketch of mechanical injection (a) and atomization (b) [17].

Table 2. Characteristics of different injection modes

Continuous stream injection enables:	Atomization enables:
• avoiding the plasma jet perturbation caused by the atomization gases [61-2]; • injection with fixed velocity, no dispersion of droplet trajectories and sizes [61]; • through the use of a precise positioning system of the injector the droplets, injection at the chosen location in plasma plume [61]; • droplets more effectively injected into the core of the plasma jet [62]; • high spray efficiency [63]; • formulation of small, uniform droplets with constant spacing of droplets in plasma (Blazdell and Kuroda [64]).	• good control of the atomization; generation of high velocities of droplets at low feeding rates [65]; • production of small, well-defined and uniform droplets influencing formation of fine splats [62]; • reduction of unmelted particles in the coating structure [33]; • designing of atomizer in a way that limits the problems associated with injector's clogging [66].

Radial vs Axial Injection

The choice between radial and axial injection results from the construction of the plasma torch. The method of radial injection of a slurry means that the feedstock material is introduced perpendicularly to the plasma jet axis. This way of introducing powder or liquid material is still the most popular in the practice due to the popularity of plasma torches which do not allow axial injection. The radial injection can be realized by: (i) atomizer; or, (ii) mechanical injector. In both cases the liquid slurry has to reach the core of plasma jet. Otherwise the feedstock will not interact sufficiently with plasma. The problem is complex because the suspension droplets have various sizes. This difference influences their momentum and the possibility of penetration into plasma jet. The injection may be optimized and there are always some droplets with different trajectories: (a) which remain on the the periphery of the plasma jet, (b) which get into the core of plasma, and (c) part of particles, which pass through the plasma plume. The described trajectories are visualized in Figure 5.

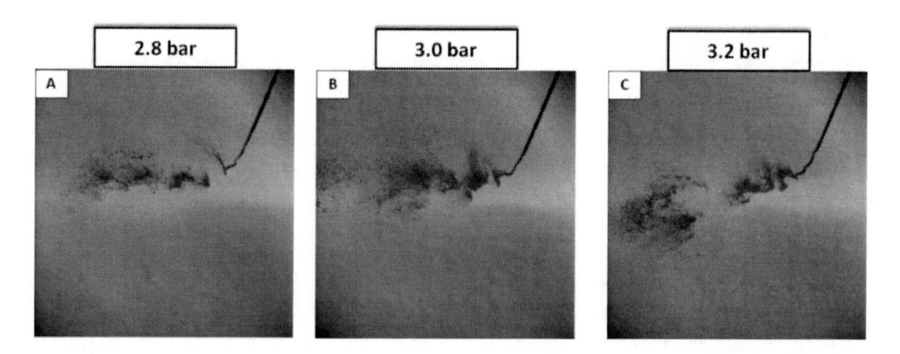

Figure 5. Trajectiories of zirconia suspension injected radially with different pressures vizualized using high-speed camera: (a) suspension remains on the periphery of the jet at pressure of 2.8 bars, (b) suspension penetrates into plasma core at pressure of 3.0 bars, (c) suspension traverses the plasma jet under pressure of 3.2 bars.

At present, the plasma torches enabling axial injection become available. The three-cathode torch Axial III of Mettech is an example. In the torch the feedstock material is injected between three separated plasma jets, which meet in a converging nozzle. This design allows introducing slurry flowing in the same direction as the plasma jet. The optimization of droplet trajectory is not necessary. Furthermore, the dispersion angle of droplets in the plasma jet is small and interaction with plasma jet is more homogeneous. When using the axial injection mode the position of injector is fixed and the parameters which can be varied are: injector diameter and suspension pressure. Moreover, the long injector having a small internal diameter resulting from the torch construction may create the problems with clogging when high viscosity suspensions are used.

The direct comparison of plasma torches having radial and axial injections was done by Ozturk et al. [69]. The authors made some mathematical modeling of SPPS processes to show that the droplets introduced axially were heated-up much faster and that the precipitation of zirconia occurs earlier. Moreover, the solid grains generated in plasma jet were more homogeneous for axial injection. The experimental work was done by the authors of the reference [62]. The authors studied the effect of injection type (radial and axial) on the coating microstructure in SPPS process. The use of axial injection mode leads to formation of fine droplets and to the creation of uniform coating microstructure.

External vs Internal Injection

The use of external or internal injection mode results from the design of plasma torch. Soma plasma torches such as SG-100 of Praxair or Axial III of Mettech give the possibility to use internal injector. Many other torches (Triplex or F4 of Oerlikon) can be used only with external injectors.

The use of internal injector allows introducing liquid directly into the hottest zones of plasma jet what enables good heat and momentum transfer to the droplets. This is especially important for refractory materials which are hard to be melted. This injection mode does not require tedious optimization of injection place. However, such mode gives no possibility to observe the beginning of liquid feedstock trajectory. Some authors, e.g., Etchart-Salas et al. [70], suggested that, when using external radial injection the nozzle clogging occurs seldom. The clogging at Axial torch spraying was discussed by the authors of reference [71]. Gotô and Atsuya [72] observed that internal injection of acid solution shorten torch lifetime. In fact, the copper electrode was partially dissolved by the aggressive liquid. This effect has to be considered when using acids for suspension stabilization. Finally, the use of internal injection of liquid feedstock limits, in general, the lifetime of the cathodes.

PLASMA TORCHES AND SPRAY PROCESS

In this section deals with various plasma torches whic are used in SPS practice. The plasma torches can be categorized in different ways. Two main parameters were chosen to a closer discussion, namely: (i) power level that the torch can generate, and, (ii) way of plasma stabilization.

The typical torch, used in plasma spraying, works with direct current, is gas-stabilized and has one cathode and one anode. The properties of generated plasma are influenced by such torch design and by working gases its flow rates and power input. The power input is usually under 50kW which is enough to reach temperatures of 15000 K. An important problem of such typical torch is arc instability, discussed elsewhere [73]. The instabilities may lead to reduction of electrodes lifetime. The fluctuations of arc are reduced in a Triplex torch (of Oerlikon Metco) in which three cathodes create arcs striking single anode. This reduces the amplitude of voltage between cathods and anode. Moreover, the torch is designed in the way that the arcs are long and the voltage between anode and cathodes is high. The power imput to Triplex is similar to that applied in conventional plasma torches. However, the liquid feedstock injection point has to be carefully optimized to take into account three arcs [33], [74]. Axial III of Mettech is another torch designed with three cathodes and three anodes. The power is delivered by three independent power sources (each one delivers up to 50 kW). The torch is one of the most powerful gas-stabilized torches designed for plasma spraying of coating. It allows achieving high deposition rates and high productivity. However, the spray system is complex and expensive. The working gases consumption is important. Moreover, the change of the cathodes is time consuming and more frequent to keep uniform and stable arc [71].

Another idea of high-power plasma torch design was achieved by to the use of water to stabilize arc and to generate the working gases. Water vortex protects internal walls of the torch against overheating and water vapors stabilize and form, after dissociation and ionization, plasma composed of oxygen and hydrogen. Plasma jet produced by water stabilized plasma torch (WSP) is characterized by very high enthalpy and high temperature. Thermodynamic and thermophysical properties of water vapor plasma allow increasing of heat transfer to the sprayed material. The plasma jet is, however, very turbulent [75-76]. Finally, the use of external, rotating anode reduces some operational capacities [77].

A new hybrid WSP (h-WSP torch) was designed with arc stabilization by water and by the gas. An important advantage of the hybrid version is the possibility to use an additional plasmaforming working gas like Ar [78-79]. A complex design and sophisticated operating principles of WSP and h-WSP torches limit their application in the industry.

The basic comparison of plasma torches that were tested by authors in Suspension Plasma Spraying process of zirconia coatings is presented in Table 3.

Table 3. Comparison of different plasma torches discussed in the chapter

Torch	Injector	Injection angle	Electric power (kW)	Arc stabilization
SG-100	Internal/External	Radial	Low (40kW)	Gas
Triplex	External	Radial	Low (45kW)	Gas
Axial III	Internal	Axial	High (120kW)	Gas
WSP/ h-WSP	External	Radial	High (150kW)	Water

Operational Spray Parameters

The use of suspension, instead of dry powder, renders the spray process more complex. The phenomena, which occur in plasma jet during SPS process, are as follows [17], [80]: (i) aerodynamic break-up of liquid stream and droplets formulation, (ii) solvent evaporation, (iii) sintering of fine solids,

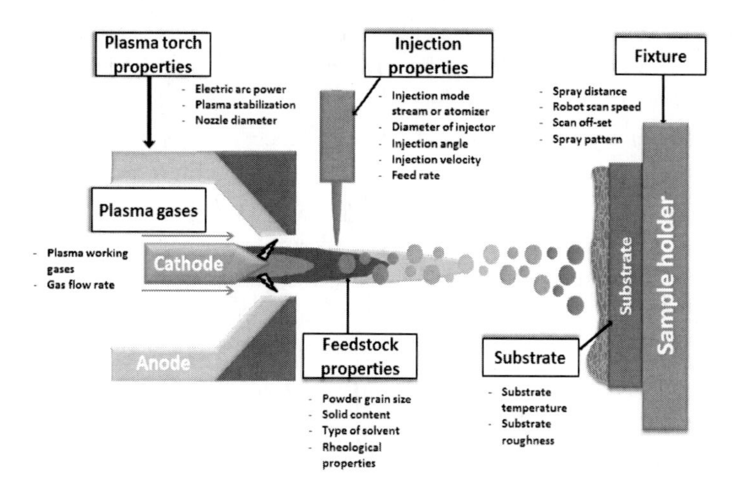

Figure 6. SPS operational deposition parameters.

(iv) melting of the sintered fines solids and powder agglomerates, (v) partial evaporation of liquid material and (vi) impact of molten particles on the substrate. The window of optimal parameters which allow obtaining desired microstructure is narrow. The process variables influencing the microstructure of SPS coatings are schematically shown on Figure 6.

STABILIZED ZIRCONIA AND ITS PROPERTIES

Stabilized zirconia coatings obtained by SPS sprayed zirconia may have two important possible applications: TBC and SOFC.

Pure ZrO_2 has three crystal phases: monoclinic, tetragonal and cubic. The monoclinic phase, M, is stable below 1170°C; the tetragonal phase, T, is stable between 1170 and 2370°C; and the cubic one, C, - above 2370°C) [81]. The phases transformations s is accompanied by the volume change, which generates stresses in the material and can cause material fracture (most critical is a rapid T → M transformation, where the volume change exceed even 4%). The doping of zirconia stabilizes zirconia in a desired phase.

Currently the most poupular is yttria-stabilized zirconia (YSZ). This material has many useful properties [82-86]:

- high melting point of about 2700°C;
- the coefficient of thermal conductivity below 2.3 $Wm^{-1}K^{-1}$ (bulk material);
- relatively high thermal expansion coefficient (~11 × 10-6°C^{-1});
- relatively high ionic conductivity (~ 0.1 Scm^{-1} at 900–1000°C);
- low electronic conductivity;
- low density of 6.4 gcm^{-3};
- hardness of around 14 GPa;
- good chemical and mechanical stability;

Other stabilizers of zirconia have been also tested. The other rare earth elements (gadolinium, lanthanum, ytterbium, cerium ...) oxides and also MgO, CaO, etc. were used. The studies of TBC's showed that use of other rare earth oxides stabilizers decreases thermal conductivity of zirconia. The use of lanthanum gadolinium oxides stabilizers improves phase stability at high temperatures and its resistance to be sintered [82], [87-88]. Another promising material is yttria with ceria-stabilized zirconia. This material has high corrosion resistance and phase stability [89-90]. The studies of SOFC showed utility of doping with scandium to increase ionic conductivity [91]. The ScSZ electrolytes allow reducing of the operating temperatures of the fuel cells. Finally, the stabilizing with copper oxide results in an improved catalytic activity of zirconia [92].

COATINGS' MICROSTRUCTURE

The described studies focused on zirconia coatings microstructure are presented based on the own experiences of authors.

Morphology

Suspension Plasma Spraying enables producing films and coatings having thickness ranging from a few micrometers to a few hundreds micrometers. The deposits can have:

- *columnar microstructure* (similar to the PVD films microstructure);
- finely grained microstructure;
- lamellar microstructure;
- microstructure being a mixture of the previous ones called *two-zone microstructure* [68].

The examples of zirconia coatings microstructures are shown in Figure 7.

An example shown in Figure 8 is helpful in characterizing lamellar microstructure of SPS sprayed coating. The figure shows homogenous, regular and well-adhering YCeSZ (yttria- and ceria-stabilized zirconia) coating sprayd using suspension formulated with fine powder obtained by milling of a coarse one. The interface does not present any delamination, cracks or discontinuities. The coating has porosity of almost 16%. The pores have submicrometer sizes and monomodal distribution of them. There is a good contact between molten lamella and no defects in coating can be observed. The electron backscattered diffraction (EBSD) image (Figure 8c) shows crystallographic orientation of individual grains. The orientation is very random. An observed particularity is a columnar form of grains near the interface with substrate. This effect results from the heat evacuation direction [44]. The sizes of the grains vary from 300 to 450 nm (Figure 8d, e). The crystal grain sizes are well within the range of initial powder particles (d_{v50}= 696nm).

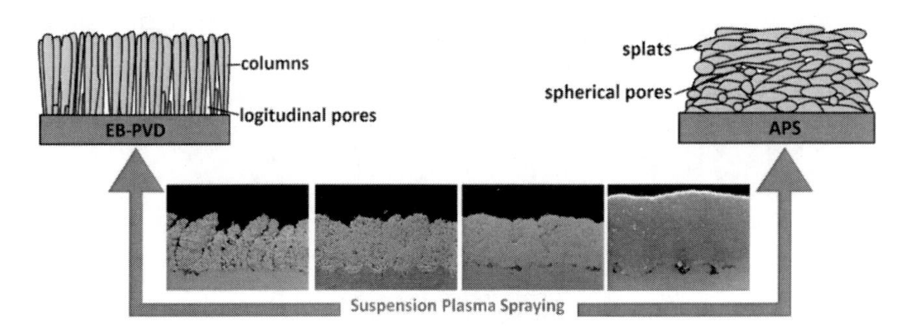

Figure 7. Examples of zirconia coatings produced by SPS technology.

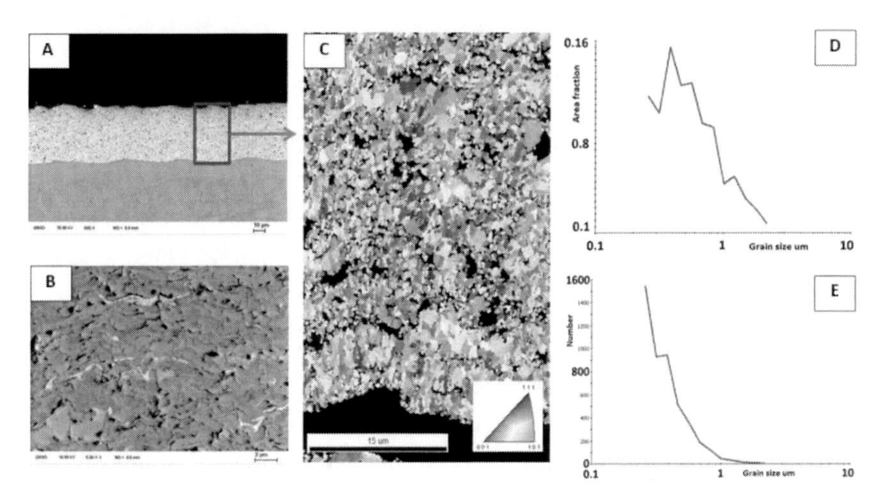

Figure 8. Microstructural images of homogeneous zirconia coatings made by: (a, b) scanning electron microscope with backscattered electrons detector; (c, d, e) electron back scattered diffraction set up [93].

Another example of zirconia microstructure is shown in Figure 9. The coating, having numerous columns, was deposited using suspension formulated using synthesized fine zirconia powder (8YSZ made by Tosoh). The heterogeneous microstructure may be resulted from low concentration of solid in suspension (only 2.5 wt.%). The coating-substrate interface has no crack or pores. The coating has high internal porosity of above 21%. The pores have nanometer or submicrometer size.

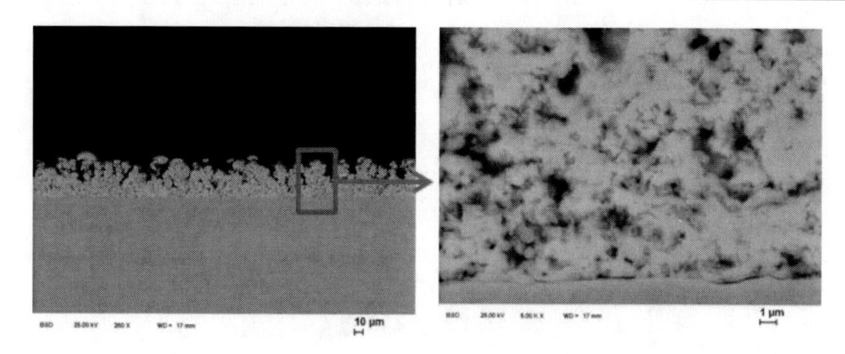

Figure 9. Scanning electron microscope (BSE) micrographs of very irregular zirconia coating [93].

Build-Up Mechanisms

The use of agglomerated and spray-dried powders used for conventional plasma spraying leads to formation of microstructure shown on the right side of Figure 10. The external layer of an agglomerated particle melts in plasma jet. This creates a coating with regularly spaced matrix including of solidified shells having inside fine particles. Suspension Plasma Spraying leads to a formation of *two-zone microstructure* (see left side of Figure 10). The microstructure includes irregularly distributed dense and well molten lamellas formed by droplets, which are transported in hot region of plasma, got agglomerated and molten and sintered and loosely bound fine grains which travelled in the plasma jet periphery. The well molten lamellas in SPS coatings are not regularly distributed and can be smaller than that deposited with agglomerated powder. This results from uncontrolled agglomeration of fine solids included in suspension droplets inside plasma jet. Moreover, the pore distribution is in SPS coatings is more uniform [94].

The growth-up mechanisms of columnar-like structure has been carefully analyzed in the recent years. Our research has brought a conclusion that the creation of columns in the SPS coatings is a superposition of various process parameters including roughness and topography of the substrate [30], [90]. The formation of columns is associated with the plasma flow which takes place on the substrate at spraying. This leads to a lateral movement of fine particles which are submitted by the following forces [30]:

- drag force, F_D, pushing fine solids parallel to the substrate surface;
- adhesion force, F_A making that the particles adhere to the substrate.

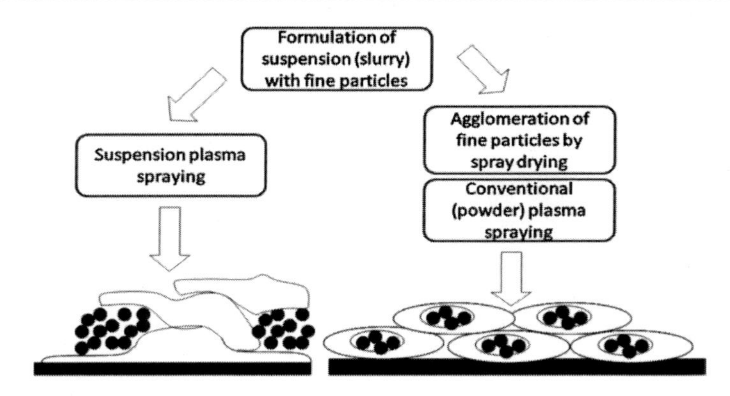

Figure 10. The comparison of formation dense and fine grained coatings: by suspension and powder plasma spraying [67].

If the particles are big enough the relation $F_A > F_D$ is valid. The small powder particles, travelling frequently on plasma jet periphery, are not adhering to the substrate directly. The balance of forces for such particles is $F_D > F_A$. These particles move laterally on the substrate's surface and may adhere to its irregularities (see Figure 11a - c).

The described effects are visible and were confirmed by EBSD studies in reference [93]. The substrate irregularities can be created by sangblasting or by laser treatment (Figure 11a). The columns can be formed on smooth substrate. This may happen if the solid phase concentration in suspension is low. The columns created in this way are less regularly distributed (Figure 11b).

The mechanism of columnar growth is supported also by the shadowing effect discussed elsewhere [95-97]. The growth of coating is an effect of the accumulation of molten lamellas on the substrate surface peaks. Then, space hidden by these lamellas may initiate a formation of long voids being perpendicular to the substrate. This leads also to the formation of columnar microstructure.

Curry et al. [42] investigated the effect of suspension preparation on columns formation. He found that the use of suspensions formulated with ethanol with low solid content promotes formation of columns. This results from and easy atomization of droplets which becomes small. The size of solids in suspension is also important and fine powder particles promote columnar microstructure formation. Bernard et al. [98] showed that the use high-enthalpy plasma which has a lot of turbulences promotes the formation of columnar microstructure. Finally, Seshadri and Sampath [99] studied the influence of particle temperature and particle velocities on generation of desired microstructure of coating.

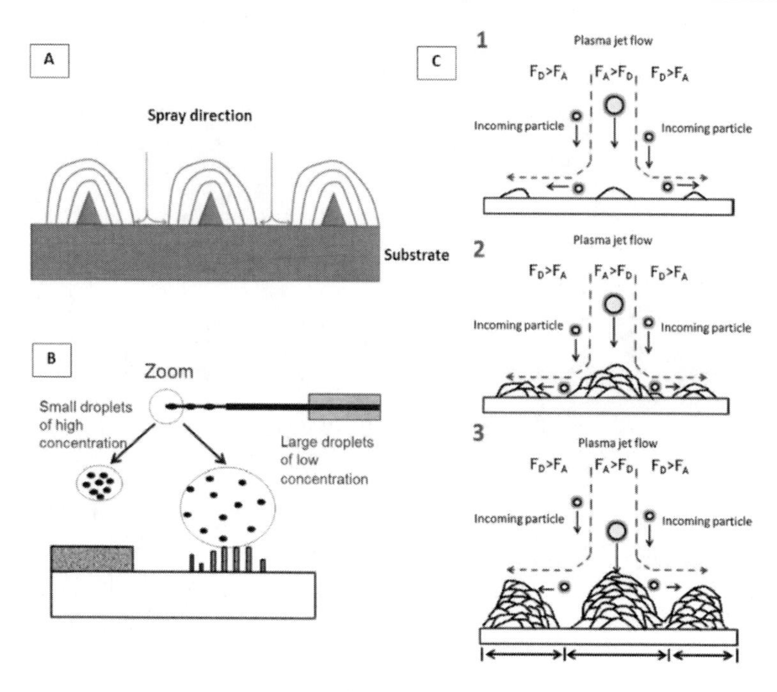

Figure 11. The formation of columnar microstructure in SPS coatings [30, 93]: (a) plasma fluctuations and shadowing effect caused by coating roughness; (b) effects related to different solid concentraton in suspension; (c) movement of particles influenced by the drag and adhesion forces.

Figure 12. The view of plasma spray set ups (installation used by our research group – a) and EB-PVD (installation at ARCI in Hyderabad in India, presented with permission – b).

The present knowledge is such that plasma spray process does not allow obtaining of regular columnar structure as in PVD films. The particles used in SPS are a few orders of magnitude greater than atoms or molecules forming the microstructure in vapor deposition. The mechanism of columns growth in PVD methods results from material evaporation and condensation onto substrate. The grains grow starting from the islands of condensation. The growth may be affected by substrate temperature, topography…[100]. The condensed films' morphology can be expressed in a simplified way as a function of melting point of coating material and temperature of substrate. If the difference between the substrate temperature and the melting point decreases then the mobility of condensed atoms is greater and there is more islands of condensation [101-102]. The EB-PVD installation is shown for a comparison to the plasma spray installation used in our research group in Figure 12.

SELECTED APPLICATIONS OF SPS ZIRCONIA COATINGS

TBC and SOFCs are complex products and information presented in this section is limited and shows only recent studies of suspension plasma sprayed zirconia coatings in comparison to other deposition technologies.

TBC – Thermal Barrier Coatings

Technology of Thermal Barrier Coatings, TBC are one of the key technologies related to gas turbines - both in aviation industry as well as in power generation plants. Two different reasons of their application can be pointed out. Firstly, the extension of turbine lifetime at given operating temperature is achieved when using thermal barrier. Secondly, an increase of operating temperature leads to better turbine efficiency [103]. Currently, the operating temperatures of turbines are above the melting point of applied substrate materials (mainly nickel superalloys). The values exceed even 1500°C-1600°C. It should be also mentioned that the change a turbine necessite to cool it down. While the turbine restarts the great jump of the temperature occurs from room temperature to 1500°C-1600°C and the thermal stresses in the turbine, dangerous for the product integrity, are generated.

Finally, the role of TBC is to protect a substrate thermally and to have high thermal shock resistance. It is difficult to develop a coating having these both characteristics. On one side, EB-PVD coatings have columnar microstructure. Due to the low in-plane modulus of such coatings the strain accommodation during cyclic compression and tension is provided and high thermal shock resistance is achieved [104]. Furthermore, EB-PVD method allows obtaining coatings having smooth surface what helps in preserving of aerodynamic design of turbine blades. However, the production cost of such coatings is relatively high, so the application of EB-PVD may be limited to the most fragile and relatively small parts of the turbine. The other parts are produced using powder plasma spraying method [105-106]. The advantage of ceramic coatings produced by APS is a brick-wall microstructure with horizontal lamellas. The defects in the microstructure as gaps, pores and horizontal voids between the splats result in reduction of heat transport and in better thermal isolation. The thermal shock resistance of plasma sprayed coatings is lower that of the coatings produced by EB-PVD.

The useful technological solution would be to add the advantages of these two mentioned methods i.e., high thermal shock resistance, low thermal conductivity, high deposition rate and inexpensive manufacturing in one process Suspension Plasma Spraying is supposed to be such a process (see Figure 14). In fact, the recent studies enabled to found out that:

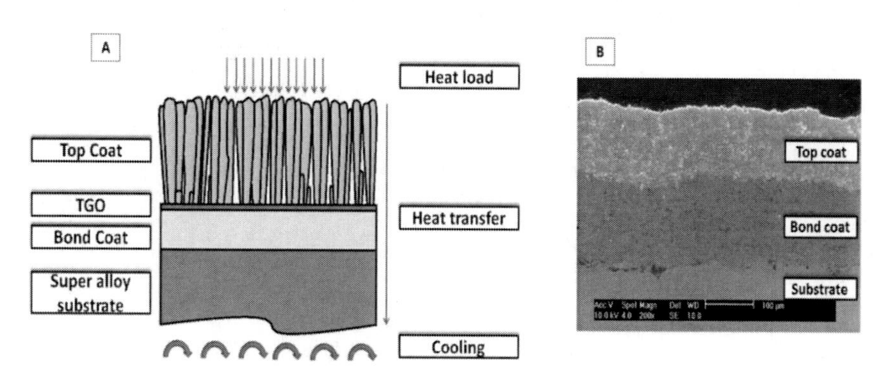

Figure 13. The example of Thermal Barrier Coating System.

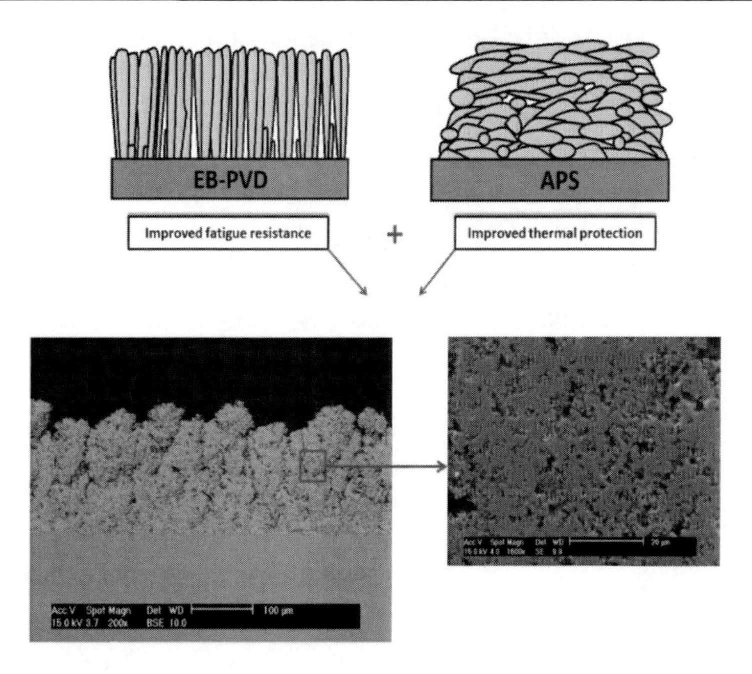

Figure 14. An alternative for EB-PVD and APS achieved by the use of SPS zirconia coatings.

- thermal conductivity of SPS sprayed zirconia coatings is lower than the coatings obtained by to EB-PVD and APS. This was explained by high porosity of coatings, very small size of pores and by finely grained microstructure [107-108];
- columnar microstructure of coatings obtained by Suspension Plasma Spraying can be achieved [30], [96],
- appropriate selection of process parameters allows controlling the columns growth mechanism in SPS method and their morphologie (shape, width, porosity, etc.) [42], [98],
- thermal shock resistance of SPS coatings columnar microstructure is improved [42].

SOFC – Solid Oxide Fuel Cells

SPS technology is also useful for manufacturing of Solid Oxides Fuel Cells, SOFC. The technology is particularly useful for production of dense electrolyte to be discussed here.

SOFC can be described as highly efficient energy conversion device that can electrochemically convert fuel into the electricity. The cells use the fuels, like biomass, alcohols; natural gases, etc. which can be easily convert into the electricity. Other important advantages are: high efficiency, lack of a typical combustion process, low emissions of harmful substances like CO_X; NO_X; SO_2, modular structure, etc. [109].

The important element of a cell is a gas tight electrolyte. The electrolyte should conduct electrical charges carried by O^{-2} ions. Electrolyte is placed between anode and cathode. Their porous microstructure allows a good access of the reacting gases to ensure the flow of reaction products [110]. The scheme of a SOFC is shown in Figure 15a.

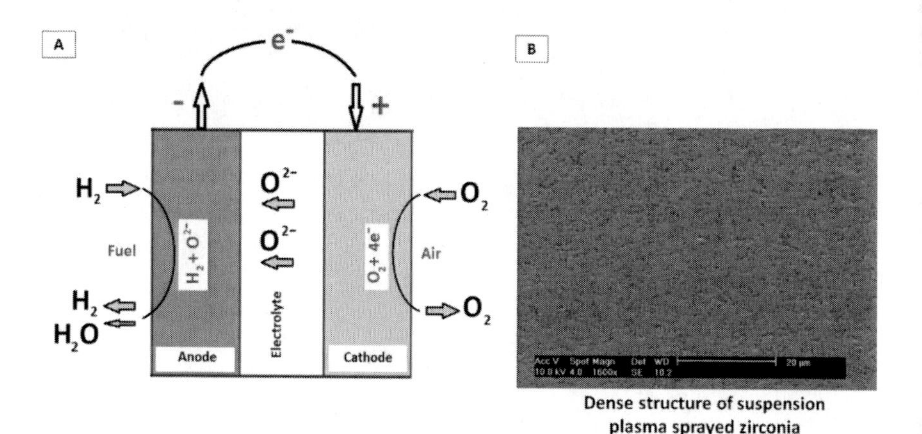

Figure 15. The scheme of Solid Oxide Fuel Cell © 2011 Taroco HA, Santos JAF, Domingues RZ, Matencio T. Published in [109] under CC BY-NC-SA 3.0 license. Available from: http://dx.doi.org/10.5772/18297 (a) and dense zirconia coating sprayed by the use of SPS that can be used as electrolyte (b).

The electrolyte is a thin, dense and strong film which must be stable in reducing and oxidizing environment. It has to have high ionic and low electronic conductivity at the temperature of operation [111]. Stabilized zirconia, in particular YSZ, fulfills all the specifications. Due to the fluorite structure yttria-stabilized zirconia is characterized by high oxide ion conductivity which is equal to 0.14 Scm^{-1} at 1000°C [112]. As the lowering of temperature is benefitial for SOFC, the other zirconia materials are tested. For instance, scandia doped zirconia, ScSZ has ion conductivity at 780°C similar to YSZ at 1000°C [86]. Ceria stabilized zirconia can also be used as the electrolyte to reduce operating temperature of a cell as it can operate at 600-

800°C. Finally, the perovskites and oxides having hexagonal structure have been considered as possible electrolyte materials [111].

There are many technologies of electrolytes manufacturing. The most popular ones, for zirconia electrolytes, are described below.

- Wet ceramic processing techniques – processes like tape casting, screen printing, spin coating, etc. The deposited layer is sintered. These fabrication techniques are multi-steps processes which render them time consuming and expensive. The high temperature sintering limits the application of some cells' electrodes [113];
- Spray pyrolysis is a method that does not need any expensive set-up. The process is simple. It allows obtaining dense and gas-tight thin films. Electrolytes can be produced at relatively low temperature and on rough substrates. The technique enables production of multilayered electrolytes [114];
- Vapor deposition methods (PVD, CVD) – able to produce high-quality electrolytes. Layers can be deposited onto porous substrate also. The typical drawbacks of vapor deposition methods are low deposition rate, high temperature of process and cost of deposition set-up [115-117];
- Atmospheric plasma spraying (APS) of dry powders allows producing electrolytes effectively. The process is not expensive. The cheap metallic electrodes can be also used. The drawback of this process is the difficulty in production of thin (< 20 μm) electrolytes [118].

The development and the use of the cells is mostly limited by durability and high manufacturing cost [119]. Suspension plasma spraying (SPS) is one of the methods which may be useful in producing the electrolytes for SOFC. The use of fine powders leads to the formation of finely grained microstructure with controlled pore distribution. There is no need to carry out any post-spray heat treatment of coating to improve their properties [43]. The coatings have the mechanical properties which meet the specification. Waldbillig and Kesler [120] investigated gas tightness, electrolyte resistance and electrodes activity of plasma sprayed SOFC. The authors found that all these properties are influenced be the coating microstructure. The measured values of open circuit voltage were slightly smaller than theoretical values for the same conditions. The measured, by Soysal et al. [121], values of peak power density were up to 0.1 Wcm^{-2} (at the temperature of 750°C). The authors of reference [122] found out that the cells electrolyte obtained using SPS had the power density in the range of 0.8 Wcm^{-2} at 800°C. Finally, the preparation of electrolytes for SOFC's by SPS should be cotinued in order to prepare high-quality fuel cells.

Acknowledgments

The authors gratefully acknowledge the help of dr Radek Musalek and colleagues from IPP Prague for their help with spraying using h-WSP torch and for the shadography observations. Prof. Per Nylen and his colleagues from University West in Sweden gave us a possibility to use the Axial III torch from their laboratory. Finally, Drs Stefan Kozerski and Leszek Łatka helped in spraying of coatings.

References

[1] Drexler, KE. "Molecular engineering: An approach to the development of general capabilities for molecular manipulation," *Proc. Natl. Acad. Sci. U. S. A.*, vol. 78, no. 9, pp. 5275–5278, Sep. 1981.

[2] Drexler, KE. *Engines of Creation*. Anchor Press/Doubleday, 1986.

[3] Bhushan, B. *Springer Handbook of Nanotechnology*. Springer Science and Business Media, 2010.

[4] Kelsall, R; Hamley, IW; Geoghegan, M. *Nanoscale Science and Technology*. John Wiley and Sons, 2005.

[5] Aliofkhazraei, M. *Nanocoatings: Size Effect in Nanostructured Films*. Springer Science and Business Media, 2011.

[6] Gerberich, Ww; Michler, J; Mook, Wm; Ghisleni, R; Östlund, F; Stauffer, Dd; Ballarini, R. "Scale effects for strength, ductility, and toughness in 'brittle' materials," *J. Mater. Res.*, vol. 24, no. 03, pp. 898–906, Mar. 2009.

[7] Ahmed, W; Jackson, MJ. *Emerging Nanotechnologies for Manufacturing*. William Andrew, 2014.

[8] Lukaszkowicz, K. "Review of Nanocomposite Thin Films and Coatings Deposited by PVD and CVD Technology," in Nanomaterials, M. Rahman, *Ed. InTech*, 2011.

[9] Pawłowski, L. *The Science and Engineering of Thermal Spray Coatings*, 2nd Edition. Wiley, 2008.

[10] Fauchais, P; Montavon, G; Bertrand, G. "From Powders to Thermally Sprayed Coatings," *J. Therm. Spray Technol.*, vol. 19, no. 1–2, pp. 56–80, Dec. 2009.

[11] Lima, RS; Marple, BR. "Thermal Spray Coatings Engineered from Nanostructured Ceramic Agglomerated Powders for Structural, Thermal Barrier and Biomedical Applications: A Review," *J. Therm. Spray Technol.*, vol. 16, no. 1, pp. 40–63, Mar. 2007.

[12] Marcinauskas, L. "Deposition of alumina coatings from nanopowders by plasma spraying," *Medziagotyra*, vol. 16, no. 1, pp. 47–51, 2010.

[13] Chen, H; Ding, CX. "Nanostructured zirconia coating prepared by atmospheric plasma spraying," *Surf. Coat. Technol.*, vol. 150, no. 1, pp. 31–36, Feb. 2002.

[14] Shaw, LL; Goberman, D; Ren, R; Gell, M; Jiang, S; Wang, Y; Xiao, TD; Strutt, PR. "The dependency of microstructure and properties of nanostructured coatings on plasma spray conditions," *Surf. Coat. Technol.*, vol. 130, no. 1, pp. 1–8, Aug. 2000.

[15] Gitzhofer, F; Bouyer, E; Boulos, MI. "Suspension plasma spray," US Patent 5609921 A, 1997.

[16] Karthikeyan, J; Berndt, CC; Tikkanen, J; Wang, JY; King, AH; Herman, H. "Preparation of nanophase materials by thermal spray processing of liquid precursors," *Nanostructured Mater.*, vol. 9, no. 1–8, pp. 137–140, 1997.

[17] Pawłowski, L. "Application of solution precursor spray techniques to obtain ceramic films and coatings", [in:] *Future Development of Thermal Spray Coatings*, N. Espallargas (ed.), Elsevier, p. 123-141, 2015.

[18] Pawłowski, L. "Suspension and solution thermal spray coatings," *Surf. Coat. Technol.*, vol. 203, no. 19, pp. 2807–2829, Jun. 2009.

[19] Jordan, EH; Xie, L; Gell, M; Padture, NP; Cetegen, B; Ozturk, A; Ma, X; Roth, J; Xiao, TD; Bryant, PEC. "Superior thermal barrier coatings using solution precursor plasma spray," *J. Therm. Spray Technol.*, vol. 13, no. 1, pp. 57–65, Mar. 2004.

[20] Lima, RS; Karthikeyan, J; Kay, CM; Lindemann, J; Berndt, CC. "Microstructural characteristics of cold-sprayed nanostructured WC–Co coatings," *Thin Solid Films*, vol. 416, no. 1–2, pp. 129–135, Sep. 2002.

[21] Berghaus, JO; Legoux, JG; Moreau, C; Hui, R; Decès-Petit, C; Qu, W; Yick, S; Wang, Z; Maric, R; Ghosh, D. "Suspension HVOF Spraying of Reduced Temperature Solid Oxide Fuel Cell Electrolytes," *J. Therm. Spray Technol.*, vol. 17, no. 5–6, pp. 700–707, Nov. 2008.

[22] Saremi, M; Valefi, Z. "The effects of spray parameters on the microstructure and thermal stability of thermal barrier coatings formed by solution precursor flame spray (spfs)," *Surf. Coat. Technol.*, vol. 220, pp. 44–51, Apr. 2013.
[23] Berndt, CC. "Materials Production for Thermal Spray Processes," in *Handbook of Thermal Spray Technology*, J. R. Davis, Ed. ASM International, 2004.
[24] Sopicka-Lizer, M. *High-Energy Ball Milling: Mechanochemical Processing of Nanopowders*. Elsevier, 2010.
[25] Tjong, SC; Chen, H. "Nanocrystalline materials and coatings," *Mater. Sci. Eng. R Rep.*, vol. 45, no. 1–2, pp. 1–88, Sep. 2004.
[26] Banerjee, S; Tyagi, DAK. *Functional Materials: Preparation, Processing and Applications*. Elsevier, 2012.
[27] Carpio, P; Rayón, E; Pawłowski, L; Cattini, A; Benavente, R; Bannier, E; Salvador, MD; Sánchez, E. "Microstructure and indentation mechanical properties of YSZ nanostructured coatings obtained by suspension plasma spraying," *Surf. Coat. Technol.*, vol. 220, pp. 237–243, Apr. 2013.
[28] TOSOH CORPORATION, "Zirconia," Tokyo, Japan.
[29] Marr, M; Kuhn, J; Metcalfe, C; Harris, J; Kesler, O. "Electrochemical performance of solid oxide fuel cells having electrolytes made by suspension and solution precursor plasma spraying," *J. Power Sources*, vol. 245, pp. 398–405, Jan. 2014.
[30] Sokołowski, P; Kozerski, S; Pawłowski, L; Ambroziak, A. "The key process parameters influencing formation of columnar microstructure in suspension plasma sprayed zirconia coatings," *Surf. Coat. Technol.*, vol. 260, pp. 97–106, Dec. 2014.
[31] Movahedi, B. "A Solid State Approach to Synthesis Advanced Nanomaterials for Thermal Spray Applications," in *Advanced Plasma Spray Applications*, H. Salimi Jazi, Ed. InTech, 2012.
[32] Koch, CC. "Synthesis of nanostructured materials by mechanical milling: problems and opportunities," *Nanostructured Mater.*, vol. 9, no. 1–8, pp. 13–22, 1997.
[33] Kaβner, H; Siegert, R; Hathiramani, D; Vassen, R; Stöever, D. "Application of Suspension Plasma Spraying (SPS) for Manufacture of Ceramic Coatings," *J. Therm. Spray Technol.*, vol. 17, no. 1, pp. 115–123, Dec. 2007.

[34] Chen, D; Jordan, EH; Gell, M. "Suspension plasma sprayed composite coating using amorphous powder feedstock," *Appl. Surf. Sci.*, vol. 255, no. 11, pp. 5935–5938, Mar. 2009.

[35] Łatka, L; Cattini, A; Chicot, D; Pawłowski, L; Kozerski, S; Petit, F; Denoirjean, A. "Mechanical Properties of Yttria- and Ceria-Stabilized Zirconia Coatings Obtained by Suspension Plasma Spraying," *J. Therm. Spray Technol.*, vol. 22, no. 2–3, pp. 125–130, Dec. 2012.

[36] Jaworski, R; Pierlot, C; Pawłowski, L; Bigan, M; Quivrin, M. "Synthesis and Preliminary Tests of Suspension Plasma Spraying of Fine Hydroxyapatite Powder," *J. Therm. Spray Technol.*, vol. 17, no. 5–6, pp. 679–684, Oct. 2008.

[37] Berghaus, JO; Marple, B; Moreau, C. "Suspension plasma spraying of nanostructured WC-12Co coatings," *J. Therm. Spray Technol.*, vol. 15, no. 4, pp. 676–681, Dec. 2006.

[38] Suryanarayana, C. "Mechanical alloying and milling," *Prog. Mater. Sci.*, vol. 46, no. 1–2, pp. 1–184, Jan. 2001.

[39] Fernandes, F; Ramalho, A; Loureiro, A; Guilemany, JM; Torrell, M; Cavaleiro, A. "Influence of nanostructured ZrO2 additions on the wear resistance of Ni-based alloy coatings deposited by APS process," *Wear*, vol. 303, no. 1–2, pp. 591–601, Jun. 2013.

[40] Movahedi, MHEB. "Thermal spray coatings of Ni–10 wt-%Al composite powder synthesised by low energy mechanical milling," *Surf. Eng.*, vol. 25, no. 4, pp. 276–283, 2009.

[41] Dong, KAKZL. "Microstructure formation in plasma-sprayed functionally graded NiCoCrAlY/yttria-stabilized zirconia coatings," *Surf. Coat. Technol.*, vol. 114, no. 2–3, pp. 181–186, 1999.

[42] Curry, N; VanEvery, K; Snyder, T; Susnjar, J; Bjorklund, S. "Performance Testing of Suspension Plasma Sprayed Thermal Barrier Coatings Produced with Varied Suspension Parameters," *Coatings*, vol. 5, no. 3, pp. 338–356, Jul. 2015.

[43] Marchand, O; Bertrand, P; Mougin, J; Comminges, C; Planche, MP; Bertrand, G. "Characterization of suspension plasma-sprayed solid oxide fuel cell electrodes," *Surf. Coat. Technol.*, vol. 205, no. 4, pp. 993–998, Nov. 2010.

[44] Fauchais, P; Etchart-Salas, R; Rat, V; Coudert, JF; Caron, N; Wittmann-Ténèze, K. "Parameters Controlling Liquid Plasma Spraying: Solutions, Sols, or Suspensions," *J. Therm. Spray Technol.*, vol. 17, no. 1, pp. 31–59, Feb. 2008.

[45] Rampon, R; Marchand, O; Filiatre, C; Bertrand, G. "Influence of suspension characteristics on coatings microstructure obtained by suspension plasma spraying," *Surf. Coat. Technol.*, vol. 202, no. 18, pp. 4337–4342, Jun. 2008.

[46] Toma, FL; Bertrand, G; Begin, S; Meunier, C; Barres, O; Klein, D; Coddet, C. "Microstructure and environmental functionalities of TiO2-supported photocatalysts obtained by suspension plasma spraying," *Appl. Catal. B Environ.*, vol. 68, no. 1–2, pp. 74–84, Oct. 2006.

[47] Chen, D; Jordan, EH; Gell, M. "The Solution Precursor Plasma Spray Coatings: Influence of Solvent Type," *Plasma Chem. Plasma Process.*, vol. 30, no. 1, pp. 111–119, Nov. 2009.

[48] Kaβner, H; Vaβen, R; Stöver, D. "Study on instant droplet and particle stages during suspension plasma spraying (SPS)," *Surf. Coat. Technol.*, vol. 202, no. 18, pp. 4355–4361, Jun. 2008.

[49] Toma, FL; Bertrand, G; Klein, D; Coddet, C; Meunier, C. "Nanostructured photocatalytic titania coatings formed by suspension plasma spraying," *J. Therm. Spray Technol.*, vol. 15, no. 4, pp. 587–592, Dec. 2006.

[50] Pawłowski, L. "Finely grained nanometric and submicrometric coatings by thermal spraying: A review," *Surf. Coat. Technol.*, vol. 202, no. 18, pp. 4318–4328, Jun. 2008.

[51] Robinson, BW; Tighe, CJ; Gruar, RI; Mills, A; Parkin, IP; Tabecki, AK; de Villiers Lovelock, HL; Darr, JA. "Suspension plasma sprayed coatings using dilute hydrothermally produced titania feedstocks for photocatalytic applications," *J Mater Chem A*, vol. 3, no. 24, pp. 12680–12689, 2015.

[52] Mueller, S; Llewellin, EW; Mader, HM. "The rheology of suspensions of solid particles," *Proc. R. Soc. Lond. Math. Phys. Eng. Sci.*, vol. 466, no. 2116, pp. 1201–1228, Apr. 2010.

[53] Wang, Z; Ni, Z; Qiu, D; Tao, G; Yang, P. "Characterization of stability of ceramic suspension for slurry introduction in inductively coupled plasma optical emission spectrometry and application to aluminium nitride analysis," *J. Anal. At. Spectrom.*, vol. 20, no. 4, pp. 315–319, Mar. 2005.

[54] Ring, TA. *Fundamentals of Ceramic Powder Processing and Synthesis*. Academic Press, 1996.

[55] Larson, RG. *The Structure and Rheology of Complex Fluids*. OUP USA, 1999.

[56] Waldbillig, D; Kesler, O. "The effect of solids and dispersant loadings on the suspension viscosities and deposition rates of suspension plasma sprayed YSZ coatings," *Surf. Coat. Technol.*, vol. 203, no. 15, pp. 2098–2101, May 2009.

[57] Greenwood, R; Kendall, K. "Selection of Suitable Dispersants for Aqueous Suspensions of Zirconia and Titania Powders using Acoustophoresis," *J. Eur. Ceram. Soc.*, vol. 19, no. 4, pp. 479–488, Apr. 1999.

[58] Wang, YH; Liu, XQ; Meng, GY. "Dispersion and stability of 8 mol.% yttria stabilized zirconia suspensions for dip-coating filtration membranes," *Ceram. Int.*, vol. 33, no. 6, pp. 1025–1031, Aug. 2007.

[59] Khan, AU; Haq, AU; Mahmood, N; Ali, Z. "Rheological studies of aqueous stabilised nano-zirconia particle suspensions," *Mater. Res.*, vol. 15, no. 1, pp. 21–26, Feb. 2012.

[60] Pawłowski, L. "Suspension and solution thermal spray coatings," *Surf. Coat. Technol.*, vol. 203, no. 19, pp. 2807–2829, Jun. 2009.

[61] Fazilleau, J; Delbos, C; Rat, V; Coudert, JF; Fauchais, P; Pateyron, B. "Phenomena Involved in Suspension Plasma Spraying Part 1: Suspension Injection and Behavior," *Plasma Chem. Plasma Process.*, vol. 26, no. 4, pp. 371–391, Apr. 2006.

[62] Killinger, A; Gadow, R; Mauer, G; Guignard, A; Vaßen, R; Stöver, D. "Review of New Developments in Suspension and Solution Precursor Thermal Spray Processes," *J. Therm. Spray Technol.*, vol. 20, no. 4, pp. 677–695, Mar. 2011.

[63] Toma, FL; Berger, LM; Jacquet, D; Wicky, D; Villaluenga, I; de Miguel, YR; Lindeløv, JS. "Comparative study on the photocatalytic behaviour of titanium oxide thermal sprayed coatings from powders and suspensions," *Surf. Coat. Technol.*, vol. 203, no. 15, pp. 2150–2156, May 2009.

[64] Blazdell, P; Kuroda, S. "Plasma spraying of submicron ceramic suspensions using a continuous ink jet printer," *Surf. Coat. Technol.*, vol. 123, no. 2–3, pp. 239–246, Jan. 2000.

[65] Vaßen, R; Kaßner, H; Mauer, G; Stöver, D. "Suspension Plasma Spraying: Process Characteristics and Applications," *J. Therm. Spray Technol.*, vol. 19, no. 1–2, pp. 219–225, Nov. 2009.

[66] Fauchais, P; Vardelle, M; Goutier, S; Vardelle, A. "Key Challenges and Opportunities in Suspension and Solution Plasma Spraying," *Plasma Chem. Plasma Process.*, vol. 35, no. 3, pp. 511–525, Nov. 2014.

[67] Kozerski, S; Łatka, L; Pawłowski, L; Cernuschi, F; Petit, F; Pierlot, C; Podlesak, H; Laval, JP. "Preliminary study on suspension plasma sprayed ZrO2 + 8 wt.% Y2O3 coatings," *J. Eur. Ceram. Soc.*, vol. 31, no. 12, pp. 2089–2098, Oct. 2011.

[68] Kozerski, S; Pawłowski, L; Jaworski, R; Roudet, F; Petit, F. "Two zones microstructure of suspension plasma sprayed hydroxyapatite coatings," *Surf. Coat. Technol.*, vol. 204, no. 9–10, pp. 1380–1387, Jan. 2010.

[69] Ozturk, A; Cetegen, BM. "Modeling of axially and transversely injected precursor droplets into a plasma environment," *Int. J. Heat Mass Transf.*, vol. 48, no. 21–22, pp. 4367–4383, Oct. 2005.

[70] Etchart-Salas, R; Rat, V; Coudert, JF; Fauchais, P; Caron, N; Wittman, K; Alexandre, S. "Influence of Plasma Instabilities in Ceramic Suspension Plasma Spraying," *J. Therm. Spray Technol.*, vol. 16, no. 5–6, pp. 857–865, Oct. 2007.

[71] Mohanty, P; Stanisic, J; Stanisic, J; George, A; Wang, Y. "A Study on Arc Instability Phenomena of an Axial Injection Cathode Plasma Torch," *J. Therm. Spray Technol.*, vol. 19, no. 1–2, pp. 465–475, Dec. 2009.

[72] Gotô, H; Atsuya, I. "Stabilization of the plasma-jet flame and determination of aluminium and boron in steel," *Fresenius Z. Für Anal. Chem.*, vol. 240, no. 2, pp. 102–110, Mar. 1968.

[73] Duan, Z; Heberlein, J. "Arc instabilities in a plasma spray torch," *J. Therm. Spray Technol.*, vol. 11, no. 1, pp. 44–51, Mar. 2002.

[74] Bobzin, K; Bagcivan, N; Zhao, L; Petkovic, I; Schein, J; Hartz-Behrend, K; Kirner, S; Marqués, JL; Forster, G. "Modelling and diagnostics of multiple cathodes plasma torch system for plasma spraying," *Front. Mech. Eng.*, vol. 6, no. 3, pp. 324–331, May 2011.

[75] Hrabovsky, M. "Thermal Plasma Generators with Water Stabilized Arc," *Open Plasma Phys. J.*, vol. 2, no. 1, pp. 99–104, 2009.

[76] Hrabovsky, M. "Water-stabilized plasma generators," *Pure Appl. Chem.*, vol. 70, no. 6, pp. 1157–1162, 1998.

[77] Raghu, S; Goutevenier, G; Gansert, R. "Comparative study of the structure of gas-stabilized and water-stabilized plasma jets," *J. Therm. Spray Technol.*, vol. 4, no. 2, pp. 175–178, Jun. 1995.

[78] Kavka, T; Gregor, J; Chumak, O; Hrabovsky, M. "Effect of arc power and gas flow rate on properties of plasma jet under reduced pressures," *Czechoslov. J. Phys.*, vol. 54, no. 3, pp. C753–C758, Mar. 2004.

[79] Jeništa, J; Bartlová, M; Aubrecht, V. "Performance of water and hybrid stabilized electric arcs: the impact of dependence of radiation losses and

plasma density on pressure," *Czechoslov. J. Phys.*, vol. 56, no. 2, pp. B1224–B1230, Oct. 2006.

[80] Delbos, C; Fazilleau, J; Rat, V; Coudert, JF; Fauchais, P; Pateyron, B. "Phenomena Involved in Suspension Plasma Spraying Part 2: Zirconia Particle Treatment and Coating Formation," *Plasma Chem. Plasma Process.*, vol. 26, no. 4, pp. 393–414, Apr. 2006.

[81] Stevens, R. *An introduction to zirconia: written for Magnesium Elektron*. Twickenham, Middx: Magnesium Elektron Ltd., 1986.

[82] Cao, XQ; Vaβen, R; Stöver, D. "Ceramic materials for thermal barrier coatings," *J. Eur. Ceram. Soc.*, vol. 24, no. 1, pp. 1–10, Jan. 2004.

[83] W. J. (Georgia I. of T. Lackey, D. P. Stinton, G. A. Cerny, A. C. (Oak R. N. L. Schaffhauser, and L. L. (Technology A. and T. Fehrenbacher, "Ceramic Coatings for Advanced Heat Engines - a Review and Projection," *Adv. Ceram. Mater.* USA, vol. 2, 1, Jan. 1987.

[84] Vaßen, R; Stöver, D. "Conventional and New Materials for Thermal Barrier Coatings," in *Functional Gradient Materials and Surface Layers Prepared by Fine Particles Technology*, M.-I. Baraton and I. Uvarova, Eds. Springer Netherlands, 2001, pp. 199–216.

[85] Cales, B; Stefani, Y. "Mechanical properties and surface analysis of retrieved zirconia hip joint heads after an implantation time of two to three years," *J. Mater. Sci. Mater. Med.*, vol. 5, no. 6–7, pp. 376–380, Jun. 1994.

[86] Yamamoto, O. "Solid oxide fuel cells: fundamental aspects and prospects," *Electrochimica Acta*, vol. 45, no. 15–16, pp. 2423–2435, May 2000.

[87] Wu, J; Wei, X; Padture, NP; Klemens, PG; Gell, M; García, E; Miranzo, P; Osendi, MI. "Low-Thermal-Conductivity Rare-Earth Zirconates for Potential Thermal-Barrier-Coating Applications," *J. Am. Ceram. Soc.*, vol. 85, no. 12, pp. 3031–3035, Dec. 2002.

[88] Krämer, S; Yang, J; Levi, CG. "Infiltration-Inhibiting Reaction of Gadolinium Zirconate Thermal Barrier Coatings with CMAS Melts," *J. Am. Ceram. Soc.*, vol. 91, no. 2, pp. 576–583, Feb. 2008.

[89] Di Girolamo, G; Blasi, C; Brentari, A; Schioppa, M. "Microstructure and thermal properties of plasma-sprayed ceramic thermal barrier coatings," *ENEA Mag.*, no. 1–2, 2013.

[90] Park, SY; Kim, JH; Kim, MC; Song, HS; Park, CG. "Microscopic observation of degradation behavior in yttria and ceria stabilized zirconia thermal barrier coatings under hot corrosion," *Surf. Coat. Technol.*, vol. 190, no. 2–3, pp. 357–365, Jan. 2005.

[91] Ivers-Tiffée, E; Weber, A; Herbstritt, D. "Materials and technologies for SOFC-components," *J. Eur. Ceram. Soc.*, vol. 21, no. 10–11, pp. 1805–1811, 2001.

[92] Dongare, MK; Dongare, AM; Tare, VB; Kemnitz, E. "Synthesis and characterization of copper-stabilized zirconia as an anode material for SOFC," *Solid State Ion.*, vol. 152–153, pp. 455–462, Dec. 2002.

[93] Sokołowski, P; Pawłowski, L; Dietrich, D; Lampke, T; Jech, D. "Advanced Microscopic Study of Suspension Plasma-Sprayed Zirconia Coatings with Different Microstructures," *J. Therm. Spray Technol.*, vol. 25, no. 1–2, pp. 94–104, Sep. 2015.

[94] Fauchais, P; Rat, V; Delbos, C; Coudert, JF; Chartier, T; Bianchi, L. "Understanding of suspension DC plasma spraying of finely structured coatings for SOFC," *IEEE Trans. Plasma Sci.*, vol. 33, no. 2, pp. 920–930, Apr. 2005.

[95] Pateyron, B; Pawłowski, L; Calve, N; Delluc, G; Denoirjean, A. "Modeling of phenomena occurring in plasma jet during suspension spraying of hydroxyapatite coatings," *Surf. Coat. Technol.*, vol. 214, pp. 86–90, Jan. 2013.

[96] VanEvery, K; Krane, MJM; Trice, RW; Wang, H; Porter, W; Besser, M; Sordelet, D; Ilavsky, J; Almer, J. "Column Formation in Suspension Plasma-Sprayed Coatings and Resultant Thermal Properties," *J. Therm. Spray Technol.*, vol. 20, no. 4, pp. 817–828, Mar. 2011.

[97] Mauer, G; Hospach, A; Zotov, N; Vaßen, R. "Process Conditions and Microstructures of Ceramic Coatings by Gas Phase Deposition Based on Plasma Spraying," *J. Therm. Spray Technol.*, vol. 22, no. 2–3, pp. 83–89, Nov. 2012.

[98] Bernard, B; Bianchi, L; Malié, A; Joulia, A; Rémy, B. "Columnar suspension plasma sprayed coating microstructural control for thermal barrier coating application," *J. Eur. Ceram. Soc.*, vol. 36, no. 4, pp. 1081–1089, Mar. 2016.

[99] Seshadri, RC; Sampath, S. "Characterization of the Deposition Formation Dynamics of Suspension Plasma Spray Coatings using In-Situ Coating Property Measurements," *Proc. Int. Therm. Spray Conf.* Shanghai, PR China May 10 - 12 2016, pp. 85–90.

[100] Rigney, DV; Viguie, R; Wortman, DJ; Skelly, DW. "PVD thermal barrier coating applications and process development for aircraft engines," *J. Therm. Spray Technol.*, vol. 6, no. 2, pp. 167–175, Jun. 1997.

[101] Dobrzański, LA; Dobrzańska-Danikiewicz, AD; "*Engineering surfaces treatment*," Open Access Libr., vol. 5, pp. 1–480, 2011 (in Polish).
[102] Zhao, H; Yu, F; Bennett, TD; Wadley, HNG. "Morphology and thermal conductivity of yttria-stabilized zirconia coatings," *Acta Mater.*, vol. 54, no. 19, pp. 5195–5207, Nov. 2006.
[103] Dimiduk, DM; Perepezko, JH. "Mo-Si-B Alloys: Developing a Revolutionary Turbine-Engine Material," *MRS Bull.*, vol. 28, no. 09, pp. 639–645, Sep. 2003.
[104] Darolia, R. "Thermal barrier coatings technology: critical review, progress update, remaining challenges and prospects," *Int. Mater. Rev.*, vol. 58, no. 6, pp. 315–348, Aug. 2013.
[105] Miller, RA. "Current status of thermal barrier coatings — An overview," *Surf. Coat. Technol.*, vol. 30, no. 1, pp. 1–11, Jan. 1987.
[106] Wortman, DJ; Nagaraj, BA; Duderstadt, EC. "Proceedings of the 2nd International Symposium on High Temperature Corrosion of Advanced Materials and CoatingsThermal barrier coatings for gas turbine use," *Mater. Sci. Eng.* A, vol. 120, pp. 433–440, Dec. 1989.
[107] Sokołowski, P; Łatka, L; Pawłowski, L; Ambroziak, A; Kozerski, S; Nait-Ali, B. "Characterization of microstructure and thermal properties of YCSZ coatings obtained by suspension plasma spraying," *Surf. Coat. Technol.*, vol. 268, pp. 147–152, Apr. 2015.
[108] Łatka, L; Cattini, A; Pawłowski, L; Valette, S; Pateyron, B; Lecompte, JP; Kumar, R; Denoirjean, A. "Thermal diffusivity and conductivity of yttria stabilized zirconia coatings obtained by suspension plasma spraying," *Surf. Coat. Technol.*, vol. 208, pp. 87–91, Sep. 2012.
[109] Taroco, HA; Santos, JAF; Domingues, RZ; Matencio, T. "Ceramic Materials for Solid Oxide Fuel Cells," in *Advances in Ceramics - Synthesis and Characterization, Processing and Specific Applications*, C. Sikalidis, Ed. InTech, 2011.
[110] Henne, R. "Solid Oxide Fuel Cells: A Challenge for Plasma Deposition Processes," *J. Therm. Spray Technol.*, vol. 16, no. 3, pp. 381–403, Aug. 2007.
[111] Ishihara, T; Sammes, NM; Yamamoto, O. "Chapter 4 - Electrolytes," in *High Temperature and Solid Oxide Fuel Cells*, Amsterdam: Elsevier Science, 2003, pp. 83–117.
[112] Mogensen, M; Lybye, D; Bonanos, N; Hendriksen, PV; Poulsen, FW. "Factors controlling the oxide ion conductivity of fluorite and perovskite structured oxides," *Solid State Ion.*, vol. 174, no. 1–4, pp. 279–286, Oct. 2004.

[113] Waldbillig, D; Kesler, O. "Effect of suspension plasma spraying process parameters on YSZ coating microstructure and permeability," *Surf. Coat. Technol.*, vol. 205, no. 23–24, pp. 5483–5492, Sep. 2011.

[114] Setoguchi, T; Sawano, M; Eguchi, K; Arai, H. "Application of the stabilized zirconia thin film prepared by spray pyrolysis method to SOFC," *Solid State Ion.*, vol. 40–41, Part 1, pp. 502–505, Aug. 1990.

[115] Seydel, J; Becker, M; Ivers-Tiffée, E; Hahn, H. "Granular nanocrystalline zirconia electrolyte layers deposited on porous SOFC cathode substrates," *Mater. Sci. Eng.* B, vol. 164, no. 1, pp. 60–64, Aug. 2009.

[116] Mingde, L; Bo, Y; Mingfen, W; Jing, C; Jingming, X; Yuchun, Z. "The Fabrication Technique of YSZ Electrolyte Film," *Prog. Chem.*, vol. 20, no. 0708, pp. 1222–1232, 2008.

[117] Nédélec, R; Uhlenbruck, S; Sebold, D; Haanappel, VAC; Buchkremer, HP; Stöver, D. "Dense yttria-stabilised zirconia electrolyte layers for SOFC by reactive magnetron sputtering," *J. Power Sources*, vol. 205, pp. 157–163, May 2012.

[118] Scagliotti, M; Parmigiani, F; Chiodelli, G; Magistris, A; Samoggia, G; Lanzi, G. "Plasma-sprayed zirconia electrolytes," *Solid State Ion.*, vol. 28–30, Part 2, pp. 1766–1769, Sep. 1988.

[119] Waldbillig, D; Kesler, O. "Electrochemical testing of suspension plasma sprayed solid oxide fuel cell electrolytes," *J. Power Sources*, vol. 196, no. 13, pp. 5423–5431, Jul. 2011.

[120] Waldbillig, D; Kesler, O. "Characterization of metal-supported axial injection plasma sprayed solid oxide fuel cells with aqueous suspension plasma sprayed electrolyte layers," *J. Power Sources*, vol. 191, no. 2, pp. 320–329, Jun. 2009.

[121] Soysal, D; Ansar, A; Ilhan, Z; Costa, R. "Nanostructured Composite Cathodes by Suspension Plasma Spraying for SOFC Applications," *ECS Trans.*, vol. 35, no. 1, pp. 2233–2241, Apr. 2011.

[122] Tsai, T; Barnett, SA. "Increased solid-oxide fuel cell power density using interfacial ceria layers," *Solid State Ion.*, vol. 98, no. 3–4, pp. 191–196, Jun. 1997.

BIOGRAPHICAL SKETCHES

Lech Pawłowski

Affiliation and Address: Professor at the University of Limoges, SPCTS, 12, rue Atlantis, 87068 Limoges, France, phone (+33) (0) 587 50 24 12, fax (+33) 587 50 23 04 E-mail: lech.pawlowski@unilim.fr

Education:

- MSc, Eng, Wrocław University of Technology, Poland, 1974, with honors;
- PhD, Wrocław University of Technology, Poland, 1978;
- DSc, The University of Limoges, France, 1985, with honors.

Research and Professional Experience and Professional Appointments:

- since 2010: University Professor at the University of Limoges (87) - Research laboratory SPCTS, UMR CNRS 7315, France;
- 1999-2010: University Professor at National High School of Engineers of Lille (59), France;
- 1995-1999: University Professor at the University of Artois (62, Bethune), France;
- 1993: Manager of H.T.I. (87, Ambazac), France;
- 1991-1992: Invited Professor at the University of Trento and scientific consultant in the company Centro Sviluppo Materiali (Trento, Italy);
- 1988-1990: Senior Research Fellow at Monash University (Melbourne, Australia);

- 1985-1988: Project manager in the company W. Haldenwanger (Berlin, Germany);
- 1984-1985: Research Fellow at the University of Stuttgart (Germany);
- 1981-1984: Research Fellow at the University of Limoges (87), France;
- 1978-1985: Research Fellow at the Wrocław University of Technology (Poland);
- 1974-1978: PhD student.

Honors:

- Hall of Fame of Thermal Spray Society ASM International in 2015;
- Doctorat honoris causa (Dr-Ing. E.h.) of the Faculty of Mechanics of Technical University of Chemnitz in Germany in 2013;
- Invited professor in China University of Petroleum in Quingdao (Shandong Province) 2011 for 5 years.

Publications Last 3 Years:

Book and chapter

1. L. Pawłowski, *Application of solution precursor spray techniques to obtain ceramic films and coatings*, [in:] “New trends and the future of thermal spray coatings and their applications", ed. Nuria Espallargas, Woodhead Publishing, Sawston, UK, 2015, ISBN 978-0-85709-769-9, 123-41.
2. L. Pawłowski and P. Blanchart, *Industrial Chemistry of Oxides for Emerging Applications*, CRC Press, 978-1-46658-712-0, Lausanne, Switzerland, 2017 (to be published).

Papers

1. B. Pateyron, L. Pawłowski, N. Calve, G. Delluc, A. Denoirjean, Modeling of phenomena occurring in plasma jet during suspension spraying of hydroxyapatite coatings, *Surface and Coatings Technology* 214 (2013) pp. 86-90.
2. L. Łatka, A. Cattini, D. Chicot, L. Pawłowski, S. Kozerski, F. Petit, A. Denoirjean, Mechanical Properties of Yttria- and Ceria-Stabilized Zirconia Coatings Obtained by Suspension Plasma Spraying, *Journal of Thermal Spray Technology* 22 (2-3) (2013) 125-130.

3. H. Podlesak, L. Łatka, D. Dietrich, L. Pawłowski, B. Wielage, T. Lampke, Electron microscopy and diffraction studies of suspension-plasma-sprayed ZrO_2 + 8 wt.% Y_2O_3 coatings, *Surface and Coatings Technology* 220 (2013) 67-73.
4. A. Cattini, L. Łatka, D. Bellucci, G. Bolelli, A. Sola, L. Lusvarghi, L. Pawłowski, V. Cannillo, Suspension plasma sprayed bioactive glass coatings: Effects of processing on microstructure, mechanical properties and in-vitro behavior, *Surface and Coatings Technology* 220 (2013) 52-59.
5. Pateyron, N. Calve, L. Pawłowski, Influence of water and ethanol on transport properties of the jets used in suspension plasma spraying, *Surface and Coatings Technology* 220 (2013) 257-260.
6. P. Carpio, E. Rayón, L. Pawłowski, A. Cattini, R. Benavente, E. Bannier, M.D. Salvador, E. Sánchez, Microstructure and indentation mechanical properties of YSZ nanostructured coatings obtained by suspension plasma spraying, *Surface and Coatings Technology* 220 (2013) 237-43.
7. L. Łatka, D. Chicot, A. Cattini, L. Pawłowski, A. Ambroziak, Modeling of elastic modulus and hardness determination by indentation of porous yttria stabilized zirconia coatings, *Surface and Coatings Technology* 220 (2013) 131-9.
8. L. Pawłowski, Strategic oxides for thermal spraying: problems of availability and evolution of prices, *Surface and Coatings Technology* 220 (2013) 14-19.
9. Z. Znamirowski, K. Nitsch, L. Pawłowski, The electric charge transport in titania–alumina composite cold cathodes made using atmospheric plasma spraying and laser engraving, *Surface and Coatings Technology* 220 (2013) 271-5.
10. P. Carpio, Q. Blochet, B. Pateyron, L. Pawłowski, M.D. Salvador, A. Borrell, E. Sánchez, Correlation of thermal conductivity of suspension plasma sprayed yttria stabilized zirconia coatings with some microstructural effects, *Materials Letters*, 107 (2013) 370-3.
11. Cattini, D. Bellucci, A. Sola, L. Pawłowski, V. Cannillo, Suspension plasma spraying of optimized functionally graded coatings of bioactive glass/hydroxyapatite, *Surface and Coatings Technology* 236 (2013) 118-26.
12. Cattini, D. Bellucci, A. Sola, L. Pawłowski, V. Cannillo, Microstructural design of functionally graded coatings composed of suspension plasma sprayed hydroxyapatite and bioactive glass,

Journal of Biomedical Materials Research - Part B Applied Biomaterials, 102 (3) (2014) 551-60.
13. Cattini, D. Bellucci, A. Sola, L Pawłowski, V. Cannillo, Functional bioactive glass topcoats on hydroxyapatite coatings: Analysis of microstructure and in-vitro bioactivity, *Surface and Coatings Technology*, 240 (2014) 110-117.
14. P. Sokołowski, S. Kozerski, L. Pawłowski, A. Ambroziak, The key process parameters influencing formation of columnar microstructure in suspension plasma sprayed zirconia coatings, *Surface and Coatings Technology* 260 (2014) 97-106.
15. Z. Znamirowski, W. Posadowski, L. Pawłowski, A. Cattini, L. Łatka, Electron emission from the zirconium coated suspension plasma sprayed bioglass, *Surface and Coatings Technology* 268 (2015) 63-9.
16. M. Winnicki, A. Małachowska, M. Rutkowska-Gorczyca, P. Sokołowski, A. Ambroziak, L. Pawłowski, Characterization of cermet coatings deposited by low-pressure cold spraying, *Surface and Coatings Technology* 268 (2015) 108-114.
17. R.T. Candidato, P. Sokołowski, L. Pawłowski, A. Denoirjean, Preliminary study of hydroxyapatite coatings synthesis using solution precursor plasma spraying, *Surface and Coatings Technology* 277 (2015) 242-50.
18. M. Winnicki, A. Małachowska, G. Dudzik, M. Rutkowska-Gorczyca, M. Marciniak, K. Abramski, A. Ambroziak, L. Pawłowski, Numerical and experimental analysis of copper particles velocity in low-pressure cold spraying process, *Surface and Coatings Technology* 268 (2015) 230-40.
19. P. Sokołowski, L. Łatka, L. Pawłowski, A. Ambroziak, S. Kozerski, B. Nait-Ali, Characterization of microstructure and thermal properties of YCSZ coatings obtained by suspension plasma spraying, *Surface and Coatings Technology* 268 (2015) 147-52.
20. P. Sokołowski, L. Pawłowski, D. Dietrich, T. Lampke, D. Jech, Advanced microscopic study of suspension plasma-sprayed zirconia coatings with different microstructures, *Journal of Thermal Spray Technology* 25 (1-2) (2016) 94-104.
21. A. Meijas, R.T. Candidato, Jr., L. Pawłowski, D. Chicot, Mechanical properties by instrumented indentation of solution precursor plasma sprayed hydroxyapatite coatings: Analysis of microstructural effect, *Surface and Coatings Technology* 298 (2016) 93-102.

22. T. Lindner, S. Bonebeau, R. Drehmann, T. Grund, L. Pawłowski, T. Lampke, Analytical methods to characterize heterogeneous raw material for thermal spray process: Cored wire Inconel 625, *IOP Conference Series: Materials Science and Engineering* 118 (1) (2016) Article 012009.
23. C. Demian, A. Denoirjean, L. Pawłowski, P. Denoirjean, R. El Ouardi, Microstructural investigations of NiCrAlY + Y_2O_3 stabilized ZrO_2 cermet coatings deposited by plasma transferred arc (PTA), *Surface and Coatings Technology* 300 (2016) 104-9.

Paweł Sokołowski

Affiliation and address: PhD student at Mechanical Engineering, Wrocław University of Technology, Faculty of Mechanical Engineering and Machine Building and at University of Limoges, SPCTS12, rue Atlantis, 87068 Limoges, France

Education Research and Professional Experience:

- PhD student at Wrocław University of Technology and University of Limoges;
- 2011-2012 MSc of Mechanical Engineering, Wrocław University of Technology, Faculty of Mechanical Engineering and Machine Building, Specialization in Machine Design and Operation;
- 2007-2011 BSc of Mechanical Engineering, Wrocław University of Technology, Faculty of Transport.

Publications Last 3 Years:

1. P. Sokołowski, S. Kozerski, L. Pawłowski, A. Ambroziak, The key process parameters influencing formation of columnar microstructure in suspension plasma sprayed zirconia coatings, *Surface and Coatings Technology* 260 (2014) 97-106
2. M. Winnicki, A. Małachowska, M. Rutkowska-Gorczyca, P. Sokołowski, A. Ambroziak, L. Pawłowski, Characterization of cermet coatings deposited by low-pressure cold spraying, *Surface and Coatings Technology* 268 (2015) 108-114.
3. R.T. Candidato, P. Sokołowski, L. Pawłowski, A. Denoirjean, Preliminary study of hydroxyapatite coatings synthesis using solution precursor plasma spraying, *Surface and Coatings Technology* 277 (2015) 242-50
4. P. Sokołowski, L. Łatka, L. Pawłowski, A. Ambroziak, S. Kozerski, B. Nait-Ali, Characterization of microstructure and thermal properties of YCSZ coatings obtained by suspension plasma spraying, *Surface and Coatings Technology* 268 (2015) 147-52.
5. P. Sokołowski, L. Pawłowski, D. Dietrich, T. Lampke, D. Jech, Advanced microscopic study of suspension plasma-sprayed zirconia coatings with different microstructures, *Journal of Thermal Spray Technology* 25 (1-2) (2016) 94-104.

In: Advances in Materials Science Research
Editor: Maryann C. Wythers
ISBN: 978-1-53610-059-4

Chapter 6

DERIVATIVES OF SILICENE: ELECTRONIC AND MECHANICAL PROPERTIES

***M. Yagmurcukardes*[1,*], *C. Bacaksiz*[1], *F. Iyikanat*[1], *E. Torun*[2], *R. T. Senger*[1,†], *F. M. Peeters*[2] *and H. Sahin*[2,3,‡]**
[1]Department of Physics, Izmir Institute of Technology, Izmir, Turkey
[2]Department of Physics, University of Antwerp, Antwerp, Belgium
[3]Department of Photonics, Izmir Institute of Technology, Izmir, Turkey

Abstract

Successful isolation of graphene from graphite opened a new era for material science and condensed matter physics. Due to this remarkable achievement, there has been an immense interest to synthesize new two dimensional materials and to investigate their novel physical properties. Silicene, form of Si atoms arranged in a buckled honeycomb geometry, has been successfully synthesized and emerged as a promising material for nanoscale device applications. However, the major obstacle for using silicene in electronic applications is the lack of a band gap similar to the case of graphene. Therefore, tuning the electronic properties of

[*]E-mail address: mehmetyagmurcukardes@iyte.edu.tr
[†]E-mail address: tugrulsenger@iyte.edu.tr
[‡]E-mail address: hasansahin.edu@gmail.com

silicene by using chemical functionalization methods such as hydrogenation, halogenation or oxidation has been a focus of interest in silicene research. In this paper, we review the recent studies on the structural, electronic, optical and mechanical properties of silicene-derivative structures. Since these derivatives have various band gap energies, they are promising candidates for the next generation of electronic and optoelectronic device applications.

PACS numbers: 62.25.-g, 73.20.At, 68.47.Gh, 78.67-n

1. Introduction

Layered bulk materials consisting of two dimensional (2D) sheets which are hold together with weak, interlayer van der Waals interaction have been the focus of interest for more than a century [1, 2, 3]. With the advancement of synthesis and characterization techniques it has been possible to isolate ultra thin films down to a monolayer of these materials which became feasible in the last decade. Monolayer forms of these layered bulk materials often exhibit different physical properties than their bulk counterparts. The first isolated 2D material is known to be graphene, a one-atom-thick carbon sheet, with extraordinary physical properties [4, 5, 6]. After the successful exfoliation of graphene by Novoselov and Geim, researchers have been searching for several other 2D materials that can exist in single layer form such as hexagonal monolayer crystals III-V binary compounds [7, 8, 9, 10, 11, 12, 13, 14, 15], transition metal dichalcogenides (TMDs) [16, 17, 18, 19] and the group IV elements (silicene, germanene, stanene) [20, 21, 22, 23, 24, 25]. Among these 2D monolayer materials, graphene and silicene are known to posses semi-metallic character while the members of TMDs family compounds generally display semiconducting behavior with a band gap of 1-2 eV. In all of these 2D materials silicene occupies an important position for the next generation of nanoscale technology which up to now is mostly based on silicon.

According to its electronic-band structure, graphene has a semi-metallic character which is not suitable for optoelectronic applications. One possible way to open a gap in the band structure of graphene is to functionalize its surface with various types of atoms such as H, F and Cl which were widely studied and successfully synthesized. It was shown that both full and partial hydrogenation of graphene leads to semiconducting materials with different band gap values [26, 27, 28]. Similar to the hydrogenation case, experimental and theoretical studies showed that the band gap of fluorinated-graphene can alter from 0 to 3 eV depending on the fluorination level. [29, 30, 31, 32, 33]

Silicene, a 2D honeycomb structure of Si atoms with a buckled geometry, has been attracting great interest due to its physical properties such as possessing massless Dirac fermions and large spin-orbit coupling resulting in an intrinsic band gap [36, 37]. The buckled structure of silicene is a consequence of sp^2-sp^3 hybridization of Si atoms. This makes the structure of silicene different from the flat structure of graphene. Another important physical property of silicene is its high surface reactivity which widens the methods of manipulating its electronic, magnetic and mechanical properties [38]. Thus, the functionalization of its surface and applying external mechanical strain are some of the widely used ways of controlling the electronic properties of silicene for its practical usage in device technology.

After the theoretical prediction and successful synthesis of silicene, researchers have focused on doping [41, 42, 43, 44, 45, 46], chemical modification [38, 47, 48, 49, 50, 51, 52, 53] and strain engineering [54, 55, 56, 57, 58, 59, 60, 61, 62, 63, 64, 65] in order to modify its electronic structure. Studies have demonstrated that fully hydrogenated silicene is a semiconductor [66, 67, 68] while half hydrogenated silicene is still a semi-metal or direct-gap semiconductor depending on the hydrogenation configuration [69]. Functionalization of silicene with halogen atoms (F, Cl, Br and I) was also considered in several studies for tuning its electronic structure [70, 71, 72, 73]. Studies on fully halogenated silicene indicated that it possesses a direct-gap semiconducting character with various band gap depending on the type of the halogen atom. Other functionalization methods like doping organic molecules on hydrogenated silicene have also been considered. [47, 48, 49, 50] Moreover, the oxidation of silicene was studied both theoretically and experimentally by several research groups which is important for the use of 2D materials in nanoscale device technology. [74, 75, 76, 77, 78, 79]

In this review we summarize the studies on the structural, electronic, optical and mechanical properties of silicene derivatives. This review is organized as follows: We first provide the physical properties of hydrogenated silicene in Section 2, the oxidized silicene in Section 3 and the halogenated silicene in Sec 4. The physical properties of silicene functionalized with organic molecules are given in Section 5 while the properties of silicene decorated with adatoms are given in Section 6. Finally we present a brief summary in Section 7

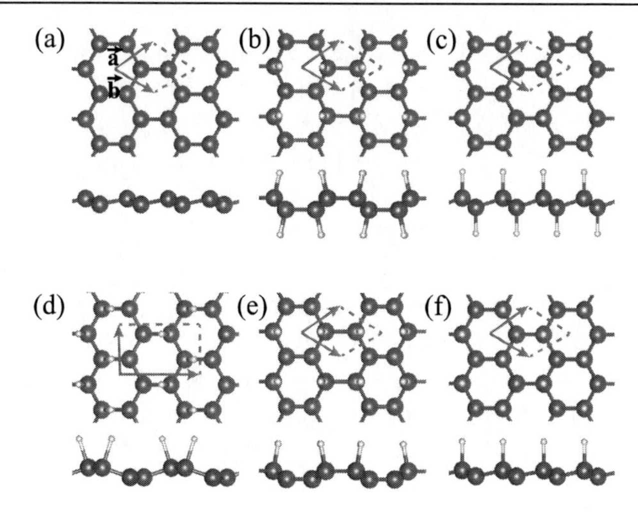

Figure 1. (Color online) Top and side views of optimized structures of (a) silicene; (b)-(c) boat-like and chair-like full-hydrogenated silicene; (d)-(f) zigzag, boat-like and chair-like half-hydrogenated silicene, respectively. The primitive unit cell of each structure is shown by red lines. Si and H atoms are shown in blue and grey, respectively.

2. Hydrogenated Silicene

In this section we review the results of studies on hydrogenation of silicene. Like C atoms in graphene, Si atoms in silicene have unpaired electrons which are suitable for possible functionalizations. Among these possible functionalizations, hydrogenation was studied extensively in the literature. [69, 80, 81, 82] It has been shown that two possible configurations exist for the hydrogenation process of silicene, fully-hydrogenation (fH), namely silicane, and single side hydrogenation, half-hydrogenation (hH), similar to the case of graphene.

In Figure 1, possible geometric structures are given for bare, fully and half hydrogenated silicene crystals. Zhang et al. investigated the structural properties of fH and hH cases of silicene by first principles calculations [69] and found that for the silicane structure the chair-like configuration (see Figure 1(c)) is the ground state and it has 30 meV/atom lower energy than the boat-like one (see Figure 1(b)) as confirmed by total energy calculations. The Si-H bond length was calculated to be 1.50 Å for the chair-like structure. For the hH silicene they

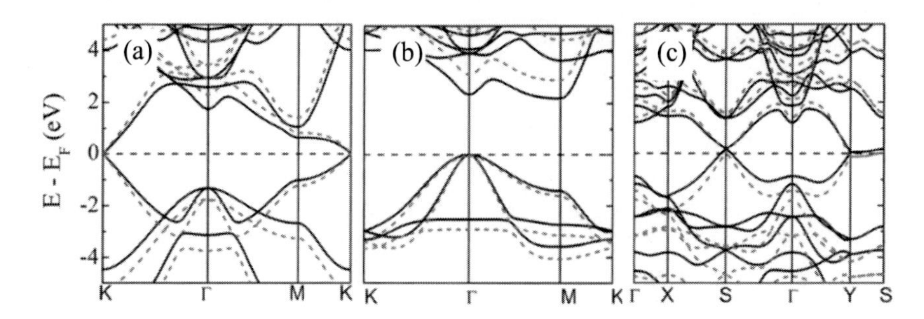

Figure 2. (Color online) Energy-band structures of (a) bare silicene, (b) chair-like silicane, and (c) zigzag HH silicene (taken from Ref. [69])

reported that the zigzag structure (Figure 1(d)) is the most stable configuration with a total energy of 33 meV/atom and 180 meV/atom lower than the boat-like and chair-like structures, respectively. In addition, Osborn et al. reported that the fH silicene structure has a higher buckling than its bare form [83]. They calculated the buckling height of fH silicene to be 0.74 Å while 0.54 Å was reported for the bare silicene case. This structural change occurs due to the interaction between Si and H atoms which widens the structure in the vertical direction. The buckling height of the hH silicene structure is reported to be less than that of silicane as expected [84].

Hydrogenation plays an important role for tuning the electronic structure of a 2D material. For instance in contrast to bare graphene, hydrogenated graphene, namely graphane, is a semiconductor [85]. The same functionalization process was studied in the case of silicene. Zhang et al. reported that the electronic-band structure of silicene can be tuned through hydrogenation. It was found that silicane is an indirect-gap semiconductor with its valence-band maximum (VBM) and conduction band minimum (CBM) residing at the Γ and M points, respectively. The band gap of silicene was found to be 2.36 eV within GGA approximation while it is reported as 3.51 eV by using HSE06 functional. These results were also predicted and supported by many other studies [66, 68, 69, 83]. In contrast to silicane, the HH silicene crystal possesses metallic character in its zigzag structure (see Figure 2(c)). However, the other two configurations, boat-like and chair-like structures, were reported to be direct-gap semiconductors [69].

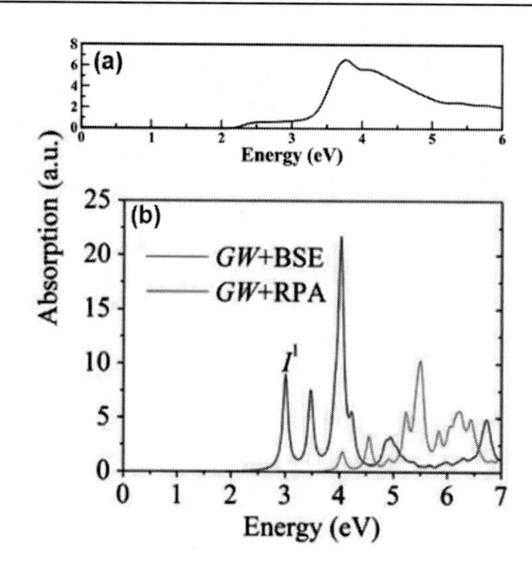

Figure 3. (Color online) Optical absorption spectra of silicane calculated with (a) GGA approach [86] and (b) GW approximations with (GW+BSE) or without (GW+RPA) considering excitonic effects (taken from Ref. [87])

Optical properties of silicane such as its optical absorption spectrum and dielectric function were investigated in the literature before [70, 86]. In addition, the optical properties of bilayer and few layer fH silicene structures were also predicted [88, 89]. Chinnathambi et al. studied the optical properties of silicane by calculating the optical absorption spectrum [86]. They reported that a transition from semi-metallic to semiconducting behavior is seen. The reason is the broken π bonds in silicene due to the saturation by H atoms. As seen in Figure 3(a), an absorption onset at 2.2 eV was predicted which is consistent with the bandgap of silicane as calculated within the GGA approximation [86]. Moreover, Wei et al. investigated the optical absorption spectra of silicane by GW approximation with random phase approximation (GW+RPA) and Bethe-Salpeter equation (GW+BSE) [70]. It was reported that the hydrogenation process removes conduction at the Dirac point and causes a finite band gap opening. It was also reported that the GW+RPA and GW+BSE methods give different absorption spectra due to the large self-energy correlations of electrons (see Fig. 3(b)). The absorption onset obtained with GW+RPA is located at about 4 eV consistent with the band gap value calculated within HSE06 functional [68, 69].

Including the excitonic correlations of electrons and holes, within the GW+BSE approximation, the excitonic effect significantly shifts the onset of the absorption spectrum towards lower energy (see Figure 3(b)).

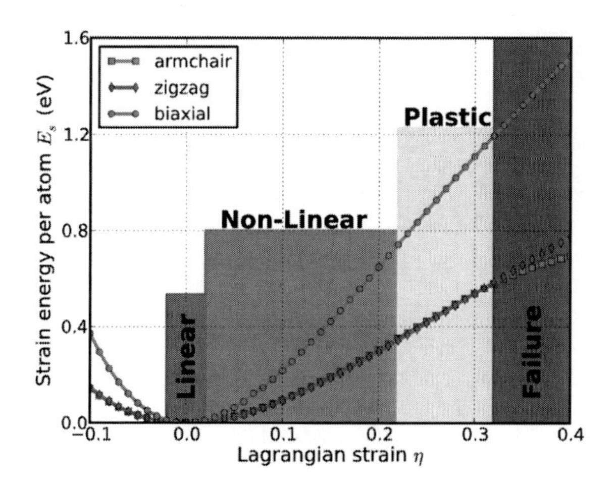

Figure 4. (Color online) The strain energy per Si-H pair as a function of applied uniaxial Lagrangian strain η along armchair and zigzag directions, and biaxial Lagrangian strain along both directions (taken from Ref. [91])

In addition to the electronic and optical properties, the mechanical properties of silicane were also investigated theoretically in the previous studies [90, 91, 92]. Peng et al. reported that the in-plane stiffness (58 N/m) and Poisson ratio (0.24) values for silicane are reduced by 16% and 26%, respectively, when compared to those of silicene [91]. The elastic limits in terms of ultimate tensile strains were found to be 0.22, 0.28, and 0.25 along armchair, zigzag, and biaxial directions, respectively. It was reported that these values increases by 9%, 33%, and 24%, respectively from silicene to silicane. Moreover, Jamdagni *et al.* reported that the band gap of silicane reduces to zero with increasing applied biaxial tensile strain leads to a semiconducting to metallic transition for silicane. Their calculations indicated that at 2% of tensile strain, the magnitude of the bandgap first increases to 2.22 eV and the indirect band-gap character of silicane changes to a direct bandgap. Then with every 2% increment of tensile strain, the band gap decreases nearly by 0.3 eV and the 22% value of the strain is the critical value for semiconducting-to-metallic transition [90].

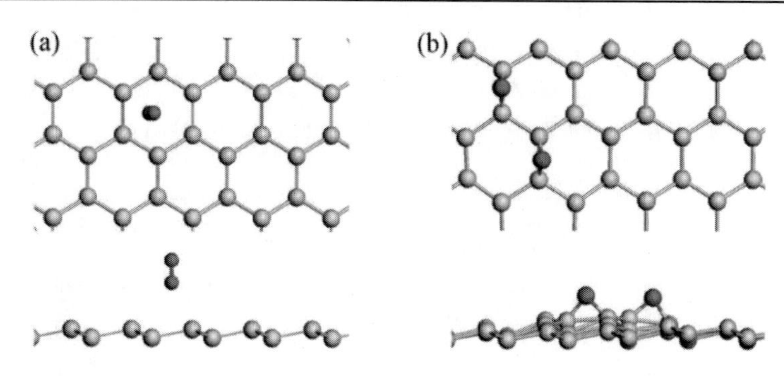

Figure 5. (Color online) Geometric structure of oxygen molecule adsorption and dissociation on pristine silicene (a) before and (b) after relaxation (taken from Ref. [79])

3. Oxidation of Silicene

Oxidation has important consequences on the usage of materials in real life device technologies. Thus, the oxidation processes of both bulk and 2D materials were widely studied and investigated in previous works. Similar to all materials, the oxidation of silicene is an important question for scientists during the fabrication of silicene-based devices. Therefore, the possibilities of silicene oxide formation and the effects of oxidation on the physical properties of silicene were studied both experimentally [53, 75, 76, 77, 78] and theoretically [74, 79, 93, 94, 95, 96].

Liu et al. investigated the oxygen adsorption and dissociation on a free-standing silicene monolayer [79]. It was reported that the O_2 molecule dissociates into O atoms on free-standing silicene and the formation of Si-O compound occurs. Also it was pointed out that the oxidation of silicene is easy because of the very low energy barrier for the O_2 molecule to dissociate into O atoms. Depending on the initial vertical distance of the O_2 molecule to the silicene layer the resultant O atoms can bind to different sites of silicene. Among these sites the lowest energy configuration is the one for which the two O atoms reside on bridge sites of silicene (see Figure 5(b)). The dissociation of O_2 molecule on free-standing silicene is confirmed by Ozcelik et al. [94] The Si-O bond lengths were calculated to be 1.71 Å and 1.73 Å for upper and lower Si atoms, respectively. The possible migration paths for an O atom from one bridge site to a

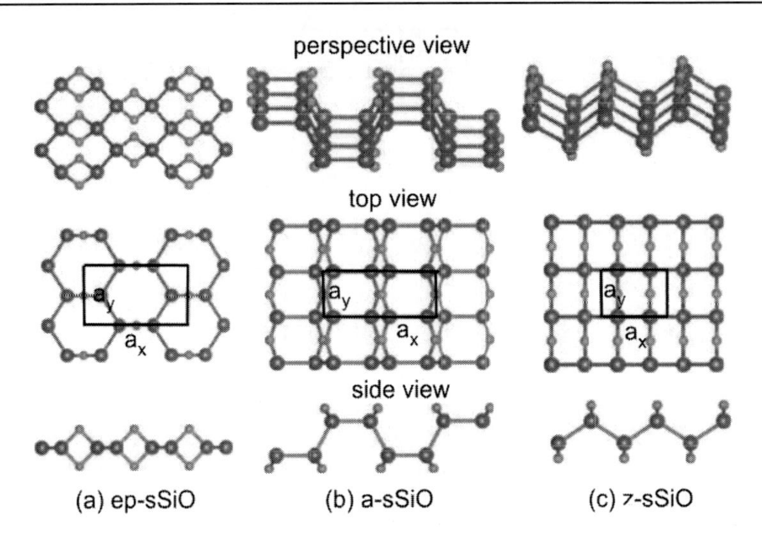

Figure 6. (Color online) Atomic structures of sSiO with (a) ep-conformation (O atoms are located as epoxy-pair groups), (b) a-conformation, and (c) z-conformation (taken from Ref. [93]).

neighboring bridge site exhibit energy barriers of 1.05 eV and 1.18 eV energy barriers. These are large values as compared to those for a graphene surface [79]. In another study, the single layer phase of silica, SiO_2, was predicted as a stable honeycomblike structure by Ozcelik et al. [95]

The effect of oxidation on the electronic properties of silicene was investigated by Wang et al. [93] They studied fully oxidized silicene with stoichiometric ratio of Si:O = 1:1. The zigzag ether-like conformation stoichiometric silicene oxide (z-sSiO) was found to be the ground state configuration (see Figure 6(c)). The z-sSiO configuration has 14 and 165 meV/atom lower energy than the a-conformation and the ep-conformation, respectively. They reported that the z-sSiO structure is a semiconductor with a direct band gap of 0.18 eV as calculated within the GGA approximation (see Figure 7(a)) while it is found to be 1.05 eV when the HSE06 functional is considered. In addition, Ozcelik et al. reported that single O adsorption on a silicene layer results in a direct-gap semiconducting structure with a band gap of 0.21 eV [94]. In the study of Ozcelik et al., the new phase of SiO_2, monolayer silica, was found to be a direct-gap semiconductor with a relatively large band gap of 3.3 eV when compared to

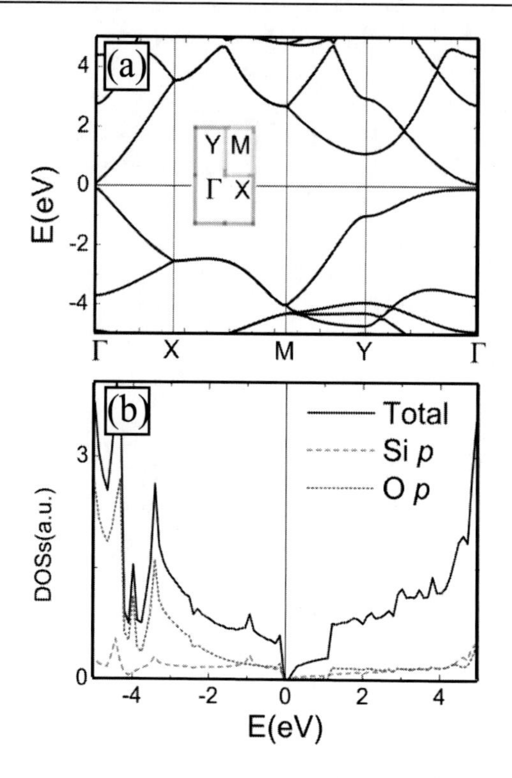

Figure 7. (Color online) (a) Energy-band structure of z-sSiO calculated within GGA and (b) the corresponding partial density of states (taken from Ref. [93])

O-doped silicene layer [95].

Wang et al. studied the mechanical properties of the stoichiometric SiO structure. They found that the z-sSiO monolayer has some prominent elastic characteristics, as negative Poisson ratios and exhibits an unconventional auxetic behavior. When these auxetic materials are stretched in one direction, they become thicker in the perpendicular direction. The reason for this auxetic behavior is the assembly of Si-O bonds into bending -O- network along the y-direction. The mechanical properties of monolayer silica were investigated in terms of in-plane stiffness and Poisson ratio [95]. It was reported that single-layer silica has an in-plane stiffness of 22.6 J/m^2 which is smaller than that of graphene. Moreover, the Poisson ratio for the monolayer phase of silica was calculated to be negative like the z-sSiO monolayer. Having negative Poisson ratio

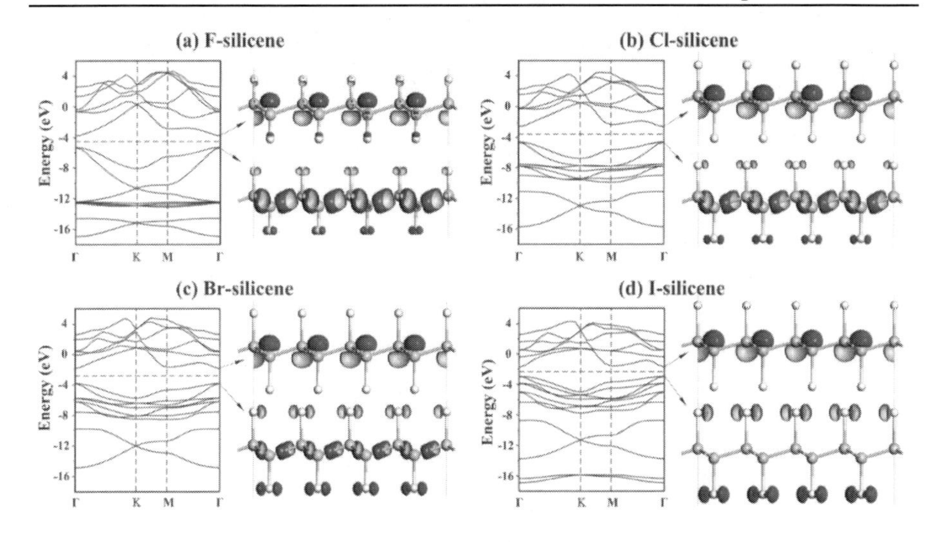

Figure 8. (Color online) The electronic band structures (at the left panel) and charge density distributions (at the right panel) of the CBM (upper) and the VBM (lower) states at the Γ point for (a) F-silicene, (b) Cl-silicene, (c) Br-silicene, and (d) I-silicene structures. Dashed line indicates the Fermi level and the isosurface value is 0.025 au (arbitrary unit) (taken from Ref. [71])

is an important mechanical property for the usage of a material in biomedical and nanosensor applications.

4. Halogenated Silicene

In order to integrate 2D monolayers into nanotechnological devices, fluorination and functionalization with other halogen atoms are promising methods similar to hydrogenation. Experimental and theoretical studies have shown that the band gap of fluorinated-graphene can be tuned from 0 to 5.0 eV by changing the fluorination level [29, 30, 31, 32, 33, 34, 35] and half-fluorinated graphene was predicted to be magnetic. [97] Similar to the case of graphene, chemical functionalization of silicene with fluorine (F) and other halogen atoms (such as Cl, Br, and I) have also been extensively studied in the literature. Gao et al. reported that halogenation of silicene opens a band gap with various gap values depending on the atomic number of the halogen atoms. [71] The obtained band gaps are 1.19, 1.47, 1.95, and 1.98 eV for I, Br, F, and Cl atoms, respectively as shown in Figure 8. They also reported that the formation energy increases with the increase in the atomic number of the halogen atom.

As in the hydrogenation case, several structural configurations (see Figure 1) were also considered for the halogenation of silicene. Ding et al. studied the structural and electronic properties of fluorinated silicene alongside with hydrogenated silicene. They reported that the band gap of the boat-like (Z-line type in the corresponding study) fluorinated silicene increases almost linearly with strain, on the other hand, the band gap of the chair-like structure has a parabolic dependence around the strain value of $\epsilon = 0.02$. [66]

Zhang et al. investigated the geometric and the electronic structure as well as the mechanical properties of halogenated silicene XSi (X = F, Cl, Br and I) in various conformers (as shown in Figure 1) by using first principles calculations within DFT. Their results indicated that halogenated silicene shows enhanced stability as compared with bare silicene and exhibits a tunable direct band gap. [72] They reported that the chair-like structure of silicene is the most favorable one for all the halogen atoms. They also showed that, consistent with the previous results [71], the formation energy increases when the atomic number of the halogen atom increases which indicates that fluorination is the most favorable one among all halogenation. In addition, as shown in Figure 9, direction dependent Poisson ratio for different conformers of the halogenated silicene were calculated and a negative Poisson ratio was predicted for the boat-like (boat2 structure in the corresponding study) structure of fluorinated silicene [72].

Moreover, Wang et al. investigated the structural, electronic and magnetic properties of half-fluorinated silicene sheets by using first principles simulation within the framework of DFT. They reported that half-fluorinated (as shown in Figure 1) silicene sheets with zigzag, boat-like or chair-like configurations were confirmed to be dynamically stable based on phonon calculations. [98] Upon the adsorption of fluorine, a band gap opening is predicted in both zigzag and boat-like conformations and they were found to be direct-gap semiconductors. Moreover, half-fluorinated silicene with chair-like configuration shows antiferromagnetic ordering which is mainly induced by the unfluorinated Si atoms.

Wei et al. investigated the optical properties of fluorinated silicene by using the many-body effects by using Greens function perturbation theory. [70] As in hydrogenation, fluorination of silicene also opens a band gap which is consistent with the previous studies. They also reported that strong excitonic effects dominate the absorption properties of hydrogenated, fluorinated silicene, and

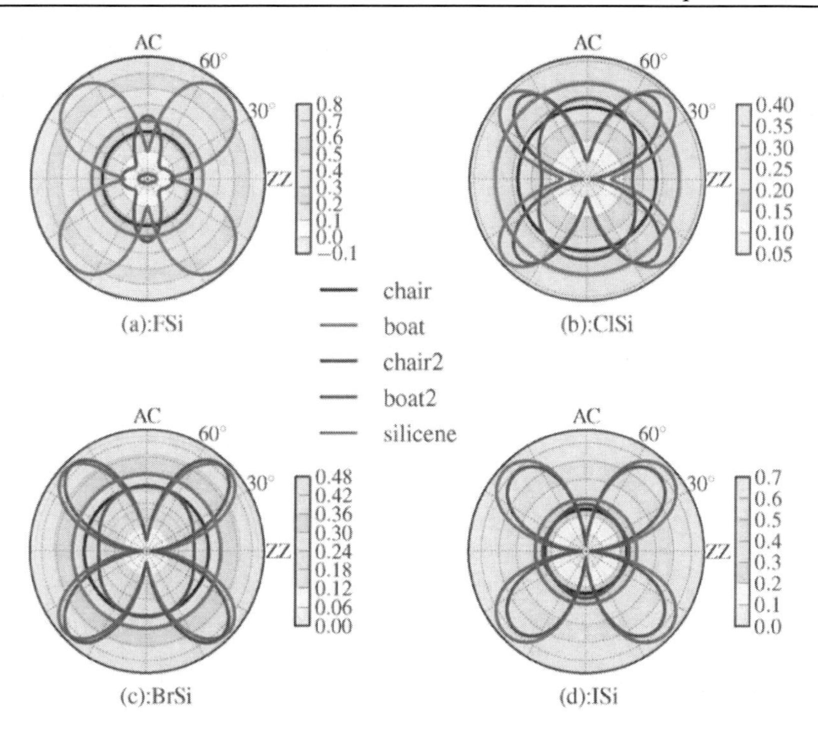

Figure 9. (Color online) Polar diagram for the Poisson ratio, ν, of halogenated silicene. (a)-(d) Represent the FSi, ClSi, BrSi and ISi systems, respectively. The angle θ identifies the extension direction with respect to the zigzag one. *ZZ* and AC represent the zigzag and armchair direction, respectively. The numerical values are represented by different background colors. Isotropic (anisotropic) behavior is associated with a circular (noncircular) shape of the ν_n plot (taken from Ref. [72])

silicene nanoribbon with high exciton binding energies.

5. Functionalization via Organic Molecule Adsorption

The adsorption of different chemical functional groups on silicene have potential applications for silicene-based nanoelectronic devices. Different from the highly stable planar structure of graphene, the buckled honeycomb structure of silicene leads to high chemical reactivity for functional groups. Thus, adsorption of functional groups could be a prominent method for tuning the electronic structure of silicene.

Hue et al. investigated the adsorption of NH_3, NO and NO_2 on silicene and found that the electronic properties of silicene are strongly depend on the type of adsorbate. Their findings revealed a significant potential of silicene for highly sensitive molecule sensors. In addition, Wen et al. found high reactivity of silicene towards NO_2, O_2 and SO_2 molecules. [100] Binding energies of these molecules on silicene are larger than 1 eV. In contrast, the binding energies of NO and NH_3 are 0.35 and 0.60 eV, respectively. While the band gap of silicene is enhanced upon adsorption of NO, O_2, NH_3, and SO_2, it becomes half-metallic when NO_2 is adsorbed. The structural and electronic properties of diverse molecules adsorbed on silicene were investigated by van der Waals included DFT. [101] Considered molecules are shown in Figure 10 and their calculated adsorption energies vary from -0.11 to -0.95 eV indicating no adsorption. Moreover, electronic structure in hydrogenated silicene as well as fluorinated silicene calculations showed that the calculated band gaps range from 0.01 to 0.35 eV for acetonitrile to acetone, respectively.

Recently, Prasongkit et al. investigated the change of the electronic and transport properties when NO_2, NO, NH_3, and CO molecules are adsorbed onto pristine and B/N-doped silicene. [102] Their results showed that NO and NO_2 can be sensitively detected by pristine silicene. On the other hand, due to the weak interaction of CO and NH_3 molecules with pristine silicene, the possibility of detection of those gases is relatively low. Increased sensitivity towards NH_3 and CO obtained when pristine silicene is doped either by B or N atoms. Quantum conductance properties of CO molecule adsorbed silicene nanoribbons were investigated by Osborn et al. [103] They showed that the quantum conduction is modified in a detectable way by weak chemisorption of a single CO molecule on a silicene nanoribbon. The adsorption of N_2 and CO_2 molecules do not affect the conductance. However, O_2 and H_2O molecules can be strongly chemisorbed and can diminish the CO detection capability of silicene. Moreover, they found that CO, O_2 and H_2O are easily detectable molecules among CO, CO_2, O_2, N_2, and H_2O. Gurel et al. investigated the interaction of H_2, O_2, CO, H_2O, and OH molecules with graphene and silicene [96] and found that H_2, O_2, and CO remain intact on both graphene and silicene. When these molecules adsorb at the vicinity of vacancy centers they can dissociate. The dissociations of other atoms are hindered by high energy barriers

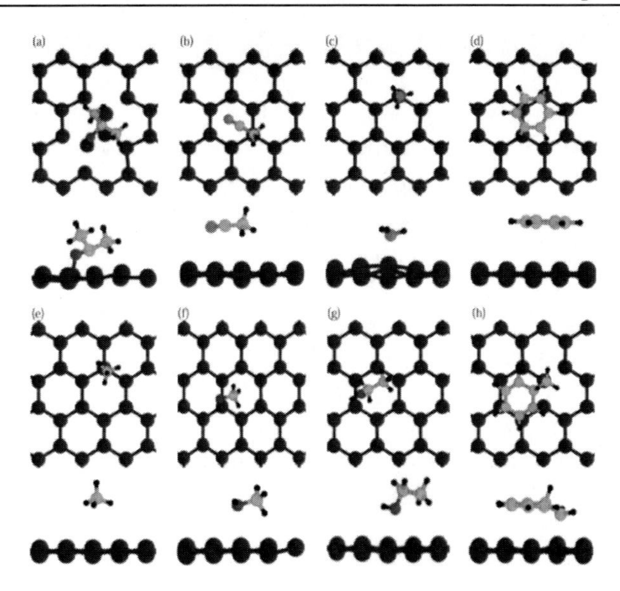

Figure 10. (Color online) Top and side views of relaxed geometric structures of (a) acetone, (b) acetonitrile, (c) ammonia, (d) benzene, (e) methane, (f) methanol, (g) ethanol, and (h) toluene molecules on silicene. Blue, green, black, violet, and red spheres are Si, C, H, N, and O atoms, respectively (taken from Ref. [101])

Stephan et al. studied adsorption of benzene molecule on (3×3) silicene which was placed on the (4×4) Ag (111) surface. [104] Their study revealed that benzene molecule can be chemisorbed on a silicene layer deposited on Ag(111) through a cycloaddition reaction. They also showed that, Si (100) and Si (111) surfaces are more reactive than the other surfaces of the structure. In addition, Stephan et al. investigated the adsorption characteristics of H_2Pc molecule on silicene above Ag (111). [105] They showed that, due to an electostatic or polarization repulsion between H_2Pc molecule and Si surface, H_2Pc molecule adopts a butterfly configuration on this surface. However, this molecule shows a planar configuration on the SiC and SiB surfaces. This study revealed the possibility of chemisorption of such large molecules on the Si/Ag system.

6. Functionalization of Silicene via Adatom Decoration

Due to the buckled honeycomb structure of silicene, it is chemically a very active material. In order to maintain and tune its electronic properties as required diverse growth mechanisms and various substrates were used. During growth processes the presence of foreign atoms and cluster formation is inevitable. The quality of fabricated silicene-based devices is strongly affected by the adsorbed foreign atoms. Therefore, the investigation of the decoration mechanisms of these atoms on silicene is quite essential.

Ni et al. investigated the geometric and electronic properties of silicene with five different transition metal atoms (Cu, Ag, Au, Pt, and Ir) adsorbed at different coverages by using first principle methods. [43] Optimized geometric structures of Cu-covered silicene with different covarages are shown in Figure 11. Similiar to Cu, favorable geometric configuration of Ag , Au, Pt, and Ir atoms is the center of silicene hegzagons. A sizable band gap can be opened without degrading the electronic properties at the Dirac point of silicene when these atoms are adsorbed. Adsorption characteristics of the metal atoms are given in Figure 12. As shown in the figure, a band gap opening occurs in all the considered coverages and the value of the gap increases from 0.03 to 0.66 eV with increasing coverage range. Using the method of the adsorption of different transition metal atoms on different regions of silicene, they designed a silicene p-i-n tunnelling field effect transistor.

The adsorption characteristics and the stability of Li atoms on silicene was investigated by first principles calculations. [106, 107] It was reported that, Li adsorbed silicene compounds are energetically favorable and fully lithiated silicene (silicel) is the most stable form among them. The stability of the silicene sheet in the presence of completely adsorbed lithium atoms on the atom-down sites of both sides (Figure 13) was confirmed by molecular dynamic simulations conducted at elevated temperatures. Lithiation can be used to tune the band gap of silicene and complete Li adsorption results in a band gap of 0.368 eV

Quhe et al. examined the gap opening in silicene in the presence of single-side adsorption of alkali atoms such as Li, Na, K, Rb, and Cs. [42] They showed that the band gap of silicene can be tuned by alteration of adsorption coverage resulting in a band gap up to 0.5 eV. Moreover, quantum transport simulation of a bottom-gated FET based on a Na-covered silicene was also conducted and a transport gap with an on/off current ratio up to 10^8 was predicted. The electronic structure, mechanical stability, and hydrogen storage capacity of strain

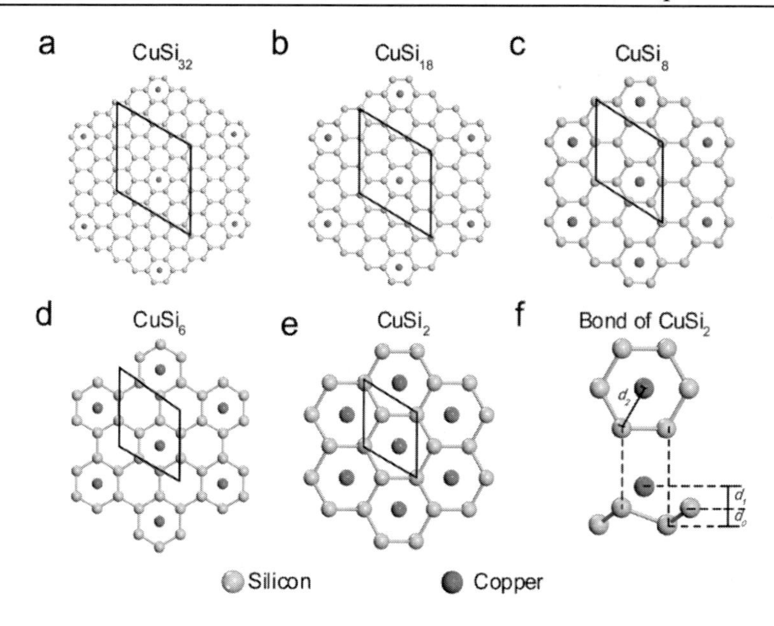

Figure 11. (Color online) Optimized geometric structures of Cu-covered silicene with a coverage of N = 3.1% (a), 5.6% (b), 12.5% (c), 16.7% (d), and 50.0% (e). (f) Top and side views of a Cu-covered silicene supercell. The parallelograms show the unit cell for each structure and the parameters d_0 and d_1 demonstrate the buckling of siilicene and the height of Cu atom to upper Si atom, respectively. The distance d_2 represents the distance from the adsorbed atom to the upper Si atom in silicene (taken from Ref. [43])

induced Mg functionalized silicene (SiMg) and silicane (SiHMg) monolayers have been investigated by Hussain et al. [108] Their results revealed that high doping concentration of Mg atom can be obtained on both monolayers by biaxial symmetric strain up to 10%. The adsorption energy of H_2 molecules on silicene was found to be ideal for the application in hydrogen storage devices. Li, Na, K, Be Mg and Ca adsorbed silicene sheets were studied to investigate their hydrogen-storage capacity. [109] It is found that Li and Na atoms have strong metal-to-substrate binding and they are suitable for high-capacity storage of hydrogen. Using DFT calculations, the effects of an external electric field on the adsorption-desorption of H_2 on a Ca-decorated silicene system was studied. [110] Doubled binding energy enhancement is observed for H_2 when

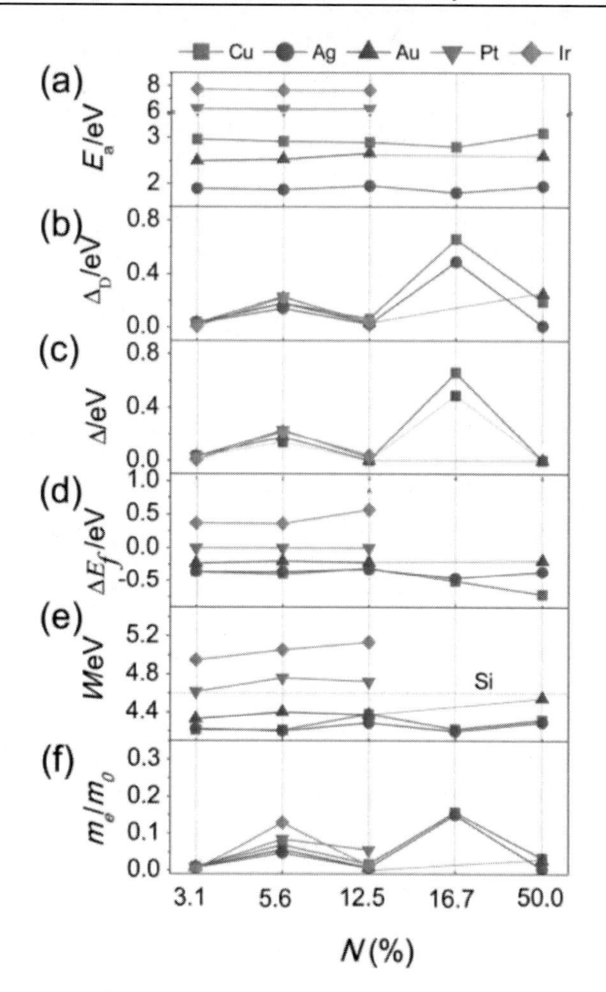

Figure 12. (Color online) Calculated adsorption energy (per atom) (a), band gap at the Dirac point (b), global band gap (c), Fermi level shift of metal covered-silicene (d), work function (the horizontal dashed line shows the work function of pure silicene) (e), and effective mass of holes of the metal covered silicene (f) as a function of coverag (taken from Ref. [43])

0.004 au external electric field was applied on the Ca-silicene system. On the other hand, when -0.004 au external electric field was applied to the system, the

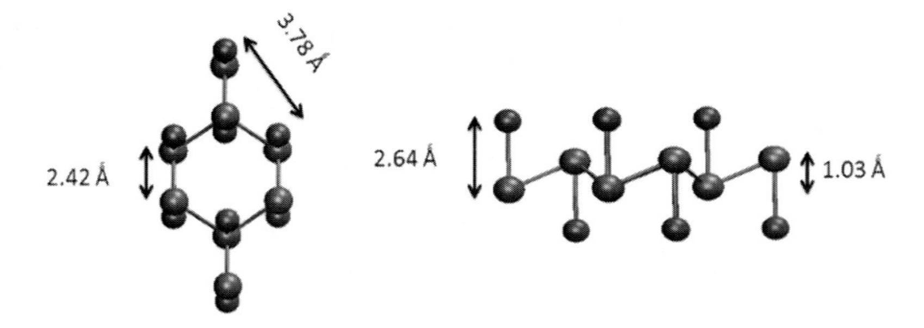

Figure 13. (Color online) Optimized geometric structure of fully lithiated silicene (silicel) (taken from Ref. [106])

binding of $9H_2$ on Ca- monolayer or bilayer silicene system is getting weaker.

Conclusion

Functionalization of 2D materials is an efficient way to tailor their electronic, optical and mechanical properties. Silicene, with its highly-reactive surface structure, is a good candidate for various functionalization techniques and have been widely studied by the researchers. Recent studies have demonstrated that hydrogenation and halogenation of silicene can tune the electronic-band structure from semi-metal to semiconductor. However, by single-side adsorption of alkali atoms, Li, Na, K, Rb, and Cs, only a relatively small band gap opens. Opening a gap in silicene is rather important for its potential usage of the material in optoelectronic applications. In addition to its electronic properties, chemical functionalization of silicene can also change the Poisson ratio from positive to negative values which is important for applications in biomedicine and nanosensors. Like other 2D monolayer materials, the electronic, optical and mechanical properties of silicene could be tuned by chemical functionalization to integrate it into nanotechnological device applications.

Acknowledgments

This work was supported by the Flemish Science Foundation (FWO-Vl) and the Methusalem foundation of the Flemish government. Computational resources were provided by TUBITAK ULAKBIM, High Performance and Grid Computing Center (TR-Grid e-Infrastructure). H.S. is supported by a FWO Pegasus Long Marie Curie Fellowship.

References

[1] B. C. Brodie, *Phil. Trans. R. Soc. Lond.* **149**, 249 (1859).

[2] R. E. Peierls, *Ann. Inst. Henri Poincare* **5**, 177 (1935).

[3] R. E. Peierls, *Helv. Phys. Acta*, **7**, 81 (1934).

[4] K. S. Novoselov, A. K. Geim, S. V. Morozov, D. Jiang, Y. Zhang, S. V. Dubonos, I. V. Grigorieva, and A. A. Firsov, *Science* **306**, 666 (2004).

[5] K. S. Novoselov, D. Jiang, F. Schedin, T. Booth, V. V. Khotkevich, S. Morozov, and A. K. Geim, Proc. Natl. Acad. Science U.S.A. **102**, 1045 (2005).

[6] K. S. Novoselov, A. K. Geim, S. V. Morozov, D. Jiang, M. I. Katsnelson, I. V. Grigorieva, S. V. Dubonos, and A. A. Firsov, *Nature* **438**, 197 (2005).

[7] H. Sahin, S. Changirov, M. Topsakal, E. Bekaroglu, E. Akturk, R. T. Senger, and S. Ciraci, *Phys. Rev. B* **80**, 155453 (2009).

[8] D. Golberg, Y. Bando, Y. Huang, T. Terao, M. Mitome, C. Tang, and C. Zhi, *ACS Nano* **4**, 2979 (2010).

[9] H. Zeng, H. Zhi, C. Zhang, Z. Wei, X. Wang, X. Guo, W. Bando, Y. Golberg, *D. Nano Lett.* **10**, 5049 (2010).

[10] L. Song, L. Ci, L. Lu, H. Sorokin, P. B. Jin, C. Ni, J. Kvashnin, A. G. Kvashnin, D. G. Lou, J. Yakobson, B. I. Ajayan, P. M. *Nano Lett.* **10**, 3209 (2010).

[11] C. Bacaksiz, H. Sahin, H. D. Ozaydin, S. Horzum, R. T. Senger, and F. M. Peeters, *Phys. Rev. B* **91**, 085430 (2015).

[12] H. L. Zhuang and R. G. Hennig, *Appl. Phys. Lett.* **101**, 153109 (2012).

[13] Q. Wang, Q. Sun, P. Jena, and Y. Kawazoe, *ACS Nano* **3**, 621 (2009).

[14] K. K. Kim, A. Hsu, X. Jia, S. M. Kim, Y. Shi, M. Hofmann, D. Nezich, J. F. Rodriguez-Nieva, M. Dresselhaus, T. Palacios, and J. Kong, *Nano Lett.* **12**, 161 (2012).

[15] M. Farahani, T. S. Ahmadi, and A. Seif, *J. Mol. Struct.* **913**, 126 (2009).

[16] Q. H. Wang, K. K. Zadeh, A. Kis, J. N. Coleman, and M. S. Strano, *Nat. Nanotechnol.* **699**, 699 (2012).

[17] J. A. Wilson and A. D. Yoffe, *Adv. Phys.* **18**, 193 (1969).

[18] S. Horzum, D. Cakir, J. Suh, S. Tongay, Y. S. Huang, C. H. Ho, J. Wu, H. Sahin, and F. M. Peeters, *Phys. Rev. B* **89**, 155433 (2014).

[19] C. Bacaksiz, S. Cahangirov, A. Rubio, R. T. Senger, F. M. Peeters, and H. Sahin, *Phys. Rev. B* **93**, 125403 (2016).

[20] S. Changirov, M. Topsakal, E. Akturk, H. Sahin, and S. Ciraci, *Phys. Rev. B* **102**, 236804 (2009).

[21] P. Vogt, P. D. Padova, C. Quaresima, J. Avila, E. Frantzeskakis, M. C. Asensio, A. Resta, B. Ealet, and G. L. Lay, *Phys. Rev. Lett.* **108**, 155501 (2012).

[22] C. L. Lin, R. Arafune, K. Kawahara, N. Tsukahara, E. Minamitami, Y. Kim, N. Takagi, and M. Kawai, *Appl. Phys. Express* **5**, 045802 (2012)

[23] A. Fleurence, R. Friedlein, T. Osaki, H. Kawai, Y. Wang, and Y. Y. Takamura, *Phys. Rev. Lett.* **108**, 245501 (2012).

[24] M. E. Davila, L. Xian, S. Cahangirov, A. Rubio, and G. L. Lay, *New J. Phys.* **16** 095002 (2014).

[25] F. F. Zhu, W. J. Chen, C. L. Gao, D. D. Guan, C. H. Liu, D. Qian, S. C. Zhang, and J. F. Jia, *Nat. Mater.* **14** 1020 (2015).

[26] D. W. Boukhvalov, M. I. Katsnelson, and A. I. Lichtenstein, *Phys. Rev. B* **77**, 035427 (2008).

[27] D. Haberer et al., *Nano Lett.* **10**, 3360 (2010).

[28] J. O. Sofo, A. S. Chaudhari, and G. D. Barber, *Phys. Rev. B* **75**, 153401 (2007).

[29] J. T. Robinson, J. S. Burgess, C. E. Junkermeier, S. C. Badescu, T. L. Rei necke, F. K. Perkins, M. K. Zalalutdniov, J. W. Baldwin, J. C. Culbertson P. E. Sheehan, and E. S. Snow, *Nano Lett.* **10**, 3001 (2010).

[30] D. K. Samarakoon, Z. Chen, C. Nicolas, and X. Q. Wang, *Small* **7**, 96 (2011).

[31] S. H. Cheng, K. Zou, F. Okino, H. R. Gutierrez, A. Gupta, N. Shen, P. C Eklund, J. O. Sofo, and J. Zhu, *Phys. Rev. B* **81**, 205435 (2010).

[32] K. J. Jeon, Z. Lee, E. Pollak, L. Moreschini, A. Bostwick, C. M. Park R. Mendelsberg, V. Radmilovic, R. Kostecki, T. J. Richardson, and E Rotenberg, *ACS Nano* **5**, 1042 (2011)

[33] J. C. Garcia, D. D. B. Lima, L. V. C. Assali, and J. F. Justo, *J. Phys Chem. C* **115**, 13242 (2011).

[34] A. L. Walter, H. Sahin, K. J. Jeon, A. Bostwick, S. Horzum, R. Koch, F Speck, M. Ostler, P. Nagel, M. Merz, S. Schupler, L. Moreschini, Y. J Chang, T. Seyller, F. M. Peeters, K. Horn, and E. Rotenberg, *ACS Nano* **8**, 7801 (2014).

[35] A. L. Walter, H. Sahin, J. Kang, K. J. Jeon, A. Bostwick, S. Horzum, L Moreschini, Y. J. Chang, F. M. Peeters, K. Horn, and E. Rotenberg, *Phys Rev. B* **93**, 075439 (2016).

[36] A. Kara, H. Enriquez, A. P. Seitsonen, L. C. L. Y. Voon, S. Vizzini, B Aufray, and H. Oughaddou, *Surf. Sci. Rep.* **67**, 1 (2012).

[37] M. Xu, T. Liang, M. Shi, and H. Chen, *Chem. Rev.* **113**, 3766 (2013).

[38] H. Sahin, J. Sivek, S. Li, B. Partoens, and F. M. Peeters, *Phys. Rev. B* **88** 045434 (2013).

[39] B. Lalmi, H. Oughaddou, H. Enriquez, A. Kara, S. Vizzini, B. Ealet, and B. Aufray, *Appl. Phys. Lett.* **97** 223109 (2010).

[40] L. Meng, Y. Wang, L. Zhang, S. Du, R. Wu, L. Li, Y. Zhang, G. Li, H Zhou, W. A. Hofer, and H. J. Gao, *Nano Lett.* **13** 685 (2013).

[41] X. Lin and J. Ni, *Phys. Rev. B* **86**, 075440 (2012).

[42] R. Quhe, R. Fei, Q. Liu, J. Zheng, H. Li, C. Xu, Z. Ni, Y. Wang, D. Yu, Z. Gao, and J. Lu, *Sci. Rep.* **2**, 853 (2012).

[43] Z. Ni, H. Zhong, X. Jiang, R. Quhe, G. Luo, Y. Wang, M. Ye, J. Yang, J. Shi, and J. Lu, *Nanoscale* **6**, 7609 (2014).

[44] Y. C. Cheng, Z. Y. Zhu, and U. Schwingenschlogl, *Europhys. Lett.* **95**, 17005 (2011).

[45] J. Sivek, H. Sahin, B. Partoens, and F. M. Peeters, *Phys. Rev. B* **87**, 085444 (2013).

[46] R. Zheng, X. Lin, and J. Ni, *Appl. Phys. Lett.* **105**, 092410 (2014).

[47] H. Okamoto, Y. Sugiyama, and H. Nakano, *Chem. Eur. J.* **17**, 9864 (2011).

[48] H. Nakano, M. Nakano, K. Nakanishi, D. Tanaka, Y. Sugiyama, T. Ikuno, H. Okamoto, and T. Ohta, *J. Amer. Chem. Soc.* **134**, 5452 (2012).

[49] H. Okamoto, Y. Kumai, Y. Sugiyama, T. Mitsuoka, K. Nakanishi, T. Ohta, H. Nozaki, S. Yamaguchi, S. Shirai, and H. Nakano, *J. Amer. Chem. Soc.* **132**, 2710 (2010).

[50] Y. Sugiyama, H. Okamoto, T. Mitsuoka, T. Morikawa, K. Nakanishi, T. Ohta, and H. Nakano, *J. Amer. Chem. Soc.* **132**, 5946 (2010).

[51] P. R. Pereda and N. Takeuchi, *J. of Chem. Phys.* **138**, 194702 (2013).

[52] M. J. S. Spencer, M. R. Bassett, T. Morishita, I. K. Snook, and H. Nakano, *New J. Phys.* **15**, 125018 (2013).

[53] Y. Du, J. C. Zhuang, H. S. Liu, X. Xu, S. Eilers, K. H. Wu, C. Peng, J. J. Zhao, X. D. Pi, K. See, G. Peleckis, X. Wang, and X. Dou, *ACS Nano* **8**, 10019 (2014).

[54] G. Liu, M. S. Wu, C. Y. Ouyang, and B. Xu, *Euro Phys. Lett.* **99**, 17010 (2012).

[55] R. Qin, C. H. Wang, W. Zhu, and Y. Zhang, *AIP Adv.* **2**, 022159 (2012).

[56] H. Zhao, *Phys. Lett. A* **376**, 3546 (2012).

[57] M. Hu, X. Zhang, and D. Poulikakos, *Phys. Rev. B* **87**, 195417 (2013).

[58] T. P. Kaloni, Y. C. Cheng, and U. Schwingenschlagl, *J. App. Phys.* **113**, 104305 (2013).

[59] A.P. Durajski, D. Szczesniak, and R. Szczesniak, *Solid State Comm.* **200**, 17 (2014).

[60] B. Mohan, A. Kumar, and P.K. Ahluwalia, *Physica E* **61**, 40 (2014).

[61] T. Hussain, S. Chakraborty, A. D. Sarkar, B. Johansson, and R. Ahuja, *J. App. Phys.* **105**, 123903 (2014).

[62] B. Wang, J. Wu, X. Gu, H. Yin, Y. Wei, R. Yang, and M. Dresselhaus, *Appl. Phys. Lett.* **104**, 081902 (2014).

[63] J. Zhu and U. Schwingenschlogl, *ACS Appl. Mater. Interfaces* **6**, 1167: (2014).

[64] C. Yang, Z. Yu, P. Lu, Y. Liu, H. Ye, and T. Gao, *Comp. Mater. Sci.* **95**, 420 (2014).

[65] G. Cao and Y. Zhang, J. Cao, *Phys. Lett. A* **379**, 1475 (2015).

[66] Y. Ding and Y. Wang, *Appl. Phys. Lett.* **100**, 083102 (2012).

[67] L. C. L. Y. Voon, E. Sandberg, R. S. Aga, and A. A. Farajian, *Appl. Phys. Lett.* **97**, 163114 (2010).

[68] M. Houssa, E. Scalise, K. Sankaran, G. Pourtois, V. V. Afanasev, and A. Stesmans, *Appl. Phys. Lett.* **98**, 223107 (2011).

[69] P. Zhang, X. D. Li, C. H. Hu, S. Q. Wu, and Z. Z. Zhu, *Phys. Lett. A* **376**, 1230 (2012).

[70] W. Wei and T. Jacob, *Phys. Rev. B* **88**, 045203 (2013).

[71] N. Gao, W. T. Zheng, and Q. Jiang, *Phys. Chem. Chem. Phys.* **14**, 25 (2012).

[72] W. B. Zhang, Z. B. Song, and L. M. Dou, *J. Mater. Chem. C* **3**, 308 (2015).

[73] X. Wang, H. Liu and S. T. Tu, *RSC Adv.* **36**, 6238 (2015).

[74] P. De Padova, C. Ottaviani, C. Quaresima, B. Olivieri, P. Imperatori, E. Salomon, T. Angot, L. Quagliano, C. Romano, A. Vona, M. M. Miranda, A. Generosi, B. Paci, and G. L. Lay, *2D Mater.* **1**, 021003 (2014).

[75] P. De Padova, C. Quaresima, B. Olivieri, P. Perfetti, and G. L. Lay, *J. Phys. D: Appl. Phys.* **44**, 312001 (2011).

[76] A. Molle, C. Grazianetti, D. Chiappe, E. Cinquanta, E. Cianci, G. Tallarida, and M. Fanciulli, *Adv. Func. Mater.* **24**, 5088 (2013).

[77] R. Friedlein, H. V. Bui, F. B. Wiggers, Y. Y. Takamura, A. Y. Kovalgin, and M. P. de Jong, *J. Chem. Phys.* **140**, 204705 (2014).

[78] X. Xu, J. Zhuang, Y. Du, H. Feng, N. Zhang, C. Liu, T. Lei, J. Wang, M. Spencer, T. Morishita, X. Wang, and S. X. Dou, *Sci. Rep.* **4**, 7543 (2014).

[79] G. Liu, X. L. Lei, M. S. Wu, B. Xu, and C. Y. Ouyang, *J. Phys. Conden. Matter.* **26**, 355007 (2014).

[80] L. B. Drissi, E. H. Saidi, M. Bousmina, and O. F. Fehri, *J. Phys. Conden. Matter.* **24**, 485502 (2012).

[81] L. B. Drissi and F. Z. Ramadan, *Physica E* **68**, 38 (2015).

[82] P. Zhang, B. B. Xiao, X. L. Hou, Y. F. Zhu, and Q. Jiang, *Sci. Rep.* **4**, 3821 (2014).

[83] T. H. Osborn, A. A. Farajian, O. V. Pupysheva, R. S. Aga, and L.C. L. Y. Voon , *Chem. Phys. Lett.* **511**, 101 (2011).

[84] C. W. Zhang and S. S. Yan, *J. Phys. Chem. C* **116**, 4163 (2012).

[85] H. Sahin and F. M. Peeters, *Phys. Rev. B* **87**, 085423 (2013).

[86] K. Chinnathambi, A. Chakrabarti, A. Banerjee, and S.K. Deb, arXiv:1205.5099v1.

[87] W. Rong, M. S. Xu, and X. D. Pi, *Chinese Phys. B* **24**, 086807 (2015).

[88] B. Huang, H. X. Deng, H. Lee, M. Yoon, B. G. Sumpter, F. Liu, S. C. Smith, and S. H. Wei, *Phys. Rev. X* **4**, 021029 (2014).

[89] Y. Liu, H. Shu, P. Liang, D. Cao, X. Chen, and W. Lu, *J. Appl. Phys.* **114**, 094308 (2013).

[90] P. Jamdagni, A. Kumar, M. Sharma, A. Thakur, and P. K. Ahluwalia, AIP Conf. Proc. **1661**, 080007 (2015).

[91] Q. Peng and S. De, *Nanoscale* **6**, 1207 (2014).

[92] C. H. Yang, Z. Y. Yu, P. F. Lu, Y. M. Liu, S. Manzoor, M. Li, and S. Zhou, Proc. SPIE **8975**, 89750K-1 (2014).

[93] Y. Wang and Y. Ding, *Phys. Stat. Solidi RRL* **7**, 410 (2013).

[94] V. O. Ozcelik and S. Ciraci, *J. Phys. Chem. C* **117**, 26305 (2013).

[95] V. O. Ozcelik, S. Cahangirov, and S. Ciraci, *Phys. Rev. Lett.* **112**, 24680[illegible] (2014).

[96] H. H. Gurel, V. O. Ozcelik, and S. Ciraci, *J. Phys. Chem. C* **118**, 2757[illegible] (2014).

[97] Y. D. Ma, Y. Dai, M. Guo, C. W. Niu, L. Yu, and B. B. Huang, *Nanoscale* **3**, 2301 (2011).

[98] X. Wang, H. Liu, and S. T. Tu, *RSC Adv.* **5**, 6238 (2015).

[99] W. Hu, N. Xia, X. Wu, Z. Li, and J. Yang, *Phys. Chem. Chem. Phys.* **16**, 6957 (2014).

[100] J. Wen, Y. J. Liu, H. X. Wang, J. X. Zhao, Q. H. Cai, and X. Z. Wang, *Comp. Mater. Sci.* **87**, 218 (2014).

[101] T. P. Kaloni, G. Schreckenbach, and M. S. Freund, *J. Phys. Chem. C* **118**, 23361 (2014).

[102] J. Prasongkit, R. G. Amorim, S. Chakraborty, R. Ahuja, R. H. Scheicher, and V. Amornkitbamrung, *J. Phys. Chem. C* **119**, 16934 (2015).

[103] T. H. Osborn and A. A. Farajian, *Nano Res.* **7 (7)**, 945 (2014).

[104] R. Stephan, M. C. Hanf, and P. Sonnet, *Phys. Chem. Chem. Phys.* **17**, 14495 (2015).

[105] R. Stephan, M. C. Hanf, and P. Sonnet, *J. Chem. Phys.* **143**, 154706 (2015).

[106] T. H. Osborn and A. Farajian, *J. Phys. Chem. C* **116**, 22916 (2012).

[107] G. A. Tritsaris, E. Kaxiras, S. Meng, and E. Wang, *Nano Lett.* **13**, 2258 (2013).

[108] T. Hussain, S. Chakraborty, A. D. Sarkar, B. Johansson, and R. Ahuja, *Appl. Phys. Lett.* **105**, 123903 (2014).

[109] T. Hussain, S. Chakraborty, and R. Ahuja, *Chem. Phys. Chem. Commun.* **14**, 3463 (2013).

[110] E. H. Song, S. H. Yoo, J. J. Kim, S. W. Lai, Q. Jiang, and S. O. Cho, *Phys. Chem. Chem. Phys.* **16**, 23985 (2014).

In: Advances in Materials Science Research
Editor: Maryann C. Wythers
ISBN: 978-1-53610-059-4

Chapter 7

ARYLAZO PYRIDONE DYES: STRUCTURE-PROPERTIES RELATIONSHIPS

Jelena M. Mirković, Bojan Đ. Božić, Gordana S. Uščumlić and Dušan Ž. Mijin*
Faculty of Technology and Metallurgy,
University of Belgrade, Belgrade, Serbia

ABSTRACT

Arylazo pyridone dyes (APDs) represent disperse dyes commonly used in textile industry due to their good technical features: excellent coloration properties, simplicity of preparation, good light and wash fastness properties. They find application in inkjet printing, liquid crystal displays and inks for heat-transfer printing. The main characteristic of APDs bearing –OH group in *ortho*- position to azo group, is existence of an azo-hydrazone tautomerism. Generally, APDs are present in the hydrazone form in the solid state and most of the solvents, while in highly dipolar solvents there is hydrazone-anion equilibrium with the predominance of the hydrazone form.

Subtle structural changes of APD molecules lead to their different physico-chemical properties and the knowledge of the origin of these changes is essential for their further application. Significant progress has been made in the area of experimental investigations of APDs using various techniques including UV-Vis, FT-IR and NMR spectroscopy, as

*E-mail: kavur@tmf.bg.ac.rs.

well as X-ray crystallography. Furthermore, the investigation of the geometry and the relative stabilities of the APD structures, as well as the charge transfer through the molecule and the possibility of forming intramolecular hydrogen bonds can be obtained using quantum chemical calculations. The transmission of electronic effects of the substituents through the molecule can be interpreted by linear free energy relationships (LFER) analysis of spectroscopic data (UV-Vis, NMR, FT-IR). Specific and nonspecific interactions (solvent-solute), which strongly influence the coloration of the APD solutions, most commonly are analyzed by linear solvent energy relationships (LSER) concept. So, in this chapter, the correlations between experimental and theoretical results have been established in order to gain complete insight into structural features which enable modeling and synthesis of novel APD molecules with potential usage in various industries.

Keywords: pyridone dyes, azo-hydrazone tautomerism, hydrazone-anion equilibrium, UV-Vis, X-ray, DFT, solvent effects, substituent effects

1. INTRODUCTION

APDs represent azo dyes having pyridone derivatives as coupling components and heterocyclic or carbocyclic compounds as diazo components. They generally find application as disperse dyes, colorants characterized by low water solubility which are, in their colloidal form, suitable for dying and printing of hydrophobic fibers and fabrics. Disperse dyes are predominantly used on polyester and polyamide fibers [1–3]. They are also utilized on nylon, cellulose acetate and acrylic fibers, but in smaller extent due to the poor wet-fastness properties on these fibers. The importance of disperse dyes have increased along with the development of new polyester fibers, polyethylene terephthalate (PET) and polytrimethylene terephthalate (PTT). Nowadays, disperse dyes constitute the second largest sector in dying industry [3]. Color of the polyester fibers after application of APDs is usually yellow or yellow-greenish [4, 5]. Azo pyridone structure is found in reactive dyes, which are used for dyeing of cellulosic and polyamide fibers [6–8], as well as in basic dyes applied to acrylic fibers and in ink-jet inks [9, 10]. Thermally stable yellow pigments obtained from azo pyridone dyes are suitable for plastics, baking varnishes and printing inks [11, 12].

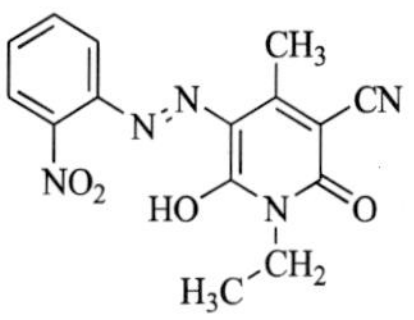

C.I. Disperse Yellow 119

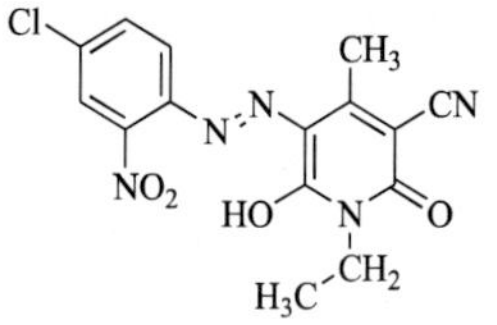

C.I. Disperse Yellow 211

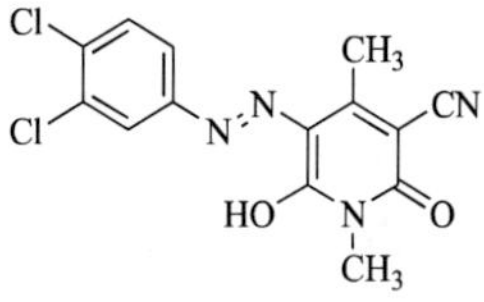

C.I. Disperse Yellow 241

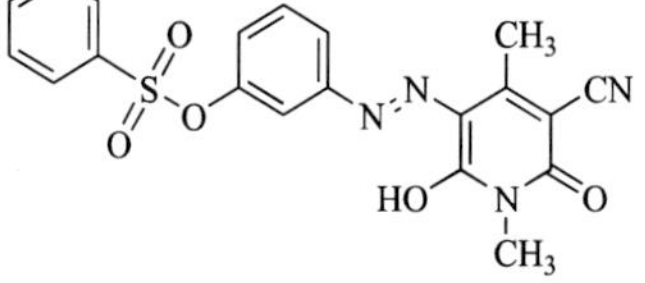

C.I. Disperse Yellow 114

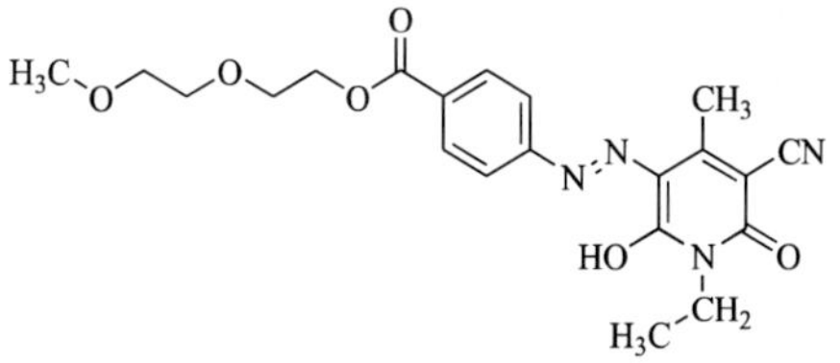

C.I. Disperse Yellow 126

Figure 1. The examples of commercial APDs.

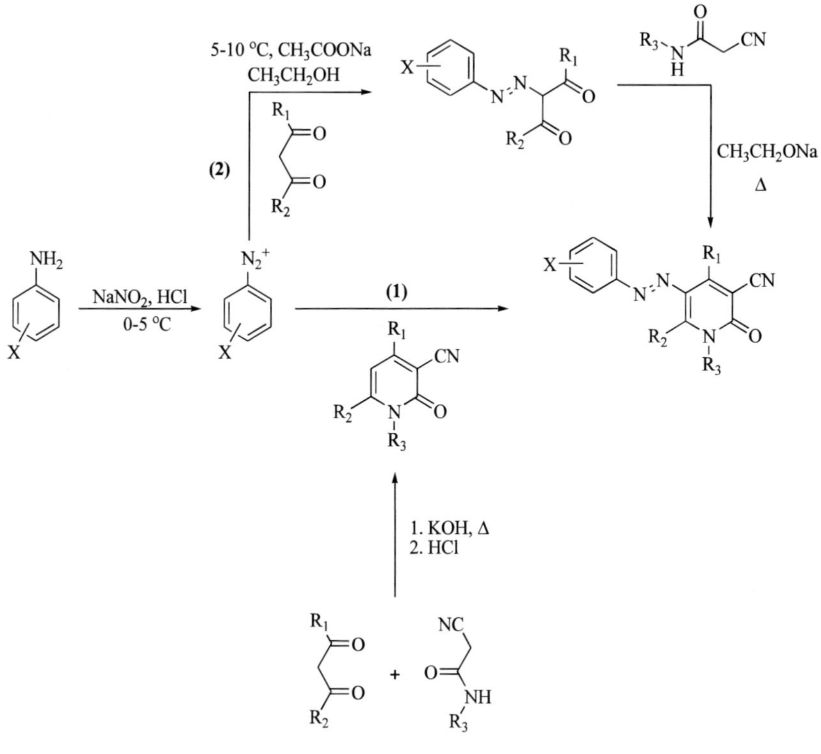

5-10 °C, CH_3COONa CH_3CH_2OH

(2)

CH_3CH_2ONa Δ

$NaNO_2$, HCl 0-5 °C

(1)

1. KOH, Δ
2. HCl

Scheme 1. Synthetic pathways of APDs.

The great significance and broad application of APDs in last few decades are due to their outstanding properties, such as good light and wash fastness properties, vivid colors and high molar extinction coefficients [1, 2]. They find application in liquid crystal displays (LCD) and in ink-jet printing [13–18]. APDs are known as typical transfer-printing yellow dyes for textiles [2]. Also, they are suitable compounds for hot melt or phase change inks [16]. The examples of commercially available APDs are presented in Figure 1.

Wide commercial application of APDs is due to the economic aspect of their preparation and to the almost innumerable possibilities for synthesis combining various diazo and coupling components [1]. APDs can be obtained following two synthetic pathways (Scheme 1). The first pathway comprises direct coupling of the pyridone moiety with diazonium salts obtained in the reaction of diazotation (Scheme 1, synthetic pathway 1) [17]. The second one includes coupling of *β*-diketones or *β*-ketoesters with diazonim salts in order to obtain aryl azo diketones or aryl azo ketoesters which are further followed by condensation to form molecule of pyridone dye (Scheme 1, synthetic pathway 2) [4, 18–21].

2. PROPERTIES OF APDs

Generally, absorption maxima of APDs usually lie in the visible region in the range of yellow to orange, but it has been reported that some dyes show deeper color strength and shades such as red or violet [22]. APDs exhibit significant fluorescence properties [23]. Moreover, APDs are characterized by good light and wash fastness properties and these properties largely depend on the structures of arylazo component and pyridone moiety as well as the type of the colored substrates. 5-(4-Arylazo)-1,4-dialkyl substituted-3-cyano-6-hydroxy-2-pyridones show high affinity towards polyester fibers, good light fastness and excellent wash fastness, which is expected due to hydrophobic nature of these dyes [24]. APDs having ethyl groups in the position 1 and 4 of pyridone ring posses good and excellent fastness to perspiration and washing, respectively, and moderate light fastness which is significantly affected by the nature of the substituents in phenyl ring where electron-accepting groups improve the fastness [25]. Introduction of methyl group in the position 4 and methyl, ethyl or propyl groups in the position 1 of the pyridone ring, can lead to APDs with better light fastness properties [26–28]. APDs obtained from 3-cyano-4,6-dimethyl-2-pyridone and 3-cyano-4-methyl-6-phenyl-2-pyridone show good light, wash, rub and perspiration fastness after dying of polyamide and polyester fabrics. The phenyl group in the position 6 upgrades stability of these dyes [29]. APDs such as 5-(2- and 4-disubstituted phenylazo)-1-alkyl-3-cyano-6-hydroxy-4-methyl-2-pyridones are not suitable for dyeing of nylon fabrics [30]. On the other hand, APDs synthesized from *N*-alkyl substituted pyridones and *p*-(β-sulphatoethyl)sulphonylaniline applied to polyester/cotton blend show excellent coloration properties [31].

Synthetic azo dyes from the textile industry (including APDs), are pollutants which represent a significant source of ground and surface waters contamination. Thus, the investigations based on the removal of the azo dyes from textile effluents have gained importance in last few decades.

Photodegradation and electrochemical methods are some of the methods that are used for the decolorization or the degradation of APDs. Wang et al. studied photostability and photodegradation of 5-(2- and 4-disubstituted phenylazo)-3-cyano-6-hydroxy-4-methyl-2-pyridones in amide solvents by 254 nm light. The photofading rate is increased in the presence of electron-accepting groups in the diazo component and also increases in the following order of the solvents: formamide < *N,N*-dimethyacetamide (DMA) < *N,N*-dimethylformamide (DMF) [32]. Further examination of photodegradation in DMF reveals that the presence of two electron-accepting groups in diazo component increases the fading rate, while the presence of alkylol group in position 1 of the coupling component decreases the rate. The nature of alkylol group does not significantly affect photodegradation of these APDs [33]. The fading rate of these dyes on the polyester substrates is more affected by the chemical constitution and the position of the substituent in the phenyl ring than its physical state. Introduction of electron-accepting groups in *para*- position increases stability to light, while β-hydroxyethyl group in the position 1 of pyridone ring causes decrease of lightfastness [34, 35].

The study of photodegradation of 5-(4-sulphophenylazo)-3-cyano-6-hydroxy-4-methyl-2-pyridone shows that this APD can be completely decolorized and mineralized in the presence of TiO_2 and simulated sun solar light after 240 min. The kinetics of degradation is affected by different parameters such as TiO_2 content, pH value, initial dye concentration as well as addition of ethanol and hydrogen peroxide [36, 37].

Electrocatalytic decolorization of the 5-(4-arylazo)-3-cyano-6-hydroxy-1-(2-hydroxyethyl)-4-methyl-2-pyridones in the presence of sodium chloride using DSA Ti/PtOx electrode in the diluted sodium hydroxide is attributed to the indirect oxidation of the investigated dyes by the electrogenerated hypochlorite formed from the chloride oxidation [38]. The kinetics of the reaction is followed by the change of the UV-Vis absorption spectra (Figure

2). After 20 min, a complete decolorization is observed. The decrease of the dye absorption on UV-Vis spectra is the indication of the decolorization. The reaction rate is dependable upon sodium hydroxide, sodium chloride and dye concentrations, current and agitation speed. The reaction is promoted when electron-donating substituents are present in phenyl ring and inhibited in the presence of electron-accepting substituents.

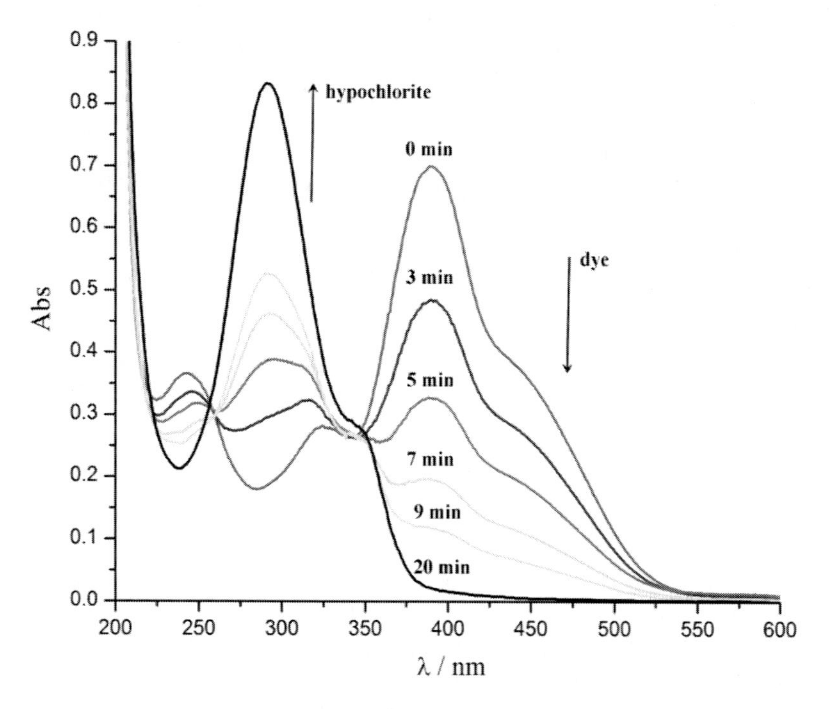

Figure 2. UV-Vis change of 5-(4-methoxyphenylazo)-3-cyano-6-hydroxy-1-(2-hydroxyethyl)-4-methyl-2-pyridone in NaOH solution during electrochemical decolorization.

The study of the electrochemical behavior of six APDs on a gold electrode in 0.1 M NaOH using cyclic and square wave voltammetry show various effects of the selected functional groups with respect to their position and the nature on their electrochemical activity. Among the different substituents in the phenyl ring (*p*-OMe, *p*-NO_2, *p*-Cl and *o*-NO_2) and in the position 1 of the pyridone ring (*N*-H, *N*-CH_2CH_3 and *N*-CH_2CH_2OH), according to the cyclic voltammetry, the most active APD is 5-(4-methoxyphenylazo)-6-hydroxy-3-cyano-4-methyl-2-pyridone. Square wave voltammetry shows apparent activity of the dyes with ethyl and hydroxyethyl groups in pyridone moiety, which is attributed to the accumulation time of this technique. Inclusion complexes of APDs with (2-hydroxypropyl)-β-cyclodextrine (1:1) show improved electrochemical activity in comparison to the pure APDs, especially in the presence of electron-accepting groups in phenyl ring indicating better inclusive abilities of these APDs towards (2-hydroxypropyl)-β-cyclodextrine [39].

Biological study of pyridone dyes show that some of them possess antimicrobial activity depending on the chemical structure of the dyes. Pyridone dyes such as 5-(4-hydroxyphenylazo)-3-cyano-6-hydroxy-4-methyl-1-propyl-2-pyridones and 5-thiazolyl-3-cyano-6-hydroxy-4-methyl-1-propyl-2-pyridones show significant activity towards Gram positive bacteria (*Staphylococcus auerus* and *Bacillus subtilis*), while APD with methyl- group in phenyl ring did not show any activity against these strains. Heteroaryl pyridone dyes obtained from 5-methylisoxazole and 5-methylsulfanyl-1,2,4-triazol as diazo components and 3-cyano-6-hydroxy-4-methyl-1-propyl-2-pyridone show moderate activity. APDs obtained from 3-cyano-6-hydroxy-4-methyl-1-propyl-2-pyridone have no antimicrobial activities against Gram negative bacteria (*Escherichia coli*, *Klebsiellae pneumoniae* and *Pseudomonas aeruginosa*) [28]. Series of 5-(4-arylazo)-3-cyano-1,4-diethyl-6-hydroxy-2-pyridones show significant antimicrobial activity. The most active compounds against Gram positive and negative bacteria are hydroxy and unsubstituted derivatives, while methoxy- and chlorine derivatives possess the most pronounced antifungal activity [25]. Disperse bisazo pyridone dyes derived from 5-amino-4-arylazo-3-methyl-1H-pyrazoles and 3-cyano-6-hydroxy-4-methyl-2-pyridone are tested against the yeast, Gram-positive and negative bacteria. *Para*-chlorine derivative of the series show promising antimicrobial activity with broad spectrum and could lead to the development of new antimicrobial drug, while *ortho*-chlorine derivative revealed a selective activity against *Streptococcus faecalis* indicating a potential use for the development of a new selective antimicrobial drug. Furthermore, most of the dyes in this series have fungicidal activity against *Candida utilis* [40].

3. Tautomerism and Equilibriums of APDs

Tautomerism represents reversible intramolecular rearrangements in organic molecules in which interconversion of structural isomers relatively rapidly occurs. Tautomers differ from each other in the electron distribution and the position of the mobile atom or group. In the great majority of examples, mobile atom is hydrogen and this tautomerism is known as prototropy [41]. APDs exhibit two types of prototrpy depending on the structure of the dyes: keto-enol tautomerism which includes pyridone ring and azo-hydrazone tautomerism involving azo group.

3.1. Keto-Enol Tautomerism of APD

Keto-enol tautomerism is the most common tautomeric equilibrium in organic chemistry. In APDs, keto-enol equilibrium includes prototropy of –NH–C(=O)– group in pyridone ring. 6-Alkyl or aryl substituted APDs exhibit this tautomerism as 2-pyridon/2-hydroxypyridine (lactam/lactim) equilibrium (Scheme 2). The infrared spectra of 5-(3- and 4-substituted phenylazo)-3-cyano-4,6-diphenyl-2-pyridones show an intense band in the region 1666–1692 cm^{-1} ascribed to carbonyl group and intense band in the region 3376–3495 cm^{-1} which originate from the N–H vibrations suggesting the existence of 2-pyridone tautomeric form in the solid state [42].

Keto-enol equilibrium in the solutions is strongly affected by the nature of the medium, as well as by the structure of the APDs. For 5-(3- and 4-substituted phenylazo)-3-cyano-4,6-dimethyl-2-pyridones and 5-(3- and 4-substituted phenylazo)-3-cyano-4,6-diphenyl-2-pyridones, it is shown that strong electron-donating and moderate electron-accepting groups stabilize keto form in protic and nonpolar aprotic solvents. Strong electron-accepting groups favor enol form and in this case, the equilibrium between two forms exists [42, 43]. The chemical shifts from N–H pyridone of 5-(3- and 4- substituted phenylazo)-3-cyano-4,6-diphenyl-2-pyridones in ^{1}H NMR spectra appear in the region 11.67–12.15 ppm in DMSO-d_6. The dyes with electron-accepting groups in phenyl ring show another signal in the region 12.08–14.04 ppm, appearing from OH proton of enol tautomeric form, supporting the fact that equilibrium exists [42].

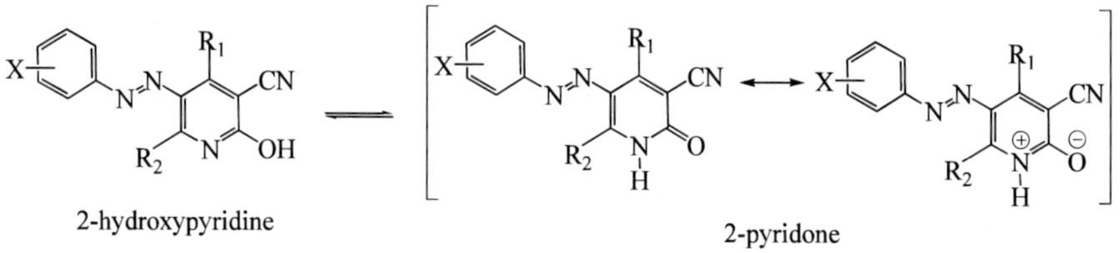

2-hydroxypyridine

2-pyridone

Scheme 2. Keto-enol tautomerism of APDs.

The study of solvent effects on keto-enol equilibrium showed that increasing of solvent polarity shifts the equilibrium towards the keto form [44]. 2-Pyridone form is more polar than 2-hydroxypyridine form due to the contribution of the charge-separated mesomeric form. Considering hydrogen-bonding ability of the solvent, hydrogen-bond donor solvents stabilize keto form, while hydrogen-bond acceptor solvents affect the stabilization of 2-hyrdoxypyridine form [42, 43]. Comparison of two series of APDs obtained from 3-cyano-4,6-dimethyl-2-pyridones and 3-cyano-4,6-diphenyl-2-pyridones showed that introducing of phenyl rings in pyridone ring instead of methyl, influences higher sensitivity of azo group to the solvent and substituent effects and these APDs are shifted to higher absorption maxima. 5-(3- and 4-substituted phenylazo)-3-cyano-4,6-diphenyl-2-pyridones exhibit color hue in the range from dark yellow to brown, while methyl analogues show light orange to brown-orange hues [42].

3.2. Azo-Hydrazone Tautomerism of APDs

The phenomenon of azo-hydrazone tautomerism is of the considerable interest, not only from the theoretical but also from the economic standpoint, since the two tautomers have different optical, physical and toxicological features, and the most important different tinctorial strengths. The hydrazone tautomer is usually shifted to higher wavelengths with respect to azo form and have higher extinction coefficients [1, 2, 45]. The main structural feature of the azo dyes, including APDs, required for azo-hydrazone tautomerism is the existence of the labile proton. This is manifested in the molecules bearing OH or NHR groups conjugated to azo bridge, i.e., in *ortho-* and *para-* position to this group (Scheme 3) [46, 47]. The tautomerism may exist both in the solutions and in the solid state. The tautomers can be identified by their distinctive spectral properties in the solution (UV-Vis, FT-IR, ^{1}H, ^{13}C and ^{15}N NMR), and by FT-IR and X-ray crystallography in the solid state [45]. In APDs, azo-hydrazone tautomerism is usually due to the presence of OH group in *ortho-* position to azo bridge, i.e., in 6-hydroxy-2-pyridones.

According to FT-IR spectra, 5-arylazo-3-cyano-6-hydroxy-2-pyridones exist in the hydrazone form in the solid state [5, 48–52]. Spectra show two characteristic intense peaks in the range 1620–1700 cm^{-1} ascribed to the vibrations of the two carbonyl groups. Additional evidence supporting the hydrazone form is broad bands in the region 3410–3500 cm^{-1} assigned to the vibrations of the imino group of the hydrazone form. Broadening of this band, as well as the low frequency indicate the existence of hydrogen bonds in the solid states [53]. Characteristic vibrations of the cyano group are in the region of 2220–2260 cm^{-1}. N–H vibrations of unsubstituted nitrogen of the pyridone ring appear in the region 3385–3388 cm^{-1}.

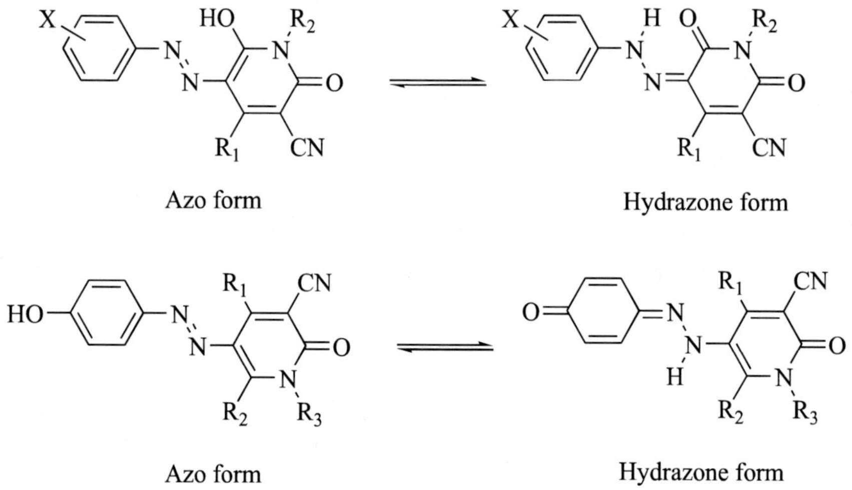

Azo form Hydrazone form

Azo form Hydrazone form

Scheme 3. Azo-hydrazone tautomerism of APDs.

Cheng et al. [54] showed that APDs derived from 3-cyano-6-hydroxy-1,4-dimethyl-2-pyridone demonstrate little variation in color from yellow to orange. UV-Vis absorption maxima in toluene were in the 423–457 nm region and are ascribed to the hydrazone form. The authors suggested that the main chromophore includes pyridone ring and the hydrazone residue and that substituents in the aryl ring exert a secondary effect. Authors also assumed that visible absorption band is consequent of the transition from the hydrazone (NH) group to the carbonyl groups. Recent investigations of 5-(4-arylazo)-3-cyano-6-hydroxy-1-(2-hydroxyethyl)-4-methyl-2-pyridones, however, have demonstrated that electron density in the molecules of APDs transfers to the cyano group as the principal acceptor [23]. Transmission mode of the electronic substituent effects in the investigated compounds is proposed according to linear free energy relationship (LFER) analysis. Scheme 4 represents the possible resonance structures of the pyridone ring with cyano and carbonyl groups as acceptors. The favored pathway is from N–N hydrazone group to cyano group (a), while the contribution of the transmission of the electron density to carbonyl group (b) on overall delocalization is almost negligible. This is in accordance with the number of the resonance structures and NBO analysis of 5-(4-methoxyphenylazo)-3-cyano-1-ethyl-6-hydroxy-4-methyl-2-pyridones [55].

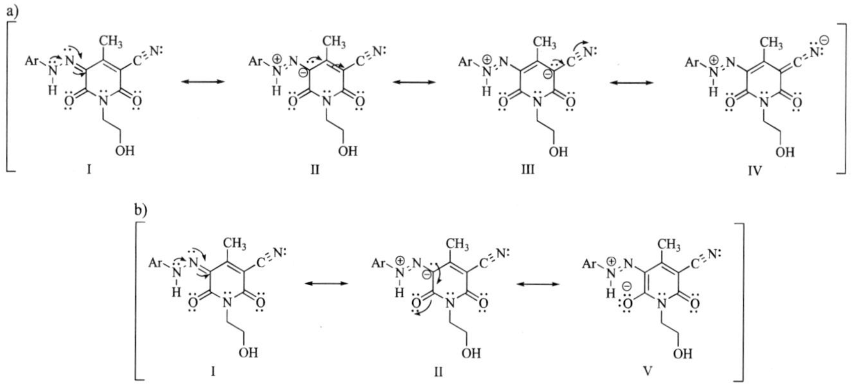

Scheme 4. Possible resonance structures of APDs with: a) cyano group as acceptor; b) carbonyl group as acceptor.

Resonance effects of electron-donating and electron-accepting groups in *para-* position of the hydrazone form of 5-(4-arylazo)-3-cyano-6-hydroxy-2-pyridones are presented in Scheme 5. According to the transmission mode, electron-donating groups in *para-* position in phenyl ring cause a significant bathochromic shift of absorption maxima of the hydrazone form due to the positive resonance effect and thus reinforcing intramolecular charge transfer (ICT) of the molecule. It is expected that electron-accepting groups in *ortho-* and *para-* positions induce hypsochromic shift, as the consequence of the destabilization of the structure due to the positive charge on nitrogen. Hypsochromic shift is observed for *meta-* substituted APDs, while groups with negative resonance effect in *para-* position, in fact, exert a small bathochromic effect due to the counteracting influence to the extended delocalization. Chlorine and bromine in *para-* position, as substituents with weak positive resonance effect, cause bathochromic shift of UV-Vis maxima even though they exert strong negative inductive effect. These substituents induce hypsochromic shift when they are present in *meta-* position. Groups in *ortho-* position show the similar trends as corresponding *para-* substituted derivatives, while the difference is caused by steric effects [33, 49, 53, 54]. The substituents in the position 1 of pyridone ring show no significant influence on the position of UV-Vis absorption spectra of the hydrazone form and chromophore properties [5].

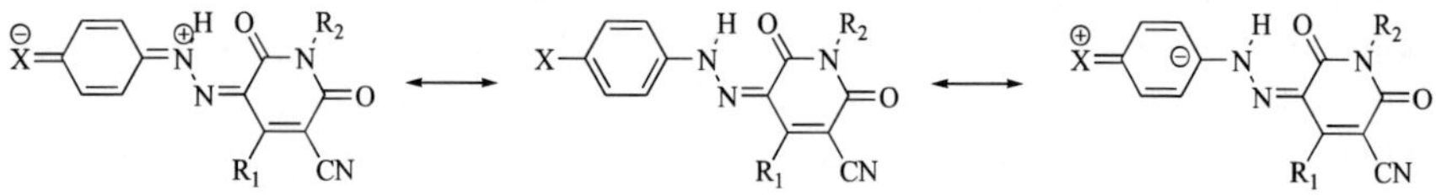

X is electron-accepting group

X is electron-donating group

Scheme 5. Resonance effects of electron-donating and electron-accepting groups in diazo component of the hydrazone form.

3.3. Hydrazone-Anion Equilibrium of APDs

Peng et al. investigated NMR spectra, relationships between structure and pH values, as well as structure and solvent effects on the dissociation of APDs obtained by coupling of 1-alkyl-3-cyano-6-hydroxy-4-methyl-2-pyridones and 1-alkyl-3-amido-6-hydroxy-4-methyl-2-pyridones with thiazoles, thiadiazoles and anilines [56–58]. ^{1}H and ^{13}C NMR were recorded in deuterated chloroform ($CDCl_3$) and deuterated dimethylsulfoxide (DMSO-d_6), and it was confirmed that these dyes exist in the hydrazone form. The peaks for the imino group are observed in the region of 14.3–16.1 ppm. Chemical shifts of the imino group for APDs obtained from 3-cyano-6-hydroxy-1,4-dimethyl-2-pyridone and different anilines are given in Table 1. NH peaks of *ortho*-substituted dyes are shifted towards the larger ppm values than the corresponding *meta-* and *para-* dyes due to the intramolecular hydrogen bond (Figure 3). After adding of piperidine as an organic base in the solutions of $CDCl_3$ and DMSO-d_6, the peak of imino group disappears. The same effect was observed after adding of anhydrous sodium carbonate [56].

Chemical shifts of the corresponding carbon atoms in both $CDCl_3$ and DMSO-d_6 show similar values. After adding of anhydrous sodium carbonate, the chemical shifts of some carbon atoms change (Table 2). The peaks of C1' moves to lower field by 13.9–14.5 ppm, C2' and C6' by 3.3–5.3 ppm, while C3,' C4' and C5' are practically unchanged. Chemical shifts of C2 and C6 have similar values (lower than 1 ppm), while after adding Na_2CO_3 this difference increases to around 6 ppm. C2 moves to lower field for 1.8, 1.9 and 2.1 ppm for X = OCH_3, H and SO_3K, respectively. C6 peaks shifts to the higher field for 3.3, 4.0 and 4.5 ppm, while C3 moves for around 15 ppm indicating an increase of electron density at these atoms.

Table 1. N–H Chemical shifts of APDs derived from 3-cyano-6-hydroxy-1,4-dimethyl-2-pyridone in DMSO-d_6

Substituents of diazo component	δ_{NH}/ppm
4-OCH_3	14.74
2-OCH_3	14.89
4-CH_3	14.53
2-CH_3	14.94
H	14.53
4-Cl	14.30
3-Cl	14.30
2-Cl	15.05
4-NO_2	14.89
2-NO_2	15.51
4-SO_3K	14.55
2-SO_3K	15.23
4-COOK	14.44
2-COOK	16.09
2-SO_3K, 4-OCH_3	15.30

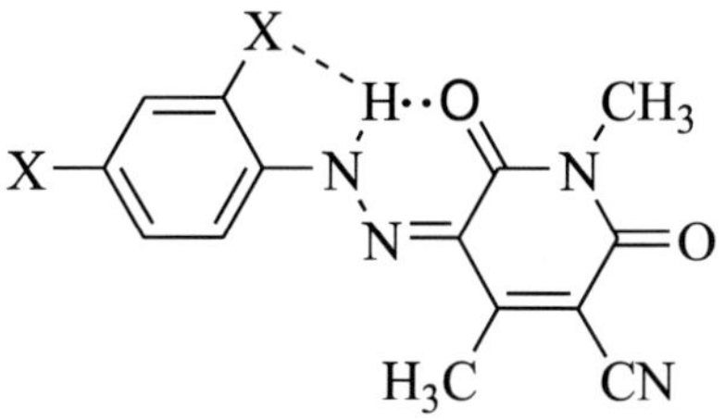

Figure 3. Intramolecular hydrogen bond in *ortho-* substituted APDs.

Therefore, on the basis of ^1H and ^{13}C NMR spectra, it was concluded that in basic solutions acid-base equilibrium exists (Scheme 6). Firstly, it was supposed that azo-hydrazone equilibrium exists, but subsequent work proved that the equilibrium is mainly dissociative process between the hydrazone form and common anion which is resonance structure of the hydrazone (structure IIIa, Scheme 6) and azo anions (structure IIIb, Scheme 6) [57–59]. The equilibrium is related to the structure of the compounds. Dyes with heterocyclic diazo component more easily undergo dissociation when compared with APDs, especially in the case of *ortho-* substituted dyes due to the intramolecular hydrogen bond (Figure 3) [57].

Table 2. ^{13}C NMR chemical shifts of 5-(4-arylazo)-3-cyano-6-hydroxy-1,4-dimethyl-2-pyridones in DMSO-d_6

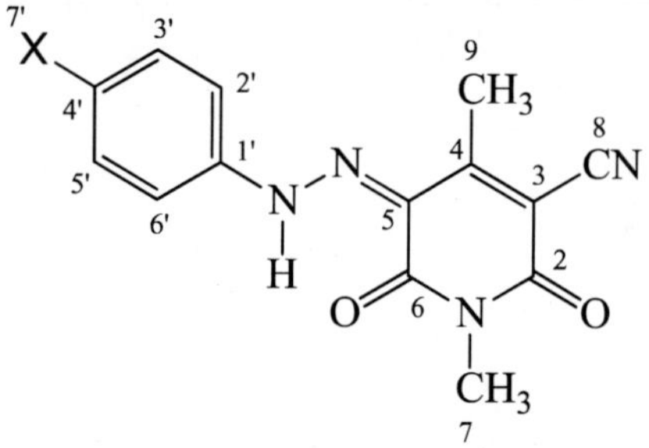

Carbon atoms	Chemical shifts/ppm					
	X = OCH_3		X = H		X = SO_3K	
	DMSO	DMSO+ Na_2CO_3	DMSO	DMSO+ Na_2CO_3	DMSO	DMSO+ Na_2CO_3
2	160.4	162.2	160.3	162.2	160.3	162.4
3	98.8	83.0	97.5	82.8	100.0	83.5
4	157.9	153.8	158.4	157.3	158.8	155.8
5	121.7	123.9	122.6	123.5	122.9	124.5
6	159.7	156.4	159.7	155.7	159.8	155.3
7	25.2	25.4	25.5	25.5	25.5	25.7
8	114.4	119.1	114.5	119.3	114.6	119.2
9	15.5	16.6	15.8	16.8	15.8	17.0
1’	134.1	148.6	140.7	154.6	140.7	154.8
2’	118.6	121.9	116.9	120.6	116.3	120.1
3’	114.8	113.8	129.3	128.4	126.8	126.0
4’	158.4	158.7	126.6	126.7	146.4	145.8
5’	114.8	113.8	129.3	128.4	126.8	126.0
6’	118.6	121.9	116.9	120.6	116.3	120.1
7’	55.2	55.1	–	–	–	–

APDs show different solution color depending on the pH value, indicating the change in a conjugated bond system. This is due to the acidic nature of these compounds and thus they can undergo proton transfer reactions. pH value at which color changes depend on the acidity of the dye. In acidic and neutral medium the hydrazone form of APDs dominates, while in a basic medium the equilibrium is shifted to anion form [57, 60]. The ratio of the hydrazone form with respect to anion form is defined as K_T [57]:

$$K_T = \frac{[Hydrazone]}{[Anion]} \quad (1)$$

The relationship between log K_T and pH is linear and from this dependence it is easy to obtain pK value at which log $K_T = 0$, i.e., the ratio of the hydrazone and anion forms equals. These values are valuable for defining the property of color change in acid-base solutions. Higher pK indicates the dominance of the hydrazone form. The equilibrium is shifted in the favor of the hydrazone form for *ortho-* substituted dyes that can form intramolecular hydrogen bond with imino group and pK values of these APDs are larger than those of *para-* and *meta-* substituted derivatives for 0.7–2.5 pH units. Dyes with thiazole and thiadiazole diazo component are more acidic than APDs due to the electron withdrawing nature of the heterocyclic residue causing the lower values of pK. Values of pK for electron-donating substituents are larger than in the APDs with electron-accepting substituents (Table 3) [57]. Strong electron-accepting substituents in pyridone ring, such as CN and NO_2 which are in *para-* position to oxygen-carrying negative charge, stabilize anion and their pK values are lower than those having amido or H groups in the same position (Table 3) [58].

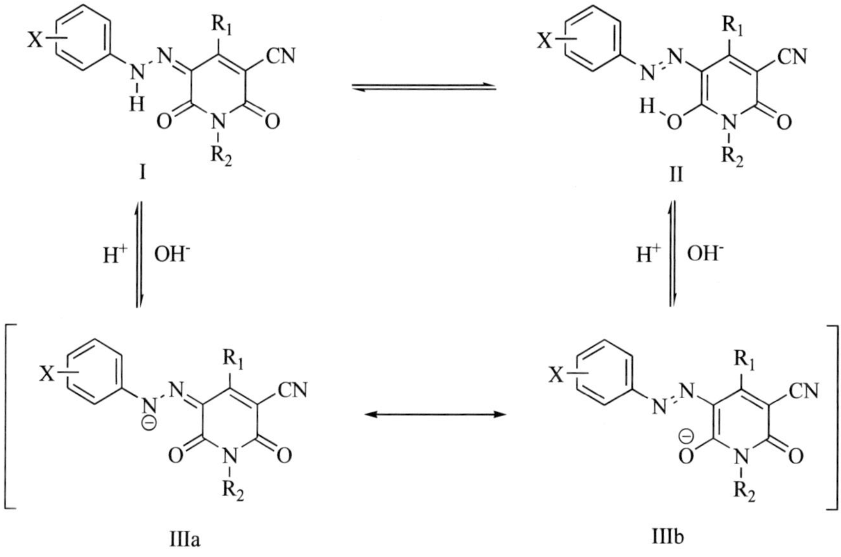

Scheme 6. Azo-hydrazone and acid-base equilibrium of APDs.

Table 3. The influence of different substituents on the hydrazone and anion UV-Vis absorption maxima and on pK values

R_1	R_2	λ_{hyd}/nm	λ_{anion}/nm	pK	R_3	R_4	λ_{hyd}/nm	λ_{anion}/nm	pK
NO_2	H	434	447	8.5	H	H	444	410	11.2
H	NO_2	441	408	9.3	H	Et	443	407	12.0
Cl	H	437	397	8.0	$CONH_2$	H	448	402	10.8
H	Cl	435	401	9.3	$CONH_2$	Et	447	401	11.3
Me	H	444	389	9.1	CN	H	466	393	7.8
H	Me	444	391	9.3	CN	Et	464	396	8.0
OMe	H	462	391	8.8	NO_2	H	462	385	7.9
H	OMe	459	390	11.3	NO_2	Et	460	390	8.8

Table 4. Percentage of the hydrazone form for APDs derived from 3-cyano-6-hydroxy-1,4-dimethyl-2-pyridone in different solvents

Substituent	Solvents					
	Chloroform	Acetone	Methanol	Ethanol	DMSO	DMF
4-OMe	100	98	94	100	71	94
2-OMe	100	100	100	100	99	100
4-Me	100	100	93	99	80	92
2-Me	100	100	99	99	75	95
H	100	99	96	98	79	91
4-Cl	100	85	88	96	46	66
3-Cl	100	97	86	97	58	37
2-Cl	100	100	100	100	75	84
4-NO_2	100	96	83	84	61	44
3-NO_2	100	94	77	88	12	35
2-NO_2	100	100	99	98	88	77
4-CN	100	97	84	95	12	58
3-CN	100	34	70	83	3	29
2-CN	100	95	70	92	14	20

According to UV-Vis spectra, hydrazone-anion equilibrium of azo pyridone dyes exists in various solvents. The thiadiazole azo pyridones easily undergo dissociation and the equilibrium depends on the solvent properties. The hydrazone form is more stabilized in hydrogen bond donor (HBD) solvents, while anion dominates in hydrogen bond acceptor (HBA) solvents. The contribution of anion form increases in the following order: acetic acid < chloroform < benzene < acetone ≈ cyclohexanone < ethanol < DMSO ≈ methanol ≈ DMF ≈ pyridine. For APDs obtained from the corresponding pyridone and different anilines, the hydrazone form is predominant in all solvents (Table 4). *Ortho-* substituted dyes that can form intramolecular hydrogen bond (Cl, NO_2, OMe) have higher contribution of the hydrazone form in HBA solvents with respect to the corresponding *meta-* and *para-* dyes. Methyl and cyano substituted dyes do not show a significant effect on the equilibrium. Dyes with acidic substituents in *ortho-* position (SO_3H, COOH) exist solely in the hydrazone form [58].

UV-Vis maxima of anion form may appear at higher or lower wavelengths with respect to the hydrazone form depending on the diazo and the coupling components [57–59]. In the case of APDs, electron-donating substituents in diazo component destabilize anion form and cause the hypsochromic shift of UV-Vis maxima, while opposite effect is present in the case of the hydrazone form (Figure 4a). Anion form of APDs with NO_2 in *para-* position appears at higher wavelengths due to the electron withdrawing nature of this group which can stabilize negative charge of anion (Table 3) and participate in extended π-delocalization throughout the whole molecule [57, 60]. If the coupling component is regarded, the substituents on the nitrogen of pyridone ring do not influence the position of UV-Vis maxima of the hydrazone and anion form, while substituent in the position 3 shows significant impact opposite to the effect of diazo component (Table 3). In fact, each substituent that increases electron density in a diazo component or decreases it in the coupling residue cause a bathochromic shift of the hydrazone form and hypsochromic shift of anion form, i.e., the greater difference between their UV-Vis maxima. Anion form of pyridone dyes with heterocyclic diazo residues (thiazoles, thiadiazoles, benzothiazoles) appears at higher wavelengths (Figure 4b) [59]. Some authors consider that cooperative intramolecular hydrogen bond in some APDs plays an important role in the stabilization of the hydrazone form, while in anion form, after deprotonation, these bonds are destroyed. Thus, the hydrazone form is more planar and has a higher tinctorial strength [61].

The hydrazone-anion equilibrium is affected by the sample concentration, where the equilibrium shifts towards anion form on decreasing concentration from 10^{-4} M to 10^{-6} M, especially in HBA solvents [48].

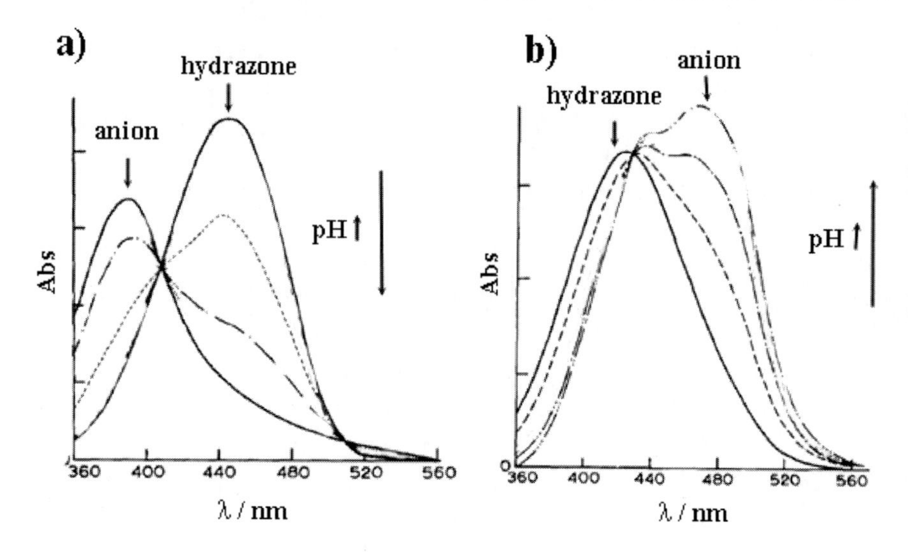

Figure 4. UV-Vis absorption spectra on different pH values. a) 5-(4-methylphenylazo)-3-cyano-6-hydroxy-1,4-dimethyl-2-pyridone; b) 5-(2-ethylthio-1,3,4-thiadiazolyl)-3-cyano-6-hydroxy-1,4-dimethyl-2-pyridone.

4. X-Ray Analysis of APDs

X-ray crystallography represents the powerful and useful tool for investigation of the molecular-level structures, as well as supramolecular interactions. Applied to APDs, this technique reveals that these dyes exist exclusively in the hydrazone form in the solid state [55, 61–67]. X-ray analysis of heteroaryl azo pyridone dye, 5-(2-nitrothiazolylazo)-1-butyl-3-cyano-6-hydroxy-4-methyl-2-pyridone, also, show the existence of the hydrazone form in the solid state [68]. A general characteristic of the crystal structures of these dyes is their planarity with dihedral angles from 0.4–16.9° depending on the functional groups.

The existence of the hydrazone form in the solid state is deduced on the basis of the bond lengths and from the position of the hydrogen atom on the hydrazone nitrogen rather than on azo oxygen. In Table 5, bond lengths of 5-(4-methoxyphenylazo)-3-cyano-1-ethyl-6-hydroxy-4-methyl-2-pyridone (1), C. I. Disperse Yellow 119 and 5-(2-nitrophenylazo)-3-cyano-1-ethyl-6-hydroxy-4-methyl-2-pyridone (2) with ethyl group in the position 1 of pyridone ring are given, while corresponding atoms numbering are given in Figure 5. In the case of 1 [55], bond lengths of C12–O3 and C8–N2 exhibit a double bond character. Another indicator of the hydrazone form of this APD is noticeable difference between the bond lengths of the N–C bonds connecting hydrazone bridge with phenyl and pyridone rings (N1–C5 and N2–C8, respectively). The length of N–N bond is shorter than single bond, but not short enough to be regarded as a double bond, thus indicating extended delocalization of electron density, which along with the intramolecular hydrogen bond, is regarded as the reason of planarity of the molecule. The same conclusions are derived from the crystal structures of the other APDs [61, 65, 66].

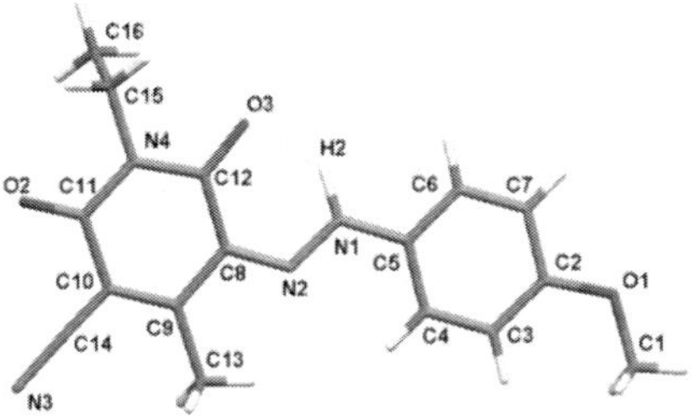

1

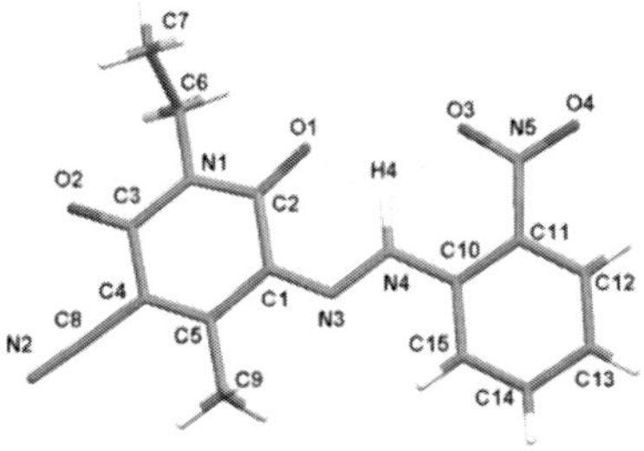

C. I. Disperse Yellow 119

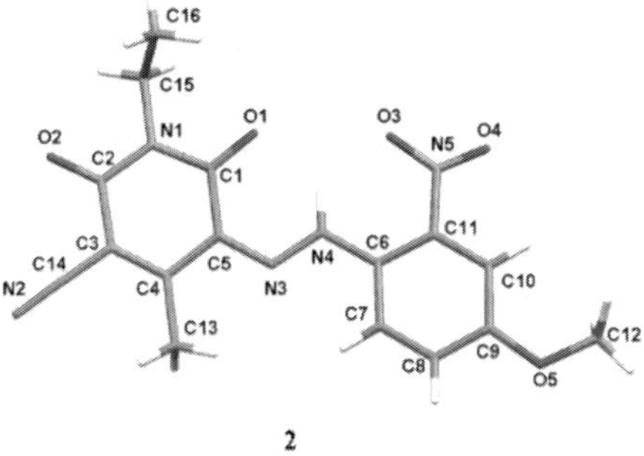

2

Figure 5. Representation of 1, C. I. Disperse Yellow 119 and 2 with atomic-numbering scheme.

Comparing the averaged bond lengths from Table 5 of APDs having the same pyridone backbone, the significant difference is noticed in –C=N–N(H)–C– fragment. Due to the presence of the nitro group in *ortho-* position in the phenyl moiety, shortening of N1–N2 bond and elongations of C5–N1 and N2–C8 are observed for molecule of 1. This is ascribed to the withdrawing nature of the nitro group having the opposite effect to the extended delocalization present in Disperse Yellow 119 and 2, as well as to the presence of additional intramolecular hydrogen bond between N–H and nitro group. The steric hindrance of the nitro group in C.I. Disperse Yellow 119 is responsible for the high dihedral angle between pyridone and phenyl rings (16.9°) when compared to 1 (3.3°).

Table 5. Corresponding bond lengths of 1, C. I. Disperse Yellow 119 and 2

Bond lengths/Å					
1		C. I. Disperse Yellow 119		2	
C5–N1	1.411(3)	C4–N10	1.388(3)	C6–N4	1.397(2)
C8–N2	1.331(3)	N3–C1	1.316(3)	C5–N3	1.322(2)
C8–C9	1.429(4)	C1–C5	1.446(3)	C4–C5	1.447(2)
C8–C12	1.461(4)	C1–C2	1.467(3)	C1–C5	1.468(2)
C9–C10	1.354(4)	C4–C5	1.348(3)	C3–C4	1.354(2)
C10–C11	1.470(4)	C3–C4	1.461(3)	C2–C3	1.463(2)
C11–O2	1.212(3)	O2–C3	1.215(3)	C2–O2	1.217(2)
C11–N4	1.396(3)	C3–N1	1.386(3)	C2–N1	1.398(2)
C12–O3	1.230(3)	C2–O1	1.223(3)	C1–O1	1.229(2)
C12–N4	1.380(3)	C2–N1	1.374(3)	C1–N1	1.382(2)
N1–N2	1.294(3)	N3–N4	1.317(2)	N3–N4	1.313(2)

For all APD structures, strong intramolecular hydrogen bond N–H···O between the hydrazone hydrogen and carbonyl oxygen are observed. These hydrogen bonds form the pseudo six-membered rings and thus stabilize the hydrazone form (Table 6). For APDs bearing nitro group in *ortho-* position of the phenyl ring (Disperse Yellow 119 and 2), bifurcated hydrogen bond is observed between NH hydrazone and oxygen of the carbonyl group, on one side, and oxygen of nitro group, on the other side, thus forming two adjacent six-membered pseudo rings additionally stabilizing the hydrazone form [61, 66].

Table 6. Intramolecular hydrogen bonding parameters for some APDs

Compound	D–H···A	*d*(D–H) /Å	*d*(H···A) /Å	*d*(D···A) /Å	∠(DHA) /°
1	N1–H2···O3	0.97	1.82	2.602(3)	136
C. I. Disperse Yellow 119	N4–H4···O1	0.86	1.93	2.582(3)	132
	N4–H4···O3	0.86	2.00	2.614(3)	127
2	N4–H4···O1	0.90	1.92	2.590(2)	132
	N4–H4···O3	0.88	2.00	2.620(2)	126

X-ray crystallography is suitable for conformational analysis of APDs. For this purpose chlorine disubstituted phenyl APDs are considered [67]. *Cis* configuration is defined for the same orientation of *N*-alkyl chain (methyl or ethyl) and chlorine in *meta-* position of the phenyl ring, while *trans* configuration is present when these groups are oriented on the opposite sides (Figure 6). X-ray analysis reveals that 3 (C.I. Disperse Yellow 241) exists in two crystallographically independent molecules in the asymmetric unit with equal amount of *cis* and *trans* isomers, while 4 exists as *cis* isomer due to the less crowding of methyl group in this form. DFT calculations show that the hydrazone form is more thermally stable than the corresponding azo form for about 150 kJ mol^{-1}. *Trans* isomers are more stable than *cis* isomers for several to ten of kJ mol^{-1}.

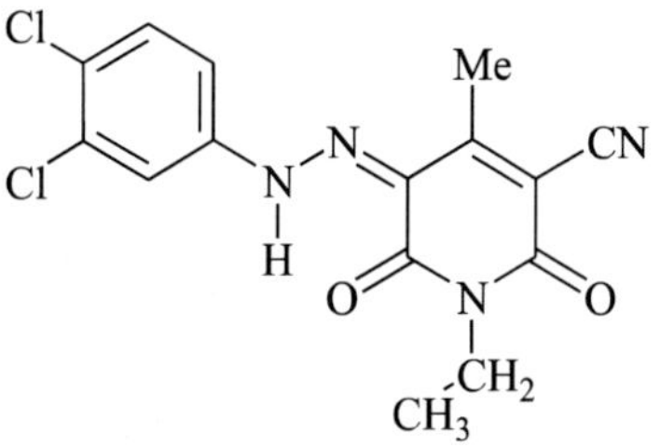

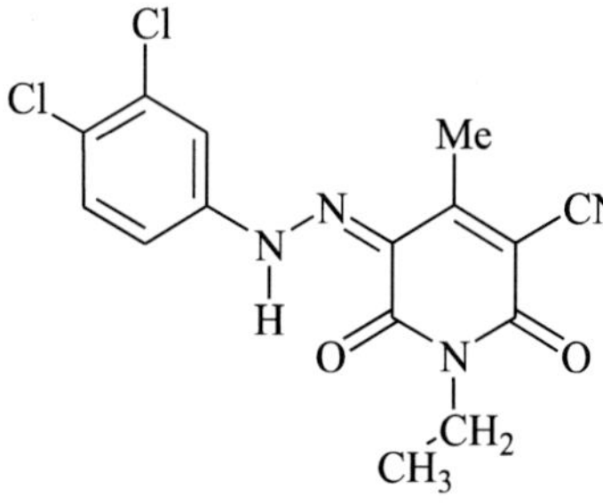

Figure 6. *Cis* and *trans* chlorine disubstituited APDs analyzed by X-ray crystallography.

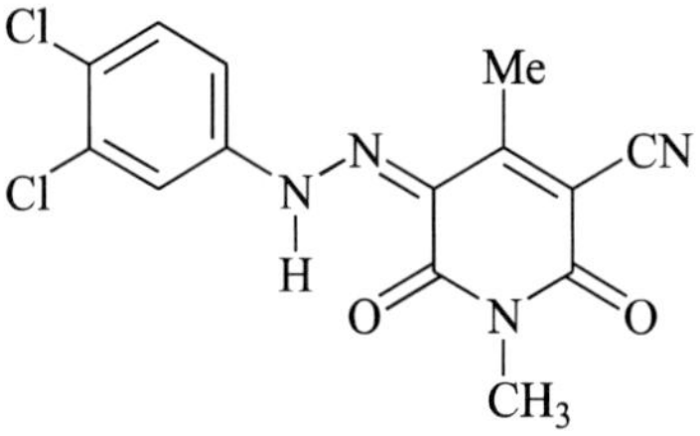

Hydrogen bonds, including C–H···O contacts, and π-π stacking interactions are responsible for the crystal packing of APDs. Different packing arrangements are observed with the subtle change of functional groups in phenyl rings [66]. Also, the different packing modes are result of the alteration of the *N*-alkyl chains. On the basis of these weak intramolecular contacts in the molecular packing of the APDs, interactions with other molecules could be predicted as well as with different substrates during the dying processes.

5. Quantum Chemical Calculations of APDs

Quantum chemical calculations are suitable for the investigation of geometry and electronic structure of the molecules. Accurately determined geometry is essential for the prediction of the physico-chemical properties, which further enable planning of modeling and synthesis of novel molecules with potential usage in various fields of industry. Density function theory (DFT) applied to APDs, so far, includes geometry optimization, vibrational and NMR analysis, electronic, NBO (Natural Bond Orbital) and MEP (Molecular Electrostatic Potential) analysis. DFT studies of APDs are often combined with experimental data. Detail theoretical study performed on 1 shows that this dye exists in the hydrazone form in the solid state [55]. The optimized geometry by B3LYP/6-311++G(d,p) method show 24 isomers, i.e., four tautomers with the corresponding number of rotational conformers. The most stable geometries belong to the hydrazone and azo form. DFT calculations of 5-arylazo-3-cyano-6-hydroxy-4-phenyl-2-pyridones show the same results [50]. In the solid state, only the hydrazone form is present, while azo form is present no more than 0.0002%. The correlations of the theoretical with experimental bond lengths show excellent agreement ($R^2 > 0.985$). Calculated harmonic vibrational frequencies of 1 are compared with experimental values (Table 7). The broad band at 3433 cm^{-1} in FT-IR spectrum assigned to N–H stretching vibration shows deviation of 273 cm^{-1} with calculated frequency value at 3160 cm^{-1}, due to the presence of the intramolecular hydrogen bond. While the C=O and C=N stretching vibrations are observed at 1672, 1627 and 1398 cm^{-1}, respectively, the calculated values are obtained at 1676, 1612 and 1386 cm^{-1}, respectively, i.e., a very good agreement is achieved between experimental and calculated data [55].

Experimental and calculated values of NMR chemical shifts of 1 are given in Table 8. The ^{1}H NMR spectrum of 1 shows characteristic singlet of methyl group in the position 4 of pyridone ring at 2.5 ppm, while calculated values are in the range 2.61–2.81 ppm. Two signals, triplet and quartet, at 1.13 and 3.80 ppm in experimental spectrum are ascribed to ethyl hydrogens, while corresponding calculated values are at 1.26–1.32 and 4.17–4.33 ppm. Characteristic doublets of aromatic hydrogen atoms appear at 7.06 and 7.71

ppm, and correlate good with theoretical values. Experimental and calculated values of N–H hydrogen are positioned at 14.81 and 14.72 ppm, respectively. In the ^{13}C NMR spectra, two carbon atoms of carbonyl group resonate at 161.2 and 161.8 ppm and they are in agreement with calculated values (172.49 and 169.59 ppm). The carbon atom of C=N group has a peak at 122.9 ppm, while corresponding calculated peak is at 129.39 ppm. These results show that both experiment and DFT calculations confirm the existence of the hydrazone form in DMSO-d_6 [55].

DFT calculations are performed in order to compare the energy differences between single-crystal structure and energy-minimized structures of azo and the hydrazone form for different APDs [55, 65, 66]. The energies and dipole moments of these structures for four commercial dyes are given in Table 9. There are large energy gaps between the single-crystal structures and energy-minimized structures for the same hydrazone form which is explained as energy compensation for the formation of the hydrogen bonds and stacking

Table 7. Experimental and calculated FT-IR bands of 1 with graphical representation of both spectra

Exp.	Calc.	Assignment[a]
3433	3160	ν N–H
3067	3109	ν C—H (arom.)
2980	3046	ν C—H (alip.)
2223	2250	ν C≡N
1672	1676	ν C2=O
1627	1612	ν C6=O
1533	1554	ν N-N
1583	1519	β N–H
1506	1490	β C—H (arom.)
1398	1386	ν C = N
1303	1288	β C—C
914	918	β N–N
897	910	γ C—H (arom.)
839	842	γ N–H
689	749	γ C2=O
684	745	γ C6=O
628	694	γ C—C
459	457	β C≡N
419	414	γ C=N

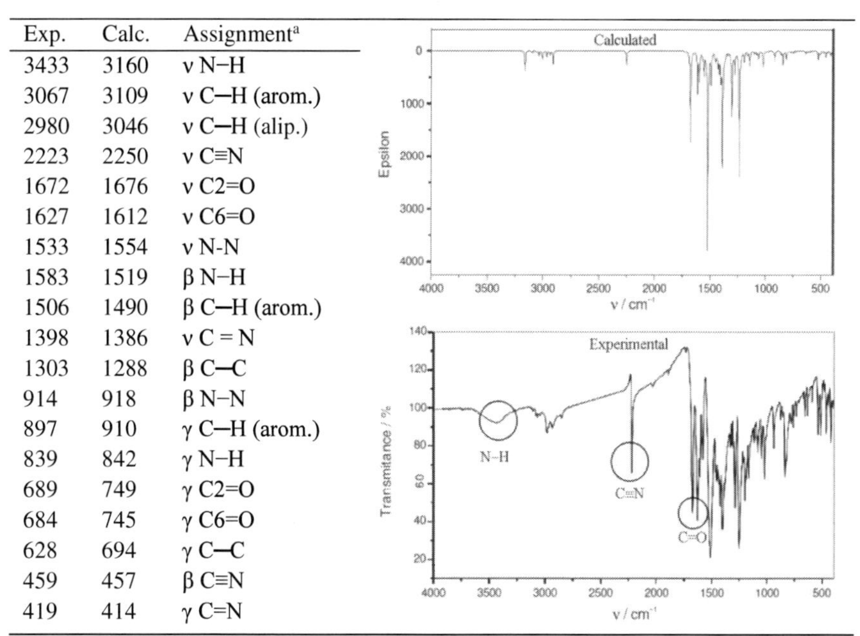

[a] ν – stretching vibrations, β – in plane bending vibrations, γ – out of plane bending vibrations.

Table 8. Experimental and calculated chemical shifts (^1H and ^{13}C NMR) of 1 in DMSO-d_6

^1H NMR δ/ppm			^{13}C NMR δ/ppm		
Exp.	Calc.	Assignments	Exp.	Calc.	Assignments
1.13	1.26	CH_3CH_2	12.9	13.9	CH_3CH_2
	1.28		16.6	19.85	CH_3
	1.32		34.5	39.6	CH_3CH_2
2.50	2.61	CH_3	55.8	58.23	OCH_3
	2.76		99.6	101.23	Py
	2.81		115.3	126.14	CN
3.80	3.96	OCH_3	115.4	117.32	Ar
	4.03		115.4	122.94	Ar
	4.15		119.5	124.8	Ar
3.89	4.17	CH_3CH_2	119.5	128.93	Ar
	4.33		122.9	129.39	C=N(Py)
7.06	7.35	Ar–H	134.8	142.05	Ar
	7.41		158.9	169.59	Ar
7.71	7.67		160.6	168.79	Py
	8.38		161.2	169.59	Py
14.81	14.72	N–H	161.8	172.49	Py

interactions between the molecules. It is obvious that the hydrazone form is thermally more stable than azo tautomer which is theoretical support for crystallization of the APDs in the hydrazone form. The ground state optimization of the dyes gives similar dipole moments for the single-crystal hydrazone form to the energy-minimized hydrazone, while higher than the energy-minimized azo forms (excluding C.I. Disperse Yellow 126). These findings are also confirmed for 1, and *meta-* and *ortho-* methoxy derivatives of 1. When all three derivatives (*ortho-* , *meta-* and *para-*) are considered, the largest energy gap between the hydrazone and azo forms is observed for *ortho-*substituted dye due to the additional stabilization of the hydrazone form through the intramolecular bond. Generally, both experimental and theoretical studies of APDs confirm the existence of the hydrazone form in the solid state. NBO performed on 1 confirms the existence of the hydrazone form in the solid state as well as intramolecular hydrogen bond causing stabilization of this tautomer [55]. NBO analysis also indicates that there is ICT from methoxy group as electron-donor to cyano group as electron-acceptor.

Table 9. Energy and dipole moments of the commercial APDs for single-crystal structure and theoretical structures

APDs		Single-crystal hydrazone	Energy-minimized structure hydrazone	Energy-minimized structure azo
C. I. Disperse Yellow 114	E (kJ mol^{-1})	–3,608,767.2	–3,609,880.7	–3,609,727.2
	$\mu \times 10^{-29}$ (C m)	3.8	3.8	2.6
C. I. Disperse Yellow119	E (kJ mol^{-1})	–3,024,979.5	–3,025,618.4	–3,025,450.3
	$\mu \times 10^{-29}$ (C m)	2.6	2.6	1.9
C. I. Disperse Yellow 126	E (kJ mol^{-1})	–3,891,007.6	–3,893,616.8	–3,893,458.9
	$\mu \times 10^{-29}$ (C m)	3.3	3.1	3.5
C. I. Disperse Yellow 211	E (kJ mol^{-1})	–3,062,610.5	–3,063,247.9	–3,063,082.1
	$\mu \times 10^{-29}$ (C m)	1.7	1.7	1.5

DFT calculations of 5-(4-arylazo)-3-cyano-6-hydroxy-4-methyl-2-pyridone show that the hydrazone form is more stable than azo form in the gas phase for 5.5–6.9 kcal mol^{-1} [60]. The theoretical UV-Vis absorption maxima positions of the hydrazone and anion form of these APDs obtained by time dependent density functional theory (TD-DFT) are given in Table 10. The results show that introduction of strong electron-withdrawing groups in phenyl ring causes higher stability of anion form which is in accordance with published experimental data [57]. Structural analysis of the hydrazone, azo and anion forms in gas phase reveals high planarity of all molecules enabling conjugation throughout the whole molecules. The bond lengths between substituent and phenyl ring are similar in azo and the hydrazone from, but differs in anion form due to the different participation of π-conjugation in these forms. In APDs with electron-donating groups, this bond length is longer in anion form due to the negative charge on oxygen atom in pyridone ring that weakens the electron-donating power of substituents, which results in decreased conjugation between substituent and phenyl ring, and thus in lengthening of this bond. In APDs with electron-accepting groups, the conjugation between substituent and phenyl ring is increased and thus bond length shortens in anion form. According to the theoretical calculations, C–N bond, connecting azo bridge and pyridone ring, is the most sensitive to electronic effects of the substituents in all three forms, indicating the stronger impact on conjugation between N–N and pyridone ring than on conjugation between N–N and benzene ring. When considering these three forms, analysis of sensitivity of the bonds reveals that C–N–N–C fragment in anion form is more affected by the nature of the substituents than in azo and the hydrazone form [60].

Table 10. UV-Vis spectral data of the hydrazone and anion form of 5-(4-arylazo)-3-cyano-6-hydroxy-4-methyl-2-pyridones obtained using TD-DFT in aqueous solution

Substituent	λ_{max}, hydrazone form/nm	λ_{max}, anion form/nm
OMe	397	367
OH	400	367
Me	386	366
H	377	365
Cl	380	370
Br	381	372
COOH	367	386
Ac	380	389
SO_3H	369	384
CN	375	386
NO_2	378	415

Theoretical analysis of 5-(4- and 3-substituted phenylazo)-3-cyano-6-hydroxy-4-methyl-1-phenyl-2-pyridones shows that these dyes exist in the hydrazone form [69]. The difference between energies of azo and the hydrazone tautomers is around 95 kJ mol^{-1}, indicating strong intramolecular hydrogen bond in the hydrazone form. The existence of the hydrogen bond is confirmed by structural and topology analysis. The introduction of either electron-donating or electron-accepting substituents causes the shortening of the hydrogen bond lengths. Electron-donating and electron-accepting groups have different effects on the geometry of the molecules. Donors support ICT from phenyl group to pyridone moiety causing the high planarity of the molecule and increase of the N–N bond length and C–N bond connecting the hydrazone unit with pyridone ring. This is supported by X-ray analysis of 1. The low deviation from planarity and increase of the N–N and C–N bonds are observed for molecules with electron-accepting groups in phenyl ring indicating two opposite electron accepting effects in the molecule: electron-accepting phenylazo group and pyridone core cause appropriate geometrical adjustment as a response to electronic demand of the electron deficient environment. MEP analysis is used for determining and visualization of the charge distribution over the 5-arylazo-3-cyano-6-hydroxy-4-methyl-1-phenyl-2-pyridones. According to this study, negative regions, i.e., regions of high electron density favorable for electrophilic attack, are localized over the cyano and carbonyl groups in pyridone ring as well as over the nitro substituted

APD. The regions of low electron density favorable for nucleophilic attack are localized over the phenyl and pyridone rings. The investigation of highest occupied molecular orbitals (HOMO) and lowest unoccupied molecular orbitals (LUMO) results that variation of the different substituents highly affects the ICT mechanism of these APDs [69]. The inspection of the HOMO and LUMO orbitals show that phenyl group in position 1 do not participate in absorption process. This is also obvious from the geometry of the molecules where phenyl ring deviates from planarity and thus do not take a part in ICT. It should be emphasized that both experimental data and DFT calculations support the fact that *N*-substituents of the pyridone ring do not affect the positions of the UV-Vis absorption maxima of APDs. Different substituents in the position 1 of pyridone ring influence different physical properties of these compounds, such as solubility, melting points etc.

6. Solvent Effects: LSER Analysis of APDs

For quantitative assessment of the solvent effects on the UV-Vis maxima of APDs, linear solvation energy relationship is often applied using Kamlet-Taft and Catalán solvent parameters. Position, shape and intensity of UV-Vis absorption bands change in the solvents of different polarity as a result of differential solvation in the ground and excited states of APDs. This phenomenon is known as solvatochromism. The extent of solvation depends on intermolecular solute-solvent interactions which include nonspecific and specific forces. Nonspecific forces comprise purely electrostatic interactions (ion/ion, ion/dipole, dipole/dipole) and polarization interactions (ion/induced dipole, dipole/induced dipole, dispersion forces). Specific interactions include HBD and HBA interactions. The degree and direction of solvatochromism depend on charge distribution, and consequently dipole moments, of the ground and the electronically excited state of the molecule. If ground state is more dipolar than the excited state, than increase of solvent polarity stabilize more the former state leading to a hypsochromic shift. Conversely, if the ground state is less dipolar than excited state a bathochromic shift is observed with increasing solvent polarity [70–72].

Correlation of UV-Vis absorption frequencies of the hydrazone tautomer of 5-(4-arylazo)-3-cyano-6-hydroxy-4-(4-methoxyphenyl)-2-pyridones with dispersive interaction function, $f(n)$ given as follows:

$$f(n) = \frac{n^2 - 1}{n^2 + 1} \quad (2)$$

where n is refractive index of the solvent, proves that dispersion forces strongly affect the position of the absorption maxima. Also, the plot of the absorption frequencies of these APDs against relative permittivity, ε_r, shows deviation from linearity indicating that specific solvent/solute interactions, besides relative permittivity, play an important role in solvatochromisim of these dyes [49].

Kamlet and Taft have developed multiparameter equation in order to describe manifold solvent/solute interactions [73]. Applied to UV-Vis absorption frequencies, ν_{max}, Kamlet-Taft equation is given as follows (3):

$$\nu_{max} = \nu_0 + a\alpha + b\beta + s\pi^* \quad (3)$$

Each parameter corresponds to a certain type of the interaction of the overall solvation capability. The Kamlet-Taft π^* parameter denotes dipolarity/polarizability of the solvent, α and β describe solvent hydrogen bond donor capacity (HBD) and hydrogen bond acceptor capacity (HBA), respectively. Solvent-independent correlation coefficients a, b and s reflect individual contribution of solvent effects on the UV-Vis shifts ν_{max}. Coefficient ν_0 is the absorption frequency of the solute in the reference system.

Kamlet-Taft equation applied on 5-arylazo-3-cyano-6-hydroxy-4-methyl-2-pyridones and 5-arylazo-3-cyano-6-hydroxy-4-phenyl-2-pyridones show that the solvent effects on UV-Vis absorption spectra are very complex and strongly dependable on the nature of the substituent on the phenyl nucleus. The electronic behavior of the nitrogen of the hydrazone group is influenced by electron-donating and electron-accepting ability of the substituents as a consequence of different conjugational or migrating ability of the electron lone pairs in the presence of the different groups. Most of the solvatochromism of APDs with electron-donating groups is due to solvent basicity and acidity rather than on the solvent polarity. On the other hand, the effect of electron attracting substituents show that dominant solvent effect on these dyes is solvent polarity rather than solvent acidity and basicity due to the positive charge on nitrogen atom in the hydrazone tautomer (Scheme 5), and stabilization of this form rather due to non-specific solute–solvent interactions than through HBD and HBA properties [48, 50].

For APDs derived from 4-(4-nitrophenyl)-3-cyano-6-hydroxy-2-pyridones, 4-(4-methoxyphenyl)-3-cyano-6-hydroxy-2-pyridones and 3-cyano-

6-hydroxy-1-(2-hydroxyethyl)-4-methyl-2-pyridones, analysis by Kamlet-Taft equation gives negative sign of independent coefficients *s* and *a,* indicating bathochromic shift with increasing of solvent dipolarity/polarizabilty and HBD ability of the solvent, respectively, suggesting stabilization of the excited state relative to the ground state. The coefficient *b* is positive for these APDs indicating hypsochromic shift with increasing of the HBA ability of the solvent. The percentage contribution of solvatochromic parameters shows that most of the solvatochromism is due to nonspecific interactions [23, 49, 51].

The effects of solvent polarity of 5-(3- and 4-substituted phenylazo)-3-cyano-4,6-dimethyl-2-pyridones exhibiting keto-enol tautomerism are analyzed by Kamlet-Taft solvent parameters. Regression results give negative signs for *s* and *b* and positive for *a*, indicating bathochromic shift with increase of solvent polarity and HBA capability, and hypsochromic shift with increase of HBD capability of the solvent, respectively. The percentage solvatochromic contribution for APDs with electron-donating and moderate electron-accepting substituents shows dominant effects of solvent dipolarity/polarizability and basicity over the solvent acidity which is in accordance with resonance structures of keto form (Scheme 2). For electron acceptors in phenyl ring, solvatochromism is mainly affected by solvent dipolarity/polarizability rather than solvent basicity and acidity. This is due to the positive charge on azo group in the dyes with strong electron-accepting, and to the decrease of hydrogen-bond accepting solvent effect when more dipolar tautomeric form $HN^{+}–CO^{-}$ transforms in then N=C–OH tautomeric form (keto-enol tautomerism) [43].

Although Kamlet-Taft approach has a considerable success in quantitative interpretation of solvation effects, there are some drawbacks of its application. Firstly, π^* empirical scale represents combination of solvent dipolarity and polarizability. Secondly, the empirical solvent parameter scales are obtained as average experimental values of numerous solvatochromic probes. Catalán et al. have introduced four empirical solvent parameter scales (Eq. (4)) [74]. The novelty of this approach is splitting of general interactions into polarizability (*SP*) and dipolarity (*SdP*) terms. Scales *SA* and *SB* correspond to Kamlet-Taft parameters α and β.

$$\nu_{max} = \nu_0 + aSA + bSB + cSP + dSdP \tag{4}$$

Applied to APDs, Catalán model provides better regressions taking into account number of the included solvents and correlation coefficients when compared to Kamlet-Taft model. Also, general conclusion is that the solvent polarizability is the most important solvent effect affecting solvatochromism of APDs. The sign of this parameter (*c*) is always negative indicating bathochromic effect with increasing of the solvent polarizability. HBD and HBA abilities have moderate impact, while dipolarity of the solvent has almost negligible impact on the position of the UV-Vis absorption maxima [23, 49, 51]. Graphical representations of correlations obtained by Kamlet-Taft and Catalán models for the series of 5-(4-arylazo)-3-cyano-6-hydroxy-1-(2-hydroxyethyl)-4-methyl-2-pyridones is given in Figure 7.

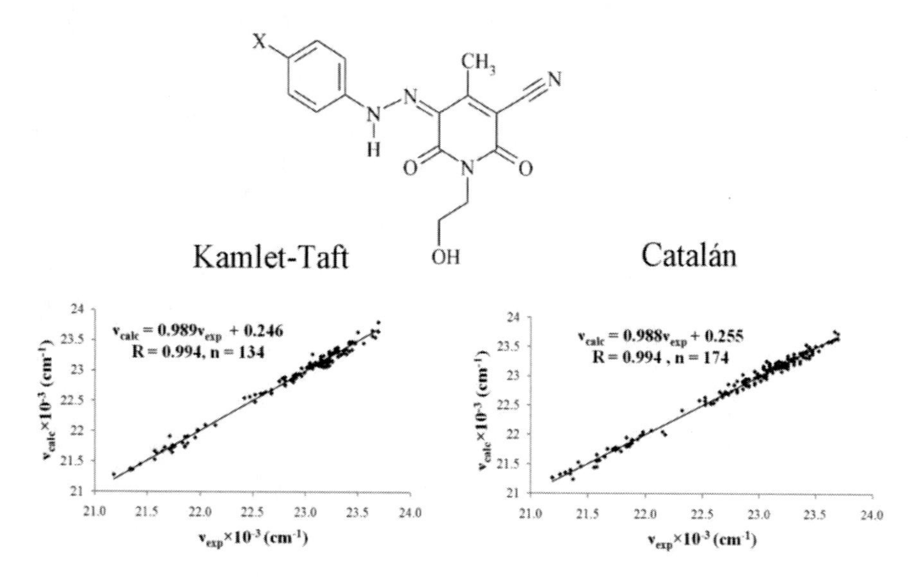

Figure 7. Experimental vs. calculated values of ν_{max} from the Eq. (3) and (4) obtained for investigated APDs (X = OH, MeO, Me, H, Cl, Br, $COCH_3$, COOH, CN, NO_2; n denotes number of the solvents included in the correlation: R is correlation coefficient).

7. Substituent Effects: LFER Analysis of APDs

The transmission mode of the electronic substituent effects of APDs is often interpreted using linear free energy relationships by correlating spectral data with Hammett substituent constants given as following equation:

$$s = \rho\cdot\sigma + h \qquad (5)$$

where s is substituent dependent value, ρ is the proportionality constant reflecting the sensitivity of the physical property to the substituent effect and h is intercept. Substituent constant σ measures the electronic effect of the substituent, σ_p corresponds to the substituents in *para-* position to the reaction site and σ_p^+ (electrophilic constant) to the substituent in *para-* position that conjugate with the electron-withdrawing reaction site [75–78].

Applied to the absorption frequencies of different APDs, Hammett equation gives a better correlations with σ_p^+ than with σ_p constants indicating extensive delocalization in aryl azo group [23, 48, 50, 51]. For 5-(4-arylazo)-3-cyano-6-hydroxy-1-(2-hydroxyethyl)-4-methyl-2-pyridones, the highest values of proportionality constant are obtained for electron-donating substituents in highly dipolar solvents (DMF, DMA, DMSO, formamide) indicating the significant impact on ICT process, while for electron-accepting substituents, ρ is negative implying counteracting effect of these groups to the extended delocalization. The increase of the polarity of the protic solvents cause higher sensibility of substituents on electronic transmission, while in aprotic solvents opposite trend is observed [23].

Since the ρ value in Eq. (5) of the dyes obtained from 4-(4-nitrophenyl)-3-cyano-6-hydroxy-2-pyridones is sensitive not only the solvent polar properties but also to HBA and HBD capabilities of the solvents, this parameter is correlated with solvent parameters given in Eq. (3) and (4). The results show that increase of the solvent dipolarity/polarizability cause the decrease of the substituents' influence on absorption maxima, while this influence increases with increase of the HBA and HBD abilities of the solvent [51].

Moreover, mono and dual substituent parameter (MSP and DSP) LFER models can be used for correlation of ^1H and ^{13}C NMR substituent chemical shifts (SCS) since the NMR shifts are highly sensitive to electron charge distribution [78]. MSP model includes Eq. (5), while DSP are extended Hammett (6) [79] and Swain-Lupton (7) equations [80]:

$$SCS = \rho_I \cdot \sigma_I + \rho_R \cdot \sigma_R + h \tag{6}$$

$$SCS = f \cdot F + r \cdot R + h \tag{7}$$

where σ_I and σ_R in Eq. (6) represent polar and resonance electronic effects, F and R in Eq. (7) are substituent parameters reflecting the field and resonance electronic effects. Values of ρ_I, ρ_R, f and r denotes sensitivity of the SCS to the corresponding substituent electronic effect. Parameter h is the intercept and describes the unsubstituted dye.

Table 11. [1]H and ^{13}C SCS data of 5-(4-arylazo)-3-cyano-6-hydroxy-1-(2-hydroxyethyl)-4-methyl-2-pyridones with respect to unsubstituted parent compound (H) in DMSO-d_6

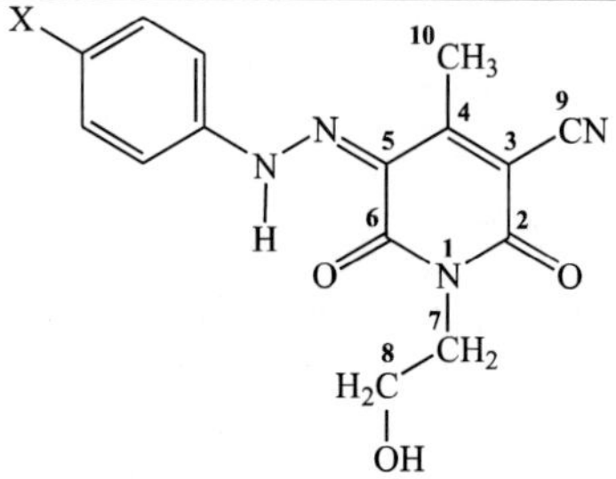

X	N–H	C2	C3	C5	C6	C7	C9
OH	–0.3	–0.1	2	1.2	–0.2	0.1	–0.5
OMe	–0.21	0	1.4	0.9	–0.2	0.1	–0.3
Me	–0.05	0	0.6	0.4	–0.1	0	–0.1
H	14.54	160.9	100.7	123.1	160.3	41.6	115.3
Cl	0.11	0.2	–0.1	–0.6	0	0.1	0.2
Br	0.13	0.2	–0.6	–0.5	0	0	0.1
CH_3CO	0.13	0.3	–1.4	–1.4	0.1	–0.1	0.3
COOH	0.15	0.2	–1.3	–1.2	0.1	–0.1	0.2
CN	0.26	0.4	–2.1	–1.8	0.2	–0.2	0.4
NO_2	0.22	0.4	–2.6	–2.5	0.2	–0.2	0.5

SCS data of N–H, C2, C3, C5, C6, C7 and C9 atoms for a series of 5-(4-arylazo)-3-cyano-6-hydroxy-1-(2-hydroxyethyl)-4-methyl-2-pyridones are given in Table 11 [23]. Electron-donating substituents in *para-* position of the phenyl ring cause the increase of electron density at N–H (upfield shift), while electron-accepting influence the decrease of electron density at this atom (downfield shift). The investigation of the substituent effect on ^{13}C SCSs show that electron-accepting substituents cause upfield shift of C3, C5 and C7 peaks, while C2, C6 and C9 are moved downfield. Electron-donating groups have opposite effect on these SCSs, excluding C2 shifts where peaks are almost unaltered by electron-donating properties of the substituents.

Application of LFER analysis to SCSs presented in Table 11 using Eqs. (5)–(7) yields excellent results indicating that the substituent impacts on the

NMR chemical shifts originate from electronic effects [23]. From the values of ρ_I and ρ_R in Eq. (6) and *f* and *r* in Eq. (7), it is obtained that resonance effect is dominant at N–H hydrazone and all carbons relative to the inductive and field electronic effects, respectively. The reverse effect is observed for the carbonyl C atom at the position 2 of the pyridone ring indicating that this atom does not participate in ICT. The most sensitive carbon atoms to electronic substituent effects are carbons in the position 3 and 5 of pyridone ring. Also, carbon atom of cyano group is also affected by the nature of the substituents. On the other hand, the lowest sensitivity is observed for methylene group of *N*-ethylhydroxy group. Moderate substituent effects are noted for carbonyl atoms. This study supports the transmission mode given in the Scheme 4.

8. METAL COMPLEXES OF APDS

Metal complexes of pyridone dyes are used in design of optical recording medium having recording layer with improved light stability capable of recording and regeneration of high-density optical information by short-wavelength laser beams. These layers consist of Ni, Co, Fe, Zn, Cu and Mn complexes of the dyes obtained from 6-hydroxy-2-pyridone structure as coupling component and isoxazole, 1,2,4-triazole or pyrazole rings as diazo components [81].

APD molecules are usually used as ligands in complexation with chromium, copper and nickel [53, 61, 67, 82, 83]. The purple chromium complexes of APDs obtained from *N*-substituted-3-cyano-6-hydroxy-4-methyl-2-pyridones and 5-substituted-2-hydroxyanilines are obtained by refluxing the mixture of ammonium chromium sulfate, sodium acetate and the appropriate APD in formamide solution. These complexes show improved stability to washing and light with respect to the pure APDs [82]. Elemental analysis show that stoichiometry corresponds to the ligand: metal ratio of 2:1. FT-IR spectra of chromium complexes show lower frequency of the cyano group vibrations when compared to FT-IR spectra of APDs. Only one strong carbonyl peak in the region 1639–1670 cm^{-1} is observed, while the peak of the second carbonyl peak in spectra of APDs disappears due to the metal ion-carbonyl oxygen bonding. UV-Vis spectra of chromium complexes show bathochromic shift of 40–55 nm in DMF. The exception is observed for nitro derivatives, in which case a hypsochromic shift is observed of 40–70 nm as a consequence of the influence of the solvent polarity on the ionic system of the complexes.

Table 12. UV-Vis absorption maxima of the dyes and complexes derived from 3-cyano-6-hydroxy-4-methyl-2-pyridones and values of log β in methanol

Substituent in diazo component	λ_{dye}/nm	$\lambda_{complex}/nm$	log β
H	452	438	10.92
4-OEt	488	451	11.15
4-Me	462	437	11.25
4-Cl	461	443	10.39
4-COOH	461	458	10.80
4-NO_2	451	461	10.43
3-OEt	466	439	11.44
3-Me	463	435	10.91
3-Cl	460	437	10.87
3-COOH	456	439	10.54
3-NO_2	447	442	10.36

Copper (II) complexes of APDs derived from 3-cyano-6-hydroxy-4-methyl-2-pyridones and *para*- and *meta*- substituted anilines are obtained in methanol [53]. FT-IR spectra of complexes show lower frequencies of cyano groups with respect to the pure APDs, one carbonyl vibration and new bands appearing from C–O and Cu–O vibrations in the range of 1281–1255 cm^{-1} and 476–421 cm^{-1}, respectively. UV-Vis absorption maxima of these complexes are shifted hypsochromically with respect to the pure APDs (Table 12), except for *p*-NO_2 derivative. The equilibrium of the complex formation is given as:

$$M + 2L = ML_2 \tag{8}$$

while overall formation constant β is given as:

$$\beta = \frac{[ML_2]}{[M][L]^2} \tag{9}$$

where [M], [L] and [ML_2] are molar equilibrium concentrations of metal ion, ligand APD and complex, respectively. The formation constants are evaluated spectrophotometrically in methanol (Table 12) [53]. The values of log β of chlorine, carboxy and nitro derivatives are lower than in the case of unsubstituted complex (H) since these groups reduce the basicity of donor atoms (N and O) and thus decrease the chelating ability of the ligand.

Introducing donors in diazo component results higher log β values and thus more stable complexes are formed.

The structures of some complexes of APDs are confirmed by X-ray single crystal diffraction studies [61, 67]. The neutral dinuclear copper complex $Cu_2(L_2)_4$ is easily produced by mixing $Cu_2(CH_3COO)_2$ and ligand L_2 (compound 2) in the absence of base, where the deprotonation of ligand happens and each Cu^{2+} is countered by two dianionic ligands and adopts six-coordinate elongated octahedral configuration (Figure 8). The basal coordination takes place over the nitrogen atoms of azo and cyano groups (N2 and N7) and over two phenol oxygen atoms (O4 and O9). Basal Cu–O and Cu–N bond lengths are in the range 1.921(2)–1.970(4) Å. The apical positions belong to oxygen of nitro group (O1) and nitrogen atom of cyano group (N5) with much longer Cu–O and Cu–N bond lengths: 2.538(4) and 2.641(5) Å, respectively. There are two types of coordination modes for four ligands, where two side ligands serve as bidentate capping ligands and the middle ones act as quadridentate bridging ligands connecting adjacent Cu^{2+} in reverse modes. Due to the fixation of coordinative bonds and geometrical requirements of the central copper polyhedron, the ligands are no more planar with dihedral angles between pyridone and phenyl rings of 22.8° and 26.4°, while in ligand this value is 6.9°. It is interesting that N–N and C–C (between carbon of carbonyl group and carbon bonded to azo bridge) are shortened with respect to the corresponding bond atoms of the ligand, exhibiting π-bond characters. On the other hand, C–N bonds (connecting azo bridge with pyridone and phenyl rings) in complex are longer than corresponding bond atoms in ligand displaying predominantly σ-bond character. Thus, the authors assumed that ligand exist in azo form [61] but it is more likely that this is azo anion form. The same observations are noted for mononuclear Ni^{2+} complex of *trans*-dichloro ligand (L_3- compound 3), [Ni(*trans*-L_3)$_2$(CH_3OH)$_2$] · $2H_2O$ [67]. Nickel adopts the same geometry as copper in previous case where equatorial plane is consisted of two oxygen and two nitrogen atoms from two ligands with Ni–O and Ni–N bond lengths of 1.975(6) and 2.071(9) Å, respectively. A bond length of Ni–O where oxygen originates from methanol molecules is 2.116(9) Å. Ligand after metal-ion complexation adopts solely *trans*- configuration (Scheme 7).

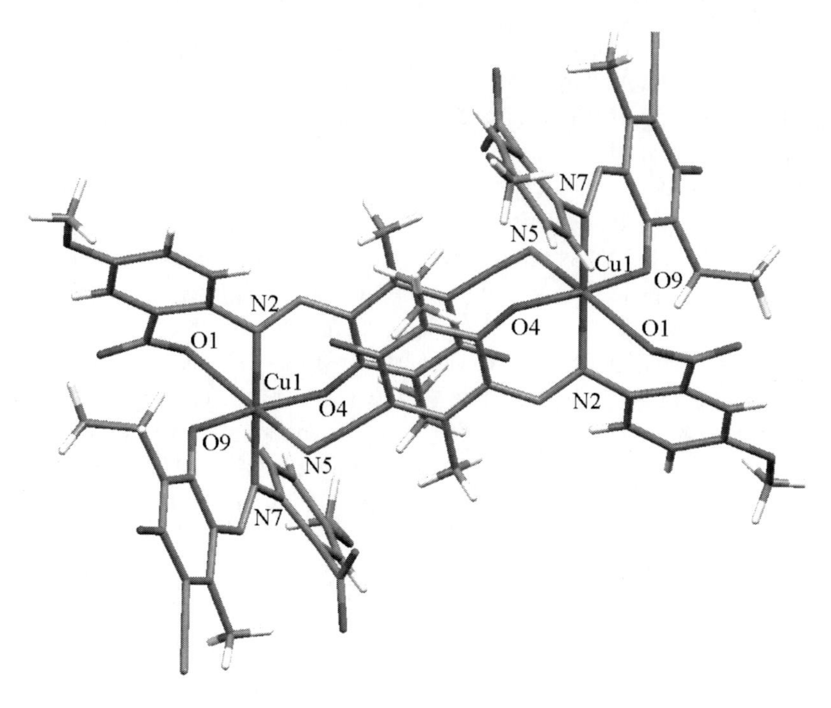

Figure 8. Drawing of copper complex of 2.

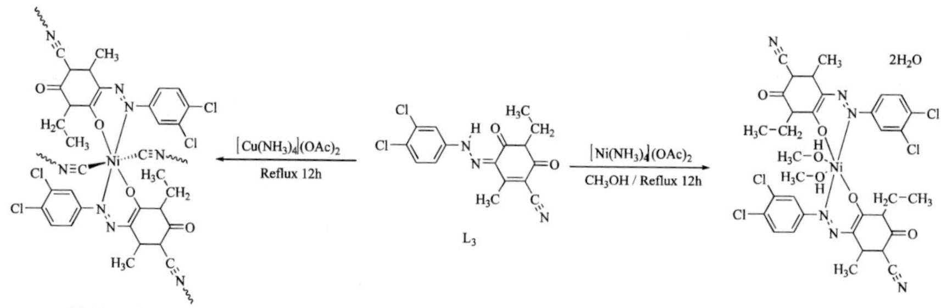

$[Cu(NH_3)_4](OAc)_2$
Reflux 12h
L_3
$[Ni(NH_3)_4](OAc)_2$
CH_3OH / Reflux 12h
$[Cu(cis\text{-}L_3)_2]_n$
$[Ni(trans\text{-}L_3)_2(CH_3OH)_2]\cdot 2H_2O$

Scheme 7. Schematic representation of the preparation of copper and nickel complexes of 3.

Starting from *trans*-dichloro ligand, a cyano-extended one-dimensional (1D) Cu^{2+} coordination polymers, $[Cu(cis\text{-}L_3)_2]_n$ (Scheme 7) and $[Cu(L_{3\text{-}OH})(H_2O)]$, are reported. The second one is obtained from *in situ* Cu^{2+} ion catalysis and complexation with H_2O_2 oxidation of *trans*-dichloro and this ligand $L_{3\text{-}OH}$ contains additional phenolic group and it coordinates with the central Cu^{2+} as a tetradentate bridging ligand [67]. TG-DTA analysis shows that thermal stabilities of these complexes are significantly improved in comparison to the pure dyes.

CONCLUSION

APDs represent disperse dyes that have gained significant interest due to their outstanding properties, such as good light and wash fastness properties, excellent coloration features and simplicity of preparation. APDs exhibit two types of prototropy depending on the structure of the dyes: keto-enol (6-alkyl or aryl substituted APDs) and azo-hydrazone. Azo-hydrazone tautomerism is usually due to the presence of OH group in *ortho-* position to azo bridge, i.e., in 6-hydroxy-2-pyridone structure. According to FT-IR data and X-ray crystallography, these dyes exist in the hydrazone form in the solid state, while NMR spectra show the existence of the hydrazone form in DMSO-d_6 and $CDCl_3$. UV-Vis spectra show the existence of hydrazone-anion equilibrium of azo pyridone dyes in various solvents, especially when the diazo component is heterocyclic compound. For APDs having carbocyclic diazo component, the hydrazone form is predominant. The hydrazone form is more stabilized in HBD solvents, while anion dominates in HBA solvents. Relative positions of the hydrazone and anion forms highly depend on the structure of the dyes. Electron-donating substituents in phenyl ring of the hydrazone form cause a significant bathochromic shift of UV-Vis absorption maxima, while electron-accepting substituents induce a small bathochromic or hypsochromic shift. The substituents in the position 1 of pyridone ring show no significant influence on the position of UV-Vis absorption spectra of the hydrazone form. DFT calculations applied to APDs confirm the existence of the hydrazone form in the solid state and the overall results are in accordance with experimental data. LSER analysis of absorption maxima of APDs lead to general conclusion that the most of the solvatochromism of these dyes is due to the nonspecific interactions rather than specific interactions. LFER analyses of UV–Vis, ^{1}H and ^{13}C NMR spectra give excellent results indicating that the substituent impacts on the spectral shifts originate from electronic effects. Due to the nature of APD molecules, they are used as ligands in complexation with chromium, copper and nickel and detail overview of these complexes is given in this chapter.

ACKNOWLEDGMENTS

The authors are grateful to the Ministry of Education, Science and Technological Development for financial assistance (project 172013).

REFERENCES

[1] Zollinger, H., (1987). Colour Chemistry. Weinheim: VCH.
[2] Hunger, K., (2003). Industrial Dyes Chemistry, Properties, Applications. Weinheim: VCH.
[3] Clark, M., (2011). Handbook of textile and industrial dyeing Volume 1: Principles, processes and types of dyes. Cambridge: Woodhead Publishing Limited.
[4] Mijin, D., Ušćumlić, G., Perišić-Janjić, N., Trkulja, I., Radetić, M., Jovančić, P., (2006). Synthesis, properties and color assessment of some new 5-(3- and 4-substituted phenylazo)-4,6-dimethyl-3-cyano-2-pyridones. *J. Serb. Chem. Soc.* 71, 435–444.
[5] Chen, C. C., Wang, I. J., (1991). Synthesis of some pyridone azo dyes from 1-substituted 2-hydroxy-6-pyridone derivatives and their color assessment. *Dyes Pigm.* 15, 69–82.
[6] Doswald, P., Wald, R., (1997). Reactive azo dyes, their preparation and their use. *WO 9730125.*
[7] Dannheim, J., (2000). Brilliant yellow reactive azo dyes containing a fluorotriazine group, their production and their use for dyeing materials containing hydroxy and amide groups. *EP 1054041.*
[8] Vines, D. R., Pedemonte, R. P., (1998). Pyridone-based disazo reactive dyes. *US 5811529.*
[9] Greve, M., (1978). Basic azo dyes free of sulfonic acid groups. *DE* 2752282.
[10] Kenworthy, M., Kenyon, R. W., (1999). Monoazo dyes and ink-jet inks containing them. *WO 9943754.*
[11] Muzik, F., Ruzicka, J., Marhan, J., Prikryl, J., Vyskocil, F., (1989). Thermally stable yellow azo pigment manufacture. *CS 258820.*
[12] Mueller, W., Lienhard, P., (1975). Azo pigments. *CH 72-6615.*
[13] Oberholzer, M., (2005). Concentrated storage-stable aqueous dye solution without any solubilizer content. *WO 2005116143.*

[14] Seto, N., Kato, Y., Fujimori, T., (2006). Colored curable compositions containing phthalocyanine and pyridone azo dyes and manufacture of color filters using them with excellent light and heat resistance. *JP* 2006071822.

[15] Tanaka, M., Kawamura, M., Murai, Y., Hirose, M., Miyazak, T., (2012). Colorant compound and ink including the colorant compound. *US 8211221.*

[16] Mayo, J. D., Duff, J. M., (2010). Multi-chromophoric azo pyridne colorant. *US 7754862.*

[17] Mijin, D. Ž., Ušćumlić, G. S., Valentić, N. V., Marinković, A. D., (2011). Synthesis of azo pyridone dyes. *Hem. Ind.* 65, 517–532.

[18] Elgemeie, G. E. H., El-Zanate, A. M., Mansour, A. K. E., (1993). Reaction of (cyano)thioacetamide with arylhydrazones of β-diketones: novel synthesis of 2(1H)-pyridinethiones, thieno[2,3-b]pyridines, and pyrazolo[3,4-b]pyridines. *Bull. Chem. Soc. Jpn.* 66, 555–561.

[19] Dostanić, J., Valentić, N., Ušćumlić, G., Mijin, D., (2011). Synthesis of 5-(substituted phenylazo)-6-hydroxy-4-methyl-3-cyano-2-pyridones from ethyl 3-oxo-2-(substituted phenyldiazenyl)butanoates. *J. Serb. Chem. Soc.* 76, 499–504.

[20] Bahreini, Z., (2009). Synthesis and characterisation of some new arylazopyridone dyes. *Pigm. Resin Technol.* 38, 298–304.

[21] Mijin, D. Ž., Baghbanzadeh, M., Reidlinger, C., Kappe, C. O., (2010). The microwave-assisted synthesis of 5-arylazo-4,6-disubstituted-3-cyano-2-pyridone dyes. *Dyes Pigm.* 85, 73–78.

[22] Yen, M. S., Wang, I. J., (2004). A facile syntheses and absorption characteristics of some monoazo dyes in bis-heterocyclic aromatic systems. Part I. Syntheses of polysubstituted 5-(2-pyrido-5-yl and 5-pyrazolo-4-yl)azo-thiophene derivatives. *Dyes Pigm.* 62, 173–180.

[23] Mirković, J. M., Božić, B. Đ., Mutavdžić, D. R., Ušćumlić, G. S., Mijin, D.Ž., (2014). Solvent and structural effects on the spectral shifts of 5-(substituted phenylazo)-3-cyano-6-hydroxy-1-(2-hydroxyethyl)-4-methyl-2-pyridones. *Chem. Phys. Lett.* 615, 62–68.

[24] Mohammed, S. A. M., El-Apasery, M. A., Elnagdi, M. H., (2014). Arylazoazines and arylazoazoles as interesting disperse dyes: Recent developments with emphasis on our contribution laboratory outcomes. *Eur. J. Chem.* 5, 192–200.

[25] Al-Etaibi, A., El-Apasery, M. A., Mahmoud, H., Al-Awadi, N., (2014). Synthesis, characterization and antimicrobial activity, and applications

of new azo pyridone disperse dyes on polyester fabric. *Eur. J. Chem.* 5, 321–327.

[26] Matusi, M., Joglekar, B., Ishigure, Y., Shibata, K., Muramatsu, H., Murata, Y., (1993). Synthesis of hydroxy-5-[2-(perfluoroalkyl) phenylazo]-2-pyridones and their application for dye diffusion thermal transfer printing. *B. Chem. Soc. Jpn.* 66, 1790–1794.

[27] Sakoma, K. J., Bello, K. A., Yakubu, M. K., (2012). Synthesis of Some Azo Disperse Dyes from 1-Substituted 2-Hydroxy-6-pyridone Derivatives and Their Colour Assessment on Polyester Fabric. *OJAppS.* 2, 54–59.

[28] Ashkar, S. M., El-Apasery, M. A., Touma, M. M., Elnagdi, M. H., (2012). Synthesis of Some Novel Biologically Active Disperse Dyes Derived from 4-Methyl-2,6-dioxo-1-propyl-1,2,5,6-tetrahydropyridine-3-carbonitrile as Coupling Component and Their Colour Assessment on Polyester Fabrics. *Molecules.* 17, 8822–8831.

[29] Helal, M. H., (2004). Synthesis and characterisation of a new series of pyridinone azo dyes for dyeing of synthetic fibers. *Pigm. Resin Technol.* 33, 165–171.

[30] Okada, Y., Hihara, T., Hirose, M., Morita, Z., (2010). Relationship between photofading and chemical structure of disperse azo dyes on nylon fabric. *Color. Technol.* 126, 215–230.

[31] Lee, J. J., Han, N. K., Lee, W. J., Choi, J. H., Kim, J. P., (2003). One-bath dyeing of a polyester/cotton blend with reactive disperse dyes from 2-hydroxypyrid-6-one derivatives. *Color. Technol.* 119, 134–139.

[32] Wang, I. J., Wang, P. Y., (1990). Photofading of azo pyridone dyes in solution. Part I. Photofading kinetics and thermodynamics of 3-(*p*- and *o*-substituted arylazo)-2-hydroxy-4-methyl-5-cyano-6-pyridone dyes in amide solvents. *Text. Res. J.* 60, 297–300.

[33] Wang, P. Y., Wang, I. J., (1990). Photofading of azo pyridone dyes in solution. Part II. Substituent effects on the UV absorption spectra and photostability of 3-(mono- and di-substituted arylazo)-2-hydroxy-4-methyl-5-cyano-6-pyridone in *N,N*-dimethylformamide. *Text. Res. J.* 60, 519–524.

[34] Wang, P. Y., Wang, I. J., (1991). Effects of substituent and aggregation on the photofading of some azo pyridone dyes on polyester substrates. *Text. Res. J.* 61, 162–168.

[35] Wang, P. Y., Wang, I. J., (1992). Photolytic Behavior of Some Azo Pyridone Disperse Dyes on Polyester Substrates. *Text. Res. J.* 62, 15–20.

[36] Dostanić, J. M., Lončarević, D. R., Banković, P. T., Cvetković, O. G., Jovanović, D. M., Mijin, D. Ž., (2011). Influence of process parameters on the photodegradation of synthesized azo pyridone dye in TiO_2 water suspension under simulated sunlight. *J. Environ. Sci. Health, Part A: Toxic/Hazard. Subst. Environ. Eng.* 46, 70–79.

[37] Dostanić, J., Lončarević, D., Rožić, Lj., Petrović, S., Mijin, D., Jovanović, D. M., (2013). Photocatalytic degradation of azo pyridone dye: Optimization using response surface methodology. *Desalin. Water Treat.* 51, 2802-2812.

[38] Mirković, J. M., Prlainović, N. Ž., Uščumlić, G. S., Grgur, B. N., Mijin, D. Ž., (2014). Optimization of electrochemical decolorization of certain arylazopyridone dyes. *J. Serb. Chem. Soc.* 79, 1523–1536.

[39] Mirković, J., Lović, J., Avramov Ivić, M., Mijin, D., (2014). Electrooxidative Behavior of Arylazo Pyridone Dyes and Their Inclusion Complexes on Gold Electrode in 0.1 M NaOH. *Electrochim. Acta.* 137, 705–713.

[40] Karci, F., Karci, F., Demirçali, A., Yamaç, M., (2014). Synthesis, solvatochromic properties and antimicrobial activities of some novel pyridone-based disperse disazo dyes. *J. Mol. Liq.* 187, 302–308.

[41] Sykes, P., (1985). A guidebook to mechanism in organic chemistry. New York: John Wiley and sons.

[42] Alimmari, A. S., Marinković, A. D., Mijin, D. Ž., Valentić, N. V., Todorović, N., Uščumlić, G. S., (2010). Synthesis, structure and solvatochromic properties of 3-cyano-4,6-diphenyl-5-(3- and 4-substituted phenylazo)-2-pyridones. *J. Serb. Chem. Soc.* 75, 1019–1032.

[43] Mijin, D. Ž., Uščumlić, G. S., Perišić-Janjić, N. U., Valentić, N. V., (2006). Substituent and solvent effects on the UV/vis absorption spectra of 5-(3- and 4-substituted arylazo)-4,6-dimethyl-3-cyano-2-pyridones. *Chem. Phys. Lett.* 418, 223–229.

[44] Beak, P., Covington, J. B., White, J. M., (1980). Quantitive model of solvent effects on hydroxypyridine-pyridone and mercaptopyridine-thiopyridone equilibriums: correlation with reaction-field and hydrogen-bonding effects. *J. Org. Chem.* 45, 1347–1353.

[45] Christie, R. M., (2001). Colour Chemistry. Cambridge: Royal Society of Chemistry.

[46] Hassanzadeh, A., Zeini-Isfahani, A., Habibi, M. H., Poor Heravi, M. R. A., Abdollahi-Alibeik, M., (2006). ^{1}H, ^{13}C, N–H, H–H, C–H COSY, H–H NOESY NMR and UV–vis studies of Solophenyl red 3BL dye azo-

hydrazone tautomerism in various solvents *Spectrochim. Acta A*. 63, 247–254.

[47] Adegoke, O. A., (2011). Relative predominance of azo and hydrazone tautomers of 4-carboxyl-2,6-dinitrophenylazohydroxynaphthalenes in binary solvent mixtures. *Spectrochim. Acta A*. 83, 504–510.

[48] Ušćumlić, G. S., Mijin, D. Ž., Valentić, N. V., Vajs, V. V., Sušić, B. M., (2004). Substituent and solvent effects on the UV/Vis absorption spectra of 5-(4-substituted arylazo)-6-hydroxy-4-methyl-3-cyano-2-pyridones. *Chem. Phys. Lett.* 397, 148–153.

[49] Alimmari, A., Božić, B., Mijin, D., Marinković, A., Valentić, N., Ušćumlić, G., (2015). Synthesis, structure and solvatochromic properties of some novel 5-arylazo-6-hydroxy-4-(4-methoxyphenyl)-3-cyano-2-pyridone dyes: Hydrazone-azo tautomeric analysis. *Arab. J. Chem.* 8, 269–278.

[50] Alimmari, A., Mijin, D., Vukićević, R., Božić, B., Valentić, N., Vitnik, V., Vitnik, Ž., Ušćumlić, G., (2012). Synthesis, structure and solvatochromic properties of some novel 5-arylazo-6-hydroxy-4-phenyl-3-cyano-2-pyridone dyes. *Chem. Cent. J.* 6, 1–8.

[51] Božić, B. Đ., Alimmari, A. S., Mijin, D. Ž., Valentić, N. V., Ušćumlić, G. S., (2014). Synthesis, structure and solvatochromic properties of novel dyes derived from 4-(4-nitrophenyl)-3-cyano-2-pyridone. *J. Mol. Liq*. 196, 61–68.

[52] Cee, A., Horakova, B., Lyčka, A., (1988). Structural analysis of substituted 3-arylazo-2-hydroxy-6-pyridones. *Dyes Pigm.* 9, 357–369.

[53] Ertan, N., Gurkan, P., (1997). Synthesis and properties of some azo pyridone dyes and their Cu(II) complexes. *Dyes Pigm.* 33, 137–147.

[54] Cheng, L., Chen, X., Gao, K., Hu, J., Griffiths, J., (1986). Color and constitution of azo dyes derived from 2-thioalkyl-4,6-diaminopyrimidines and 3-cyano-1,4-dimethyl-6-hydroxy-2-pyridone as coupling components. *Dyes Pigm.* 7, 373–388.

[55] Mirković, J., Rogan, J., Poleti, D., Vitnik, V., Vitnik, Ž., Ušćumlić, G., Mijin, D., (2014). On the structures of 5-(4-, 3- and 2-methoxyphenylazo)-3-cyano-1-ethyl-6-hydroxy-4-methyl-2-pyridone: An experimental and theoretical study. *Dyes Pigm.* 104, 160–168.

[56] Peng, Q., Li, M., Gao, K., Cheng, L., (1990). Hydrazone-azo tautomerism of pyridone azo dyes: Part 1- NMR spectra of tautomers. *Dyes Pigm.* 14, 89–99.

[57] Peng, Q., Li, M., Gao, K., Cheng, L., (1991). Hydrazone-azo tautomerism of pyridone azo dyes. Part II: Relationship between structure and pH values. *Dyes Pigm.* 15, 263–274.
[58] Peng, Q., Li, M., Gao, K., Cheng, L., (1992). Hydrazone-azo tautomerism of pyridone azo dyes. Part III- Effect of dye structure and solvents on the dissociation of pyridone azo dyes. *Dyes Pigm.* 18, 271–286.
[59] Ertan, N., Eyduran, F., (1995). The synthesis of some hetarylazopyridone dyes and solvent effects on their absorption spectra. *Dyes Pigm.* 27, 313–320.
[60] Dostanić, J., Mijin, D., Ušćumlić, G., Jovanović, D. M., Zlatar, M., Lončarević, D., (2014). Spectroscopic and quantum chemical investigations of substituent effects on the azo-hydrazone tautomerism and acid–base properties of arylazo pyridone dyes. *Spectrochim Acta A*. 123, 37–45.
[61] Chen, X., Tao, T., Wang, Y., Peng, Y., Huang, W., Qian, H., (2012). Azo-hydrazone tautomerism observed from UV-vis spectra by pH control and metal-ion complexation for two heterocyclic disperse yellow dyes. *Dalton Trans*. 41, 11107–11115.
[62] Black, S. N., Davey, R. J., Slawin, A. M. Z., Williams, D. J., (1992). Structure of 1-Butyl-5-[(4-chlorophenyl)azo]-1,2,5,6-tetrahydro-4-methyl-2,6-dioxo-3-pyridinecarbonitrile. *Acta Crystallogr. Sect. C: Cryst. Struct. Commun.* c42, 323–325.
[63] Prikryl, J., Kratochvil, B., Ondracek, J., Maixner, J., Klicnar, J., Huml, K., (1993). The crystal structure and molecular structure of charge-transfer complex of azo-dyestuff 5-(4-chloro-2-nitrophyl)-azo-6-hydroxy-1-ethyl-3-cyano-4-methyl-2-pyridone with napthalene. *Collect. Czech. Chem. Commun.* 58, 2121–2127.
[64] Qian, H. F., Wei, H., (2006). An azo dye molecule having a pyridine-2,6-dione backbone. *Acta Crystallogr. Sect. C: Cryst. Struct. Commun.* C62, o62–o64.
[65] Huang, W., Qian, H., (2008). Structural characterization of C.I. Disperse Yellow 114. *Dyes Pigm.* 77, 446–450.
[66] Huang, W., (2008). Structural and computational studies of azo dyes in the hydrazone form having the same pyridine-2,6-dione component (II): C. I. Disperse Yellow 119 and C. I. Disperse Yellow 211. *Dyes Pigm.* 79, 69–75.
[67] You, W., Zhu, H., Huang, W., Hu B., Fan Y., You, X., (2010). The first observation of azo-hydrazone and cis–trans tautomerisms for disperse

yellow dyes and their nickel(II) and copper(II) complexes. *Dalton Trans.* 39, 7876–7880.

[68] Özbey, S., Kendi, E., Ertan, N., (1997). Crystal Structure of Hydrazone Form of l-Butyl-3-cyano-6-hydroxy-4-methyl-5-(2-nitrothiazolylazo)-2(1H)-pyridone. *Dyes Pigm.* 33, 251–258.

[69] Ajaj, I., Assaleh, F., Markovski, J., Rančić, M., Brković, D., Milčić, M., Marinković A., (2015). Solvatochromism and azo-hydrazone tautomerism of novel arylazopyridone dyes: experimental and quantum chemical study. *Arab. J. Chem.* in press.

[70] Reichardt, C., (2003). Solvents and Solvent Effects in Organic Chemistry. Weinheim: VCH.

[71] Reichardt, C., (1994). Solvatochromic Dyes as Solvent Polarity Indicators. *Chem. Rev.* 94, 2319–2358.

[72] Ušćumlić, G. S., Krstić, V. V., (2000). Rastvarači i njihov značaj u organskoj hemiji. Beograd:TMF.

[73] Kamlet, M. J., Abboud, J. M., Abraham, M. H., Taft, R. W., (1983). Linear Solvation Energy Relationships. 23. A Comprehensive Collection of the Solvatochromic Parameters, π^*, α, and β, and Some Methods for Simplifying the Generalized Solvatochromic Equation. *J. Org. Chem.* 48, 2877–2887.

[74] Catalán, J., (2009). Toward a Generalized Treatment of the Solvent Effect Based on Four Empirical Scales: Dipolarity (SdP, a New Scale), Polarizability (SP), Acidity (SA), and Basicity (SB) of the Medium. *J. Phys. Chem. B.* 113, 5951–5960.

[75] Hammett, L. P., (1937). The Effect of Structure upon the Reactions of Organic Compounds. Benzene Derivatives. *J. Am. Chem. Soc.* 59, 96–103.

[76] Krygowski, T. M., Stepien, B. T., (2005). Sigma- and Pi-Electron Delocalization: Focus on Substituent Effects. *Chem. Rev.* 105, 3482–3512.

[77] Brown, H., Okamoto, Y., (1958). Electrophilic Substituent Constants. *J. Am. Chem. Soc.* 80, 4979–4987.

[78] Hansch, C., Leo, A., Taft, W., (1991). A survey of Hammett substituent constants and resonance and field parameters. *Chem. Rev.* 91, 165–195.

[79] Taft, R. W., Lewis, C. I., (1958). The General Applicability of a Fixed Scale of Inductive Effects. II. Inductive Effects of Dipolar Substituents in the Reactivities of m- and p-Substituted Derivatives of Benzene. *J. Am. Chem. Soc.* 80, 2436–2443.

[80] Swain, C. G., Unger, S. H., Rosenquist, N. R., Swain, M. S., (1983). Substituent effects on chemical reactivity. Improved evaluation of field and resonance components. *J. Am. Chem. Soc.* 105, 492–502.

[81] Miyazawa, T., Kurose, Y., (2006). Optical recording medium showing improved light stability suitable for blue laser light, metal complex compound and organic dye compound. *WO 2006104196.*

[82] Wang, I. J., Hsu, Y. J., Tian, J. H., (1991). Synthesis and properties of some pyridone chromium complex azo dyes, *Dyes Pigm.* 16, 83–91.

[83] Song, H. F, Chen, K. C., Wu, D. Q., Tian, H., (2004). Synthesis and absorption properties of some new azo-metal chelates and their ligands. *Dyes Pigm.* 60, 111–119.

In: Advances in Materials Science Research ISBN: 978-1-53610-059-4
Editor: Maryann C. Wythers

Chapter 8

COMPARISON OF ANALYTICAL POSSIBILITIES OF SCINTILLATION ATOMIC EMISSION SPECTROMETRY AND AUTOMATED MINERALOGY FOR STUDYING OF GOLD-BEARING SAMPLES

Irina E. Vasil'eva **[*]** ***and Elena V. Shabanova***
Vinogradov Institute of Geochemistry Siberian Branch
of Russian Academy of Sciences, Irkutsk, Russia

ABSTRACT

The article considers the analytical capabilities of two methods for studying the composition of gold-bearing samples: arc scintillation atomic emission spectrometry and automated mineralogy using the microprobe X-ray spectrometry. We report the results for two certified reference materials (Sukhoi Log deposit and flotation tailings gold sulfide ore) obtained by these methods. The reasons of obtaining incompatible information are discussed. The scopes of economically viable application of each method are listed.

---

[*] Corresponding Author: vasira@igc.irk.ru

Keywords: gold, silver, scintillation atomic emission spectrometry, automated mineralogy, gold-bearing certified reference material

INTRODUCTION

Unique features of precious or noble metals (PM or NM – Au, Ag and platinum group elements) specify their high demand in modern industries, medicine, energetics, etc. The increase of gold reserves provides economic stability of countries. However, only gold and silver can produce own deposits whose prospecting requires high funding, including the analysis of the composition (quantity and sizes of mineral phases that contain PM).

The used technologies of ore preconcentration or enrichment also require permanent analytical studies to find the degree of noble metal extraction. Thus, the efficiency of PM prospecting and extraction is controlled by the reliable analytical data [1]. Different analytical techniques or their combinations are applied to determine the total PM concentrations in natural and technological samples [2-4]. Nevertheless, numerous current studies indicate that these data are insufficient. When studying the gold-bearing samples, it is needed to know the concentrations of precious metals (from Clarke to industrial contents), composition and quantity of mineral phases containing PM; the size of particles keeping PM and their particle-size distribution in a certain amount of the crushed sample; the representative subsample for the analysis of each noble metal.

Two analytical methods can be regarded as perspective for obtaining the above data. They are direct methods which do not require the change of the aggregation state of powder samples, namely:

- scintillation atomic-emission spectrometry with arc discharge (scintillation AES) [5],
- automated mineralogy (MLA) with electronic scanning microscope [6].

These methods imply obtaining a great number of spectral measurements using a certain scheme and their further computer-based processing. The role of each analytical method having different physical principles and different numerical models was revealed via the data obtained by ones. For this purpose the well-studied certified reference materials (CRMs) from the collection of Vinogradov Institute of Geochemistry, Siberian Branch of Russian Academy of Sciences (IGC SB RAS) were used. This study should provide the answers to the following questions:

- How much would the information obtained by these two methods be different?
- Is it possible to use a common sample preparation for different techniques?

Subjects and Objectives

In the present study we investigated two national CRMs [7] of technogenic and natural origin, namely:

- The product from processing of quartz-sulfide ore (flotation tailings) – SZH-3 (GSO 2740-1983),
- The black shale ore sample from Sukhoi Log deposit (Irkutsk region, Russia) – SLg-1 (GSO 8550-2004).

The objectives of the present study included:

- Determination of total gold and silver contents;
- Evaluation of the of size and composition of mineral phases containing Au and Ag;
- Description of Au and Ag particle-size distribution;
- Calculation of the minimum subsample required for gold determination.

Analytical Studies of Gold-Bearing Samples

Scintillation Atomic-Emission Spectrometry with Alternative Current Arc

The device "Potok" with an electric-arc generator "Fireball" was applied to introduce the powder samples into plasma and to excite the atoms [5, 8]. The spectrograph STE-1 with a three-lens illumination system was used as a polychromator. The recording system located in a cassette part is a high-speed and high resolution (base exposition is 4 milliseconds) multichannel analyzer to measure the intensity of emission spectra in scintillation and integrated modes. The duration of a base exposition is limited by technical characteristics of the equipment and by the software ATOM [5, 8]. Spectral resolution on a line Au I 267.595 nm makes up 0.37 nm mm^{-1}, while on lines Ag I 328.06 nm and Ag I 338.289 nm it amounts to 0.47 nm mm^{-1}. The software ATOM was used to:

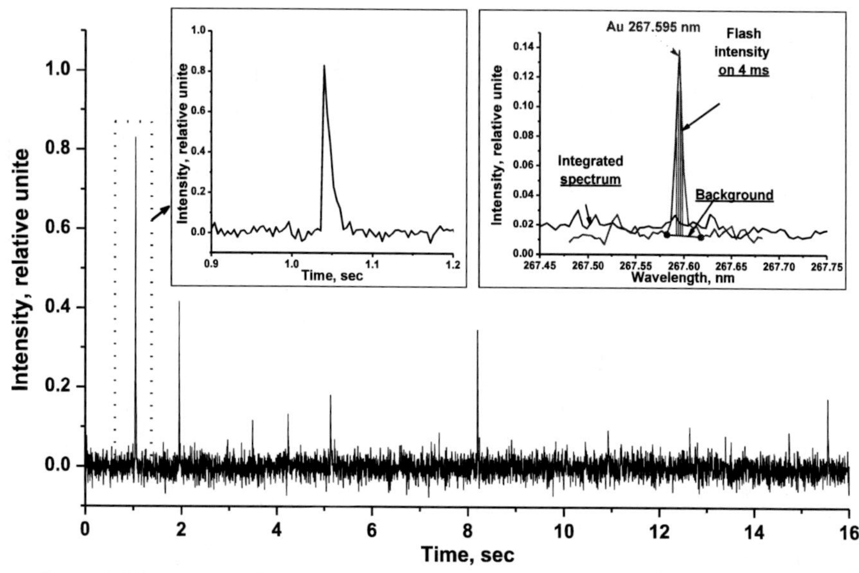

Figure 1. Scintillation spectra recorded during 14 seconds (base exposition is 4 milliseconds) of sample spectrum registration, the flash (insert 1) and the flash intensity (insert 2) of an individual gold particle.

- register the sequence of scintillation spectra during a continuous introduction of a sample into the electric arc (Figure 1);
- observe the evaporation-excitation spectra (burnout curve) for each particle as an individual emitting flash of light on a set wavelength for Au, Ag and other elements (Figure 1, insert 1);
- measure the intensity taking into account an individual background under the analyte line and the duration of this emitting flash (Figure 1, insert 2);
- record the integrated radiation on the same wavelengths;
- filter noise from the spectral background of a device and sample (Figure 2);

- count flashes from particles of the given size in the sequence of scintillation spectra during a continuous introduction of a sample into the electric arc (Figure 2);
- estimate Au and Ag particle-size distributions in each sample (Figure 2);
- construct a calibration curves for determining total Au, Ag and other element concentrations based on scintillation and integrated spectra.

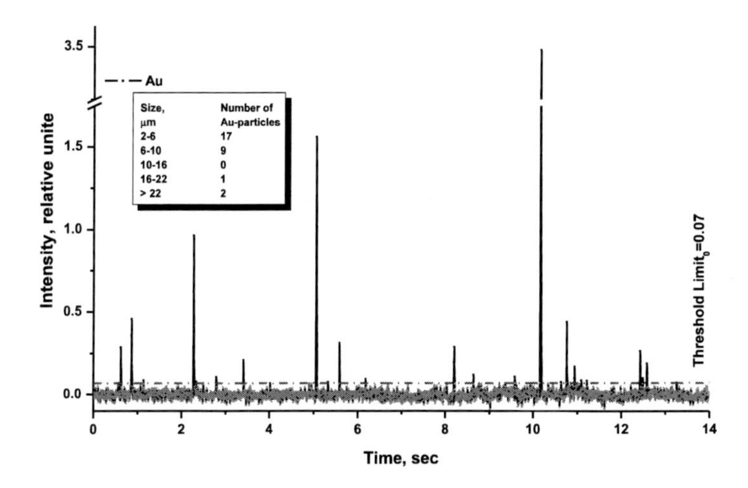

Figure 2. The sequence of the scintillation spectra at the wavelength Au I 267.5 nm and number of calculated particles containing gold in the 150 mg subsample of CRM SZH-3 (0.07 – zero threshold of intensity of flashes characterizes the equipment noise and spectral background).

The numerical model was used for defining the size of registered particles of both gold and silver [9]. The model implies that:

- the particles containing noble metals get the spherical shape in the plasma and are completely evaporated when passing through it;
- each particle in plasma of the electric arc gives only one flash;
- distribution of flash intensity from one particle corresponds to Poisson distribution;
- scintillations do not depend on each other and a simultaneous occurrence of several flashes is negligibly small.

The empirical equations of particle-size distribution for samples with unknown composition were found via the scintillation spectra of CRMs with a known Au and Ag particle-size distributions. It is known that flash intensity is depended particle weight $I^{1.4} = k \times m$, and the weight of the particle can be expressed by means of specific density and radius of this particle $m = \rho \times V = \rho \times \frac{4}{3} \times \pi \times r^3$, where I – flash intensity, k – proportionality coefficient, m – particle weight, ρ – particle specific density, V – particle volume, r – particle radius, π – constant. The dependence of the flash intensity on the diameter of gold particles is $I_{Au} = 0.067e^{0.088d}$, and the diameter of silver particles $I_{Ag} = 0.020d^{2.13}$ (Figure 3). Using these dependences we found scintillation intensities of gold and silver particles which correspond to five size intervals: 2-6; 6-10; 10-16; 16-22; > 22 μm. When analyzing the unknown samples there is an automatic Au and Ag particle-size distribution and calculation of their number in each set interval (Figure 2). The total amount of the registered scintillations is taken as 100%. This information is used for obtaining the patterns of gold and silver particle-size distributions.

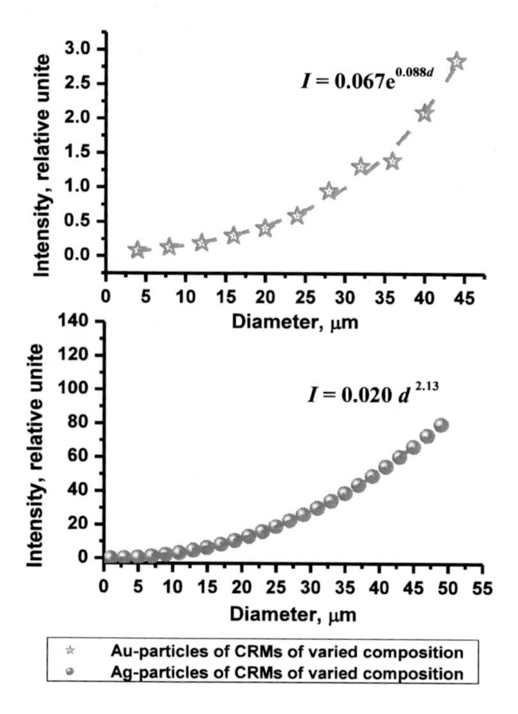

Figure 3. Dependences of the flash intensity vs. diameter of the spherical Au- and Ag-particles.

The linear calibration curves lg $I \approx$ lg C, where I is the sum of intensities of all flashes on the wavelength, were fitted by CRMs spectra with a known composition to determine the total concentration of Au and Ag. CRMs having different matrix composition (ores, soil, sediments and rocks) are used to calibrate. Figure 4 demonstrates typical calibration curves for gold and silver. The calibration curves are regarded to be optimal as a mean square deviation amounts not more than 25%, and an angle of graph slope is 40 and 42 degrees, correspondingly. It is not possible to improve the MSD value by classical ways (introduction of standard additions or of an internal standard) because the method is based on the calculation of separate flashes from particles of different size.

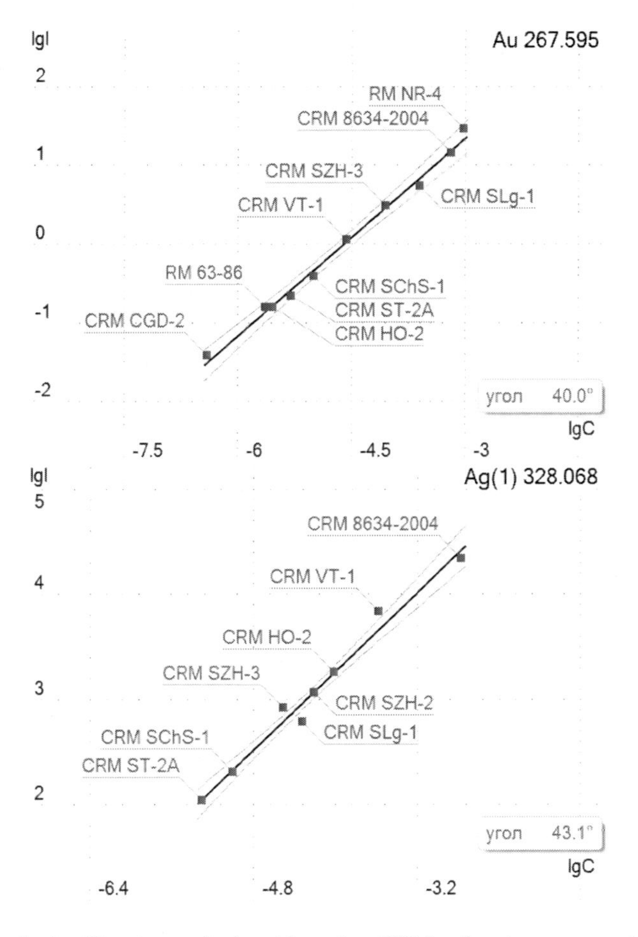

Figure 4. Typical calibrations calculated by using CRMs of various type and composition: ores, soils, sediments, and rocks for gold (a) and silver (b) analysis.

Automated Mineralogy

X-ray electronic microscope Hitachi S-570 with the CDD detector [10] and the software ImSca 2.4 was used to study the individual mineral phases containing gold and to determine the amount of gold grains and total Au concentration [6].

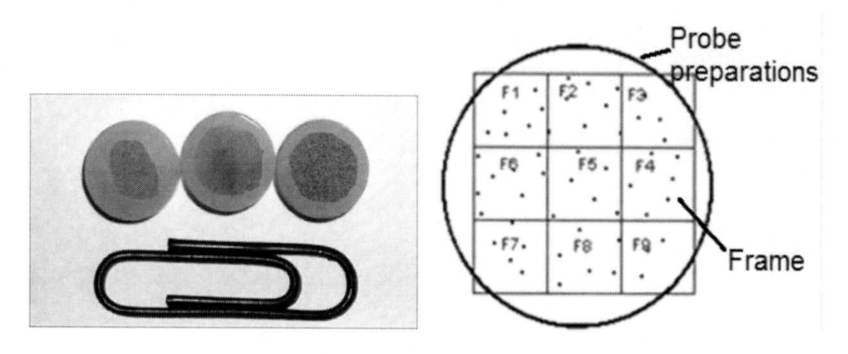

Figure 5. Probe preparations of the samples, and the separation of the study area into frames.

The probe preparations of samples were prepared after gravimetric concentration of their powders (Figure 5). Measurements were obtained in reflected electrons with the resolution of ~2.9 μm when the investigated area is divided into frames. In each frame the grains were classified by the X-ray spectrum intensity into a group of elements depending on a phase chemical composition (by a mean atomic number – density) and on the measured size (≥ 3 μm). The software ImSca 2.4 records the coordinates of each particle-carrier of PM. Following the procedure of spectral information processing the discovered grains are approximated by a cube, and the total concentration of an analyte is regarded as directly proportional to a quantity of the discovered grains of all mineral phases. As not all grains of NM minerals are extracted during the gravity concentration for calculations we use the extraction coefficient (K_{extr}), that connects the quantity of really extracted grains (n) with the quantity of theoretically present grains (n') $K_{extr} = \frac{n}{n'}$.

Thus, from the ratio of the component weight in the mixture to the sample weight with a concentration of any chemical element C, if this element occurs in n grains of different or common minerals and considering this coefficient it follows $C = \frac{\sum_{i=1}^{n}(c_i \times \rho_i \times V_i)}{M \times K_{extr}} = \frac{\sum_{i=1}^{n}(c_i \times \rho_i \times a_i^3)}{M \times K_{extr}}$, where n – quantity of the extracted grains for each i mineral; ρ_i – grain density, V_i – volume of each grain; a_i – facet of a grain cube; c_i – element concentration and M – sample weight. The results obtained by means of the MLA allow discovering rare PM mineral phases; describing the grain-size distribution of a mineral (minerals) under study; giving an element weight from phase weights.

RESULTS AND DISCUSSION

When studying CRMs SZH-3 and SLg-1 by arc scintillation atomic-emission spectrometry we analyzed Au and Ag concentrations; total number of particles-carriers of Au and Ag in a 300 mg subsamples; and amount of Au- and Ag-particles (≥2 μm) in five size groups and their size distribution (Table 1).

Table 1. Au_{total} and Ag_{total} contents obtained by scintillation AES method; repeatability of particle counting in 300 mg subsample (the number of experiments n = 45 and 56, accordingly)

CRM	Analyte	Average number of particles in 300 mg subsample	Particle-size distribution					Content, ppm	
			Intervals of particle sizes, μm						
			2-6	6-10	10-16	16-22	>22	Certified	Defined
SZH-3	Au	37 ± 4	23 ± 3	8 ± 1	4 ± 1	2	1	0.9 ± 0.1	0.96 ± 0.08
	Ag	1980 ± 73	1943 ± 75	32 ± 5	4 ± 1	1	0	0.31 ± 0.02	0.34 ± 0.01
SLg-1	Au	41 ± 3	20 ± 2	7 ± 1	7 ± 1	4 ± 1	3 ± 1	2.5 ± 0.3	2.56 ± 0.16
	Ag	1678 ± 54	1630 ± 57	36 ± 12	8 ± 1	2 ± 1	2	0.47 ± 0.08	0.37 ± 0.02

Reproducibility of particle calculations and the traceability of gold and silver contents determined in SLg-1 are given on Figure 6. Total gold and silver contents are reliably determined and well agree with the certified values despite a great scatter in a total amount of recorded particles. The particles containing only Au or Ag as well as those of the mixed composition have been discovered in SLg-1 sample (Figure 7).

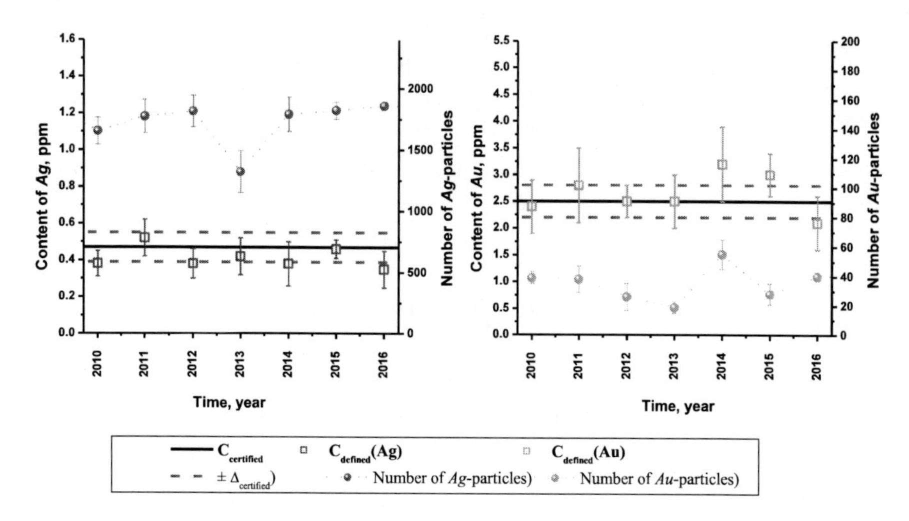

Figure 6. Repeatability of Au-, Ag-particle calculation and traceability of gold and silver total contents in CRM SLg-1.

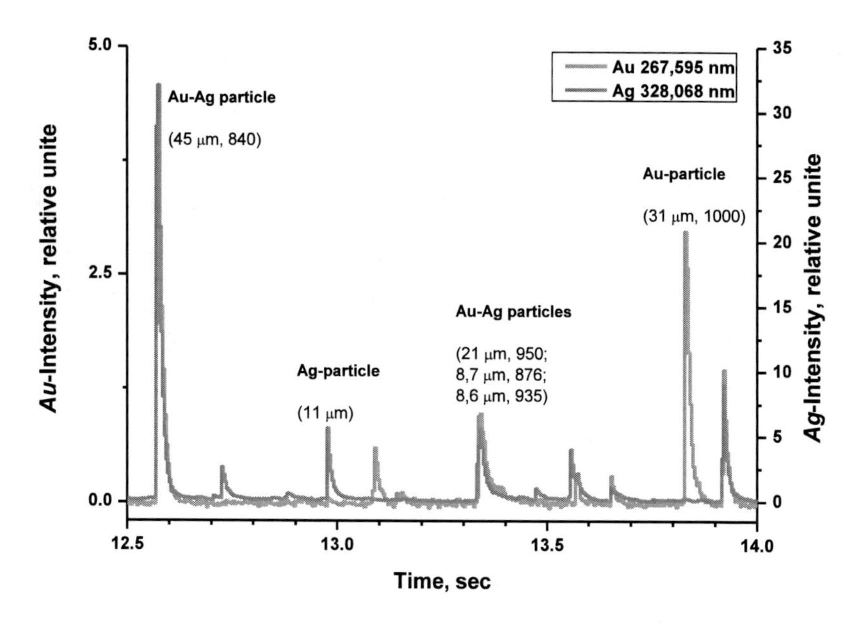

Figure 7. The results from the correlation analysis of the scintillation spectra sequences of Au 267.5 nm and Ag 328.0 nm over a time interval of 12.5-14 seconds.

We calculated the grade for over 90 gold-silver particles and found that the grade increased as the particle size grew (Figure 8). Numerous microprobe measurements done by various authors [e.g., 11, 12 etc.] show that the black shale from the Sukhoi Log deposit (Irkutsk Region) contains Au particles of different size.

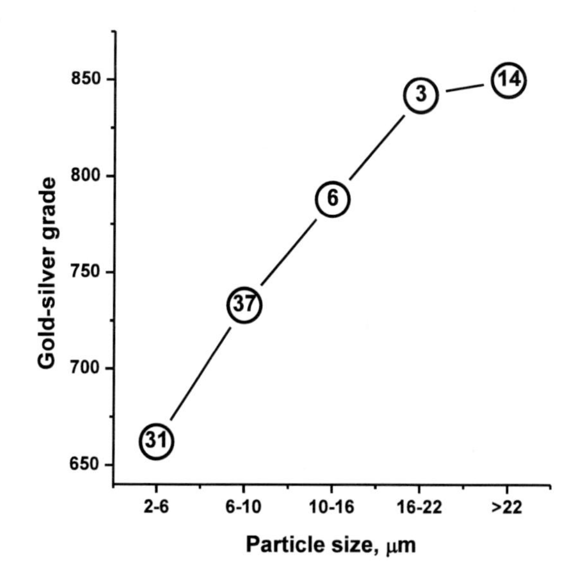

Figure 8. The average values of the gold-silver grade (number of particles for which the grade is calculated is given in circles).

Using the results of MLA (QEM*SEM – Quantitative Evaluation of Minerals by Scanning Electron Microscopy) for SGH-3 we obtained the following histogram of Au and Ag particle-size distributions and calculated the gold weight (grade) from 42 frames for each particle group. The image of one frame is given on Figure 9 (1 pixel corresponds to 2.9 μm). The classification of particles by phases is the following: yellow – Au-Ag, She, PbS, Pb-Sn, HgS, Sn, Sb, (Ce, La) OX; red – Fe, Ni, Co sulfoarsenides; green – Py. Figure 9 demonstrates the images of two Au, Ag-particles in the gravity concentration, prepared from SZH-3 powder. The QEM*SEM data (Table 2) indicate that the gold concentration in 28 discovered particles amounted to 0.051 ppm (Table 3) that is lower than the certified gold content (C_{Au} = 0.9 ppm). The largest grain is 27.7 μm (Table 2). Most of Au particles have the cube facet size as 2-10 μm. Following the solid-state polyphase calculation model [6] used in the automated mineralogy, the minimum representative subsample for SZH-3 for gold analyses made up ~ 0.5 g. Individual Ag particles of any size have not been recorded, because of it is not possible to estimate the silver content in SZH-3 powder. Note that with fine grinding of the material only a minor part of gold and silver occurs as separate grains which could enter the gravity concentration. Gold and silver are well forged, forming films on solid particles of rock-forming minerals which are stocked in tailing piles during the gravity concentration.

Table 2. Data for gold and silver in CRM SZH-3 and the probe 109ore obtained by MLA

Intervals of particle sizes, μm		Gravity concentrate					
		SZH-3			Probe 109ore		
		Number of grains (n)	Average grain size, μm (a)	Au $_{total}$ content in Au, Ag mineral, mg	Number of grains (n)	Average grain size, μm (a)	Au $_{total}$ content in Au, Ag mineral, mg
Particle-size distribution	0 - 10	21	7.1	0.141	207	7.6	1.690
	10 - 20	6	13.1	0.248	293	14.7	17.147
	20 - 30	1	27.7	0.394	70	23.8	17.612
	30 - 40	0	0	0	9	34.1	6.606
	40 - 50	0	0	0	6	44.9	10.064
	50 - 60	0	0	0	8	56.3	26.536
	60 - 70	0	0	0	15	64.9	76.216
	70 - 80	0	0	0	4	73.1	29.056
	80 - 90	0	0	0	3	82.8	31.626
	90 - 100	0	0	0	2	97.0	33.954
	100 - 110	0	0	0	1	102.5	20.033
	110 - 120	0	0	0	2	110.5	50.134
	120 - 130	0	0	0	0	0	0
Summary		28	11.5	0.782	620	30.3	320.674

The experiment with SZH-3 has shown that for the preliminary gravity concentration required for the MLA the fine powders that are prepared from the CRM material are not appropriate. Moreover, fine grinding in crushing and ball mills leads to destruction of a natural size of particles containing precious metals that could provide additional analytical errors. Therefore, SLg-1 powder was not used for assessing the analytical possibilities of the automated mineralogical analysis. The SLg-1 is the homogenized mixture of the ground material from 6 drill cores of the Sukhoi Log gold deposit [13]. For the mineralogical automated analysis we investigated the crushed sample of the duplicate core from one of ore samples, namely 109ore. The subsample of this crushed material (35.9 g) was additionally ground and classified into two size groups +63 and -63 μm with the concentration coefficients of 2 866 and 23 455, correspondingly. The gravity concentration was used to prepare the samples for microprobe analysis and to obtain measurements from 54 frames. The results are given in Table 4. The total gold concentration in the concentrate C_{Au} = 8.9 ppm; taking into account the extraction coefficient (0.8) in the crushed sample, amounted to 11.1 ppm. We discovered 620 particles of gold with the size of 110-120 μm (Table 2). The particle-size distribution has two maxima. Most particles containing gold has a size ranging from 3 to 20 μm. Individual silver particles have not been found.

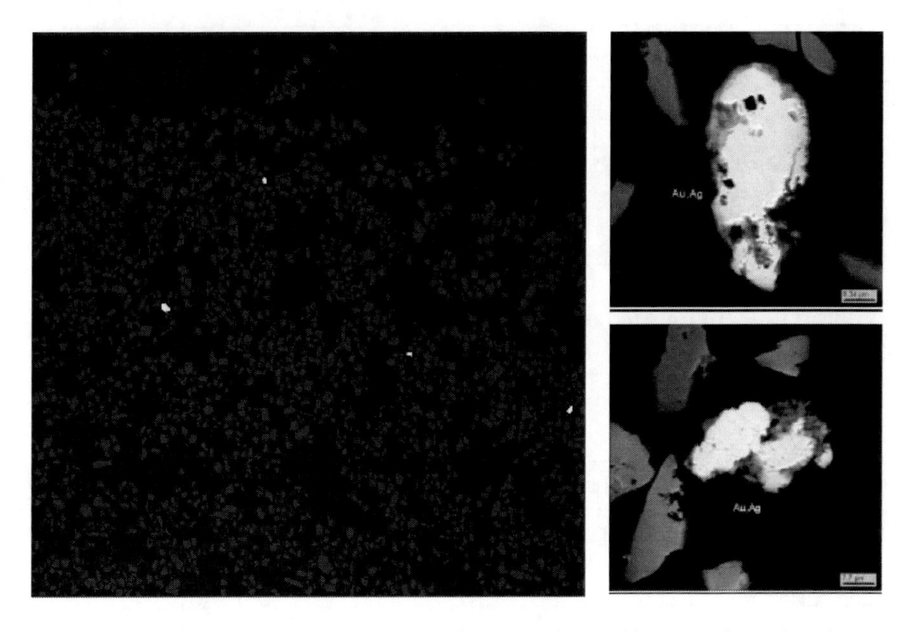

Figure 9. The image of a frame of the SZH-3 preparation and images of two Au, Ag-particles from this frame.

Table 3 presents the certified data, total gold and silver contents and number of particles containing these metals, obtained by two methods for the SZH-3. The particle-size distribution, obtained by two separate methods: scintillation AES [5] and MLA [6] were used to calculate the representative subsample [14]. Both methods demonstrate that the substance in SZH-3 is well crushed and homogenized, and gold occurs as fine particles. Moreover, most of gold particles were found in 2-10 μm interval. The total Au and Ag concentrations and amount of particles containing Au, obtained by scintillation

AES, agree well with the certified data. As regard to MLA the total Au concentration and the number of Au particles are underestimated. Ag particles have not been found. The calculated representative subsamples are in satisfactory agreement.

Table 3. Data for gold and silver in CRM SZH-3

Characteristics	Method		Certified data [7]
	SAES	MLA	
Subsample weight, g	0.3×*n*	15.2	Individually for each methods
Preliminary operations, except weighing, required for the analysis	No	Gravity preconcentration, preparation of probe drug	
Au_{total} content, ppm	0.96 ± 0.08	0.051	0.9 ± 0.1
Ag_{total} content, ppm	0.34 ± 0.01	No data	0.31 ± 0.02
Number of Au-particles in 1 g	≤120	~ 2	~ 170
Number of Ag-particles in 1 g	~ 6600	Not found	No data
Interval with the largest number of Au particles, μm	2 - 6	3 - 10	~ 5
Representative subsample, g	≥ 0.3	0.5 or a multiple of this value	1.5

Table 4. Data for gold and silver in CRM SLg-1 and the probe 109ore

Parameter	SLg-1		Sample 109ore	
	SAES	Certified [7] and literature data	MLA	Reference[13]
Subsample weight, g	0.3×*n*	Individually for each methods	35.9	Individually for each methods
Preliminary operations, except weighing, required for the analysis	No		Gravity preconcentration, preparation of probe drug	
Au_{total} content, ppm	2.6±0.2	2.5±0.3	11.1	17.5
Ag_{total} content, ppm	0.37±0.02	0.47±0.08	No data	1.7
Number of Au particles in 1 g	~ 140	No data	~ 17	No data
Number of Ag particles in 1 g	~ 5600	No data	Not found	No data
Interval with the largest number of Au particles, μm	2 - 10	0.5–10 [11]; >100 [12]	10 – 20, and 60 – 70	No data
Representative subsample, g	≥ 0.3	0.4	~ 50	~ 50
Grade of gold-silver particles	670-900	850-900 [12]	No data	~ 900

The data about the compositions of gold-bearing SLg-1 and 109ore samples obtained by different methods given in Table 4 and show the same differences in analytical data as those obtained earlier for the SZH-3.

CONCLUSION

The accuracy of analytical data obtained by two different methods corresponds to semiquantitative results. However, the obtained information about the composition of gold-bearing samples is markedly different. The amount of particles carrying Au (with the size of over 2-3 μm), discovered by MLA, is 100 times lower. It resulted in a considerable (30-50%) underestimating of total Au content. Individual Ag particles have not been discovered. The particle-size distribution well agrees for Au particles with the size of 2-3 μm. The automated mineralogy ensures a more reliable estimation of the representative subsample as compared with scintillation AES if a sample contains heavy Au particles.

We suggest that the main reasons for this disagreement of results is different sample preparation: scintillation AES requires powder samples with the size less than 80 μm while MLA requires the extraction of the gravity concentration having only heavy particles. The data are also affected by differences in numerical models of spectral information processing.

Thus, the methods under discussion are not interchangeable, and the areas of their application are different. For geological-geochemical prospecting the combination of these methods is hardly reasonable, as the cost for the MLA analysis incommensurably higher as compared with the scintillation arc AES analysis. The cost for the analysis of one sample is 100 times different. The cost of tens of sample by MLA is comparable with the expenses to analyze hundreds and thousands of samples by AES.

The use of arc scintillation atomic-emission spectrometry is quite reasonable in prospecting for express study of a great number of samples to outline the perspective areas. Prior to studies it is desirable to specify the numerical model that connects the flash intensity and size of a particle containing NM. The automated mineralogy is effective for large-scale prospecting works because it remains an irreplaceable method to study the types of noble metal and complex ores when using the technology of gold extraction after a preliminary gravity concentration.

ACKNOWLEDGMENT

The authors are very grateful V. V. Knauf for the studying of the CRM SZH-3 and 109ore sample by method MLA.

REFERENCES

[1] Coetzee, LL; Theron, SJ; Martin, GJ; van der Merwe, JD; Stanek, TA. Modern gold deportments and its application to industry. *Minerals Engineering*, 2011, vol. 24, P. 565-575.

[2] Torgov, VG; Korda, TM; Demidova, MG; Gus'kova, EA; Bukhbinder, GL. ICP AES determination of platinum group elements and gold in collective extract and strip product solution in analysis of geological samples. *J. Anal. At. Spectrom.*, 2009, vol. 24, No. 11, P. 1551-1557.

[3] Pozhidaev, Y; Vlasova, N; Vasil'eva, I; Voronkov, M. Determination of noble metals in rocks and ores using adsorbent PSTM-3T. *Advanced Science Letters*, 2013, vol. 19, No. 2, P. 615-618.

[4] Goodall, WR; Scales, PJ. An overview of the advantages and disadvantages of the determination of gold mineralogy by automated mineralogy. *Minerals Engineering*, 2007, vol. 20, P. 506-517.

[5] Shabanova, EV; Bus'ko AE; Vasil'eva, IE. Scintillation Arc Atomic Emission Analysis of Powder Samples using MAES with High Temporal Resolution. *Industrial laboratory. Diagnostics of materials*, 2012, vol. 78, No. 1(II), P. 24-33. (in Russian).

[6] Knouf, VV. NATI Research. URL: http://www.natires.com.

[7] The International Database for Certified Reference Materials COMAR. URL: http://www.comar.bam.de.

[8] Labusov, VA; Garanin, VG; Shelpakova, IR. Multichannel Analyzers of Atomic Emission Spectra: Current State and Analytical Potentials. *Russian Journal of Analytical Chemistry*, 2012, vol. 67, No. 7, P. 632-641.

[9] Raikhbaum Ya D. Fizicheskie osnovy spektral'nogo analiza [Physical Principles of Spectral Analysis], Moscow, Nauka Publ., 1980. 159 p. (in Russian).

[10] Hitachi High-Technologies EuropeGmbH. http://www.hht-eu.com/cms/4626.html.

[11] Distler, VV; Yudovskaya, MA; Mokhov, AV; Trubkin, NV; Razvozzhaeva, EA; Mitrofanov, GL; Nemerov, VK. New Data on PGE Mineralization in Gold Ores of the Sukhoi Log Deposit, Lensk Gold-Bearing District, Russia. *Doklady Earth Sciences*, 2003, vol. 393, No. 9, P. 1265-1267.

[12] Nikulin, AI; Romanchuk, AI; Pavlova, NN; Ponomarenko, VI; Zharkov, VV; Bogomolov, VA. Sukhoi Log gold deposit: Ore processing using preliminary photometric separation. *Rudy I metally* [Ores and metals], 2009, No. 2, P. 68-77 (in Russian).

[13] Petrov, LL; Kornakov, YN; Korotaeva, IYa; Anchutina, EA; Persikova, LA; Susloparova, VE; Fedorova, IN; and Shibanov, VA. Multi-Element Reference Samples of Black Shale. *Geostandards Newsletter – The Journal of Geostandards and Geoanalysis*, 2004, vol. 28, No. 1, P. 89-102.

[14] Ingamells, C. O. and Pitard, F. F. (1986). Applied Geochemical Analyses. J. Wiley and Sons, Inc. New York. 733 p.

INDEX

A

B

C

D

E

F

G

H

I

K

L

M

N

O

P

Q

R

S

T

U

V

W

X

Z